Pathogenesis of Neurodegenerative Disorders

Contemporary Neuroscience

Pathogenesis of Neurodegenerative Disorders

Edited by

Mark P. Mattson

National Institute on Aging, Baltimore, MD

Humana Press Totowa, New Jersey

© 2001 Humana Press Inc.
999 Riverview Drive, Suite 208
Totowa, New Jersey 07512

For additional copies, pricing for bulk purchases, and/or information about other Humana titles, contact Humana at the above address or at any of the following numbers: Tel.: 973-256-1699; Fax: 973-256-8341; E-mail: humana@humanapr.com, or visit our Website: http://humanapress.com

This publication is printed on acid-free paper. ∞
ANSI Z39.48-1984 (American Standards Institute) Permanence of Paper for Printed Library Materials.

Cover illustration: The cover shows a silver-strained tissue section from the hippocampus of a patient with Alzheimer's disease. The flame-shaped cell bodies of degenerating neurons (neurofibrillary tangles) and their neurites associated with amyloid deposits are stained black. The inset shows an electron micrograph of a cultured embryonic human cerebral cortical neuron that had been exposed to conditions that disrupt cellular calcium homeostasis and trigger a form of programmed cell death called apoptosis. Micrographs provided by M. P. Mattson.

Cover design by Patricia F. Cleary.

Photocopy Authorization Policy:

Printed in the United States of America. 10 9 8 7 6 5 4 3 2 1

Library of Congress Cataloging-in-Publication Data

The pathogenesis of neurodegenerative disorders / edited by Mark P. Mattson.
            p. c.m.—(Contemporary neuroscience)
Includes bibliographical references and index.
ISBN 0-89603-838-6 (alk. paper)
1. Nervous system—Degeneration—Pathogenesis. I. Mattson, Mark Paul. II. Series.

RC365 .P385 2001
616.8'0407—dc21

00-063355

Preface

As the average life expectancy of many populations throughout the world increases, so to does the incidence of such age-related neurodegenerative disorders as Alzheimer's, Parkinson's, and Huntington's diseases. Rapid advances in our understanding of the molecular genetics and environmental factors that either cause or increase risk for age-related neurodegenerative disorders have been made in the past decade. The ability to evaluate, at the cellular and molecular level, abnormalities in postmortem brain tissue from patients, when taken together with the development of valuable animal and cell-culture models of neurodegenerative disorders has allowed the identification of sequences of events within neurons that result in their demise in specific neurodegenerative disorders. Though the genetic and environmental factors that promote neurodegeneration may differ among disorders, shared biochemical cascades that will ultimately lead to the death of neurons have been identified. These cascades involve oxyradical production, aberrant regulation of cellular ion homeostasis and activation of a stereotyped sequence of events involving mitochondrial dysfunction and activation of specific proteases.

Pathogenesis of Neurodegenerative Disorders provides a timely compilation of articles that encompasses fundamental mechanisms involved in neurodegenerative disorders. In addition, mechanisms that may prevent age-related neurodegenerative disorders are presented. Each chapter is written by an expert in the particular neurodegenerative disorder or mechanism or neuronal death discussed. Chapters that consider the role of oxidative stress as a central feature of all neurodegenerative disorders and the fundamental mechanisms of neuronal apoptosis and excitotoxicity, two forms of cell death central to many different neurodegenerative disorders, open this volume. Subsequent chapters focus on specific neurodegenerative disorders. Each chapter presents information on genetic and environmental factors that may contribute to these disorders and cell death cascades involved in these disorders are detailed. Chapters focus on Parkinson's disease, trinucleotide repeat disorders (including Huntington's disease), Alzheimer's disease and Down's syndrome (two disorders that appear to involve shared mechanisms), amyotrophic lateral sclerosis, ischemic stroke, spinal cord injury, and Duchenne muscular dystrophy.

Pathogenesis of Neurodegenerative Disorders will provide a valuable working reference for graduate students and postdocs beginning their careers in this field. In addition, because each chapter presents the most up-to-date specific information in the field, this book is valuable for senior scientists in allowing them to integrate information on cellular and molecular mechanisms across the wide field of neurodegenerative disorders.

Mark P. Mattson

Contents

Contributors

STANLEY H. APPEL, *Department of Neurology, Baylor College of Medicine, Houston, TX*

MARK W. BECHER, *Department of Pathology (Neuropathology Division), University of New Mexico Health Sciences Center, Albuquerque, NM*

DAVID B. BORCHELT, *Department of Pathology (Neuropathology Division and Neuroscience), Johns Hopkins University School of Medicine, Baltimore, MD*

EDWARD A. BURTON, *Department of Human Anatomy and Genetics, University of Oxford, Oxford, England*

CARSTEN CULMSEE, *Laboratory of Neurosciences, National Institute on Aging, Baltimore, MD*

KAY E. DAVIES, *Department of Human Anatomy and Genetics, University of Oxford, Oxford, England*

ROCCO C. IANNELLO, *Centre for Functional Genomics and Human Disease, Monash Medical Centre, Clayton, Australia*

ISMAIL KOLA, *Centre for Functional Genomics and Human Disease, Monash Medical Centre, Clayton, Australia*

ISABEL KLUSMAN, *Brain Research Institute, University of Zurich, and Swiss Federal Institute of Technology, Zurich, Switzerland*

VASSILIS E. KOLIATSOS, *Departments of Pathology (Neuropathology Division), Neurology, Neuroscience, and Psychiatry and Behavioral Sciences, Johns Hopkins University School of Medicine, Baltimore, MD*

MARK A. LOVELL, *Sanders-Brown Center on Aging and Alzheimer's Disease Research Center and Department of Chemistry, University of Kentucky, Lexington, KY*

WILLIAM R. MARKESBERY, *Sanders-Brown Center on Aging and Alzheimer's Disease Research Center, University of Kentucky, Lexington, KY, and Departments of Pathology and Neurology, Vanderbilt University, Nashville, TN*

MARK P. MATTSON, *Laboratory of Neurosciences, National Institute on Aging, Baltimore, MD*

THOMAS J. MONTINE, *Department of Pathology, Vanderbilt University, Nashville, TN*

WARD A. PEDERSEN, *Laboratory of Neurosciences, National Institute on Aging, Baltimore, MD*

CARLOS PORTERA-CAILLIAU, *The Neuroscience Institute, Good Samaritan Hospital, Los Angeles, CA*

DONALD L. PRICE, *Departments of Pathology, Neurology, and Neuroscience, Johns Hopkins University School of Medicine, Baltimore, MD*

CHRISTOPHER A. ROSS, *Departments of Neuroscience, and Psychiatry and Behavioral Sciences, Johns Hopkins University School of Medicine, Baltimore, MD*

A. H. V. SCHAPIRA, *University Department of Clinical Neurosciences, Royal Free Hospital and University College Medical School and Institute of Neurology, University College London, London, UK*

GABRIELLE SCHILLING, *Department of Pathology, Johns Hopkins University School of Medicine, Baltimore, MD*

MARTIN E. SCHWAB, *Brain Research Institute, University of Zurich, and Swiss Federal Institute of Technology, Zurich, Switzerland*

M. T. SILVA, *University Department of Clinical Neurosciences, Royal Free Hospital and University College Medical School and Institute of Neurology, University College London, London, UK*

R. GLENN SMITH, *Department of Neurology, Baylor College of Medicine, Houston, TX*

JAMUNA SUBRAMANIAM, *Department of Pathology, Johns Hopkins University School of Medicine, Baltimore, MD*

SHOJI TSUJI, *Department of Neurology, Brain Research Institute, Niigata University, Niigata, Japan*

PHILIP C. WONG, *Department of Pathology, Johns Hopkins University School of Medicine, Baltimore, MD*

1

Mechanisms of Neuronal Apoptosis and Excitotoxicity

Mark P. Mattson

INTRODUCTION

Apoptosis is a form of cell death that is mediated by specific biochemical cascades involving mitochondrial changes and activation of proteases called caspases. Apoptosis provides a mechanism for cells to die without adversely affecting their neighbors (Fig. 1). This contrasts with a nonregulated form of cell death called necrosis, in which cellular organelles swell and the plasma membrane becomes permeable, resulting in release of cellular contents and massive death of groups of cells throughout a tissue. Aberrant regulation of apoptosis may be the central abnormality in many different diseases, most prominent of which are cancers wherein cells become resistant to apoptosis. Therefore, a major effort in the field of cancer research is to identify the molecular alterations responsible for such cell immortalization (Wyllie, 1997).

Cells of the nervous system differ in many ways from those in proliferative tissues. Importantly, neurons must survive for the entire lifetime of the organism in order to maintain the function of the neuronal circuits. Thus, motor neurons must maintain connections to skeletal muscles, and long-term memories require the continued survival of the neurons in the brain regions in which those memories are encoded. Many neurons undergo apoptosis during development of the nervous system, and such cell deaths occur during a time window that coincides with the process of synaptogenesis (Oppenheim, 1991). Initial overproduction of neurons, followed by the death of some, is likely an adaptive process that provides numbers of neurons sufficient to form nerve cell circuits that are precisely matched to their functional specifications. Accordingly, the decision of which neurons die is made by cellular signal transduction pathways that are "tuned" to the functionality of neuronal circuits. Two types of signaling pathways that may determine whether or not developing neurons live or die are those activated by target-derived neurotrophic factors and those activated by the excitatory neurotransmitter glutamate (Mattson and Furukawa, 1998).

Under normal conditions, many neurons remain viable and function throughout the lifetime of an individual. However, many people will not complete their lives without excessive death of one or more populations of neurons. Thus, death of hippocampal and cortical neurons is responsible for the symptoms of Alzheimer's disease (AD), death of midbrain dopaminergic neurons underlies Parkinson's disease (PD),

From: *Pathogenesis of Neurodegenerative Disorders* Edited by: M. P. Mattson © Humana Press Inc., Totowa, NJ

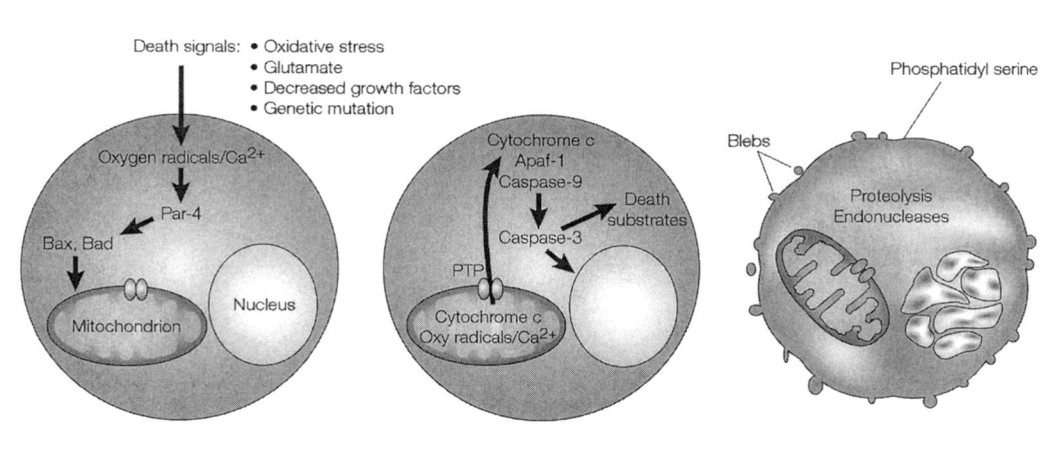

Fig. 1. Examples of morphological and biochemical features of apoptosis. During the initiation phase of apoptosis (left), the death signal activates an intracellular cascade of events that may involve increases in levels of oxyradicals and calcium, production of Par-4, and translocation of pro-apoptotic Bcl-2 family members to the mitochondrial membrane. The effector phase of apoptosis (middle) involves increased mitochondrial calcium and oxyradical levels, the formation of permeability transition pores (PTP) in the mitochondrial membrane, and release of cytochrome c into the cytosol. Cytochrome c forms a complex with Apaf-1 and caspase-9. Activated caspase-9, in turn, activates caspase-3, beginning the degradation phase of apoptosis in which various caspase and other enzyme substrates are cleaved, resulting in characteristic changes in the plasma membrane (blebbing and exposure of phosphatidylserine on the cell surface, which is a signal that stimulates cell phagocytosis by macrophages/microglia). The nuclear chromatin becomes condensed and fragmented during the degradation phase of apoptosis (right), and the cell is then at the point-of-no-return. Modified from Mattson (2000).

Huntington's disease (HD) results from death of neurons in the striatum that control body movements, and lower motor neurons in the spinal cord die in amyotrophic lateral sclerosis (ALS) patients (Fig. 2). The number of people with such neurodegenerative disorders is rapidly increasing as average lifespan increases. The present chapter, which considers the contributions of apoptosis and excitotoxicity to neurodegenerative disorders, is modified from a recent article on the same topic (Mattson, 2000).

NEURONAL APOPTOSIS

The death of neurons can be triggered by a variety of stimuli (Table 1). An intensively studied neuronal death signal is lack of neurotrophic factor support, which may trigger apoptosis during development of the nervous system and in neurodegenerative disorders (Mattson and Lindvall, 1997). A second prominent trigger of neuronal apoptosis is activation of receptors for the excitatory amino acid neurotransmitter glutamate. Calcium influx through ionotropic glutamate receptor channels and voltage-dependent calcium channels mediates glutamate-induced neuronal apoptosis and necrosis (Ankarcrona et al., 1995; Glazner et al., 2000). Such "excitotoxicity" may occur in acute neurodegenerative conditions such as stroke, trauma, and severe epileptic seizures (Choi, 1992), as well as in AD, PD, HD, and ALS (Wong et al., 1998; Mattson et al., 1999). Oxidative stress (in which free radicals such as superoxide anion radical and hydroxyl radical damage cellular lipids, proteins, and nucleic acids by attacking chemi-

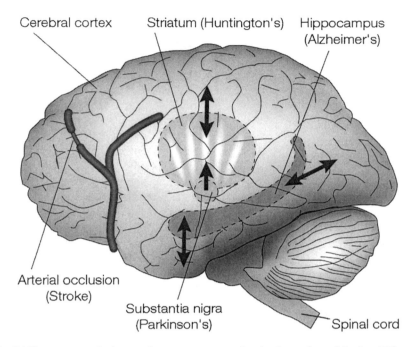

Fig. 2. Different populations of neurons are selectively vulnerable in different neuro-degenerative disorders. In AD, neurons in the hippocampus and certain regions of cerebral cortex degenerate; in PD, it is the dopaminergic neurons in the substantia nigra that undergo apoptosis; in HD, it is neurons in the striatum that die; and in ALS, spinal-cord motor neurons degenerate. Which neurons die in stroke depends on which blood vessel is affected, but often it is neurons in the cerebral cortex and striatum. Modified from Mattson (2000).

cal bonds in those molecules) is a very important trigger of neuronal death in neuro-degenerative disorders (*see* Markesbery et al., Chapter 2; Mattson, 1998; Sastry and Rao, 2000). Reduced energy availability to neurons, as occurs after a stroke or during aging, may also initiate neuronal apoptosis. Environmental toxins have been implicated in neurodegenerative disorders and can induce neuronal apoptosis; several such toxins can induce patterns of brain damage and behavioral phenotypes remarkably similar to AD, PD, and HD (Beal, 1995; Bruce-Keller et al., 1999; Duan et al., 1999b).

Although the genetic and environmental factors that trigger neuronal apoptosis may be different in physiological and pathological settings (Table 2), many of the subsequent biochemical events that execute the cell death process are highly conserved. One key focus of this death program is the mitochondrion, an organelle for which compelling evidence suggest controls the cell-death decision (Kroemer et al., 1998). Changes that occur in mitochondria in cells undergoing apoptosis include increased oxyradical production, opening of pores in their membranes, and release of cytochrome c (Fig. 1). These mitochondrial changes are central to the cell-death process because agents such as manganese superoxide dismutase and cyclosporine A, which act directly on mitochondria to suppress oxidative stress and membrane pore formation, also prevent neuronal death in various experimental models (Keller et al., 1998; Matsumoto et al., 1999).

Table 1
Examples of Proteins That Can Either Promote or Suppress Neuronal Apoptosis

Proapoptotic	
Glutamate receptor proteins	Calcium influx
Fas	Initiate death cascade
Bax, Bad	Pore formation in mitochondrial membrane— cytochrome c release
Par-4	Mitochondrial dysfunction; suppression of survival signals (NF-κB)
P53	Transcription of death genes; enhancement of Bax actions
Caspases	Cleavage of various enzyme, cytoskeletal and ion channel substrates
Antiapoptotic	
Bcl-2, Bcl-XL	Stabilize mitochondrial function; suppress oxidative stress
IAPs	Caspase inhibition
PKCz	Stimulate survival gene expression (NF-κB)
Neurotrophic factors and cytokines	Induce expression of survival genes (antioxidant enzymes, calcium-regulating proteins, IAPs, Bcl-2)
Antioxidant enzymes	Suppress oxidative stress
Calcium-binding proteins	Stabilize calcium homeostasis

Table 2
Genetic and Environmental Factors
That May Promote Apoptosis and Excitotoxicity in Neurodegenerative Disorders

Disorder	Genetic factors	Environmental factors
AD	APP, presenilin mutations, ApoE	Head trauma, low education, calorie intake
PD	α-synuclein, parkin mutations	Head trauma, toxins, calorie intake
HD	Poly-CAG expansions in huntingtin	
ALS	Cu/Zn-SOD mutations	Toxins, autoimmune response
Stroke	Cadasil mutations	Smoking, dietary calories, and fat
Trauma	Apolipoprotein E	

The events that occur upstream of the mitochondrial changes are complex and involve interactions of several types of proteins. The Bcl-2 family of proteins was originally discovered in the nematode *Caenorhabditis elegans* and includes both pro- and anti-apoptotic members (Pellegrini and Strasser, 1999). Anti-apoptotic members in neurons include Bcl-2 and Bcl-xL, while pro-apoptotic members include Bax and Bad. Bcl-2 increases resistance of neurons to death induced by excitotoxic, metabolic, and oxidative insults relevant to AD, stroke, and other disorders (Martinou et al., 1994; Guo et al., 1998). On the other hand, neurons lacking Bax are protected against apoptosis (White et al., 1998). Bcl-2 proteins may control the cell-death process by interacting with mitochondrial membranes in a manner that either promotes or prevents ion movements across mitochondrial membranes (Green and Reed, 1998).

The premitochondrial phase of apoptosis can also be regulated by other proteins including prostate apoptosis response-4 (Par-4), caspases, and telomerase. Par-4 was discovered because its expression is markedly increased in prostate tumor cells undergoing apoptosis. A series of studies subsequently showed that Par-4 has an essential role in developmental and pathological neuronal death (Guo et al., 1998; Duan et al., 1999b, 2000; Pedersen et al., 2000). In neurons, Par-4 levels increase rapidly in response to various apoptotic stimuli through enhanced translation of Par-4 mRNA (Duan et al., 1999a). A leucine zipper domain in the C-terminus of Par-4 mediates its pro-apoptotic function; Par-4 interactions with protein kinase C (PKC)ζ and Bcl-2 may be central to the mechanism whereby Par-4 induces mitochondrial dysfunction (Camandola and Mattson, 2000). Cysteine proteases of the caspase family are evolutionarily conserved effectors of apoptosis (Chan and Mattson, 1999). Caspases can act during the premitochondrial phase (e.g., caspases 2 and 8) or postmitochondrial phase (e.g., caspases 3 and 9) of apoptosis. A variety of substrate proteins are cleaved by caspases and may regulate the cell-death process. Caspase substrates include: enzymes such as poly-ADP-ribose polymerase and ataxia-telangiectasia mutated (ATM) kinase; ion channels including subunits of the AMPA subtype of neuronal glutamate receptor; and cytoskeletal proteins such as actin and spectrin (Glazner et al., 2000; Chan and Mattson, 1999).

Telomerase is a protein–RNA complex that adds a six-base DNA sequence (TTAGGG) to the ends of chromosomes, thereby preventing their shortening and protecting them during chromosome segregation in mitotic cells (Liu, 1999). Telomerase consists of a catalytic reverse transcriptase subunit called TERT and an RNA template. Telomerase activity is increased during cell immortalization and transformation, and is thought to contribute to the pathogenesis of many cancers. TERT protein and telomerase activity are present in many tissues during development, including the brain, but are downregulated during late embryonic and early postnatal development (Fu et al., 2000). Telomerase activity and expression of TERT are associated with increased resistance of neurons to apoptosis in experimental models of developmental neuronal death and neurodegenerative disorders (Fu et al., 1999, 2000; Zhu et al., 2000). The cell-survival-promoting action of TERT in neurons is exerted at an early step in the cell-death pathway prior to mitochondrial alterations and caspase activation.

ANTI-APOPTOTIC SIGNALING

Because neurons are postmitotic and not easily replaced, it is essential that signaling mechanisms are present that guard against neuronal death. The consequences of the death of neurons can be devastating, as in the cases of neurodegenerative disorders such as AD, PD, and ALS. There are several prominent anti-apoptotic signaling pathways in neurons. Activation of neurotrophic factor receptors can protect neurons against apoptosis by activating receptors linked via kinase cascades to production of cell-survival-promoting proteins (Mattson and Lindvall, 1997). Brain-derived neurotrophic factor (BDNF), nerve growth factor (NGF), and basic fibroblast growth factor (bFGF) can prevent death of cultured neurons, in part by stimulating production of antioxidant enzymes, Bcl-2 family members, and proteins involved in regulation of Ca^{2+} homeostasis (Mattson and Lindvall, 1997; Tamatani et al., 1998). Tumor necrosis factor-α (TNF-α), ciliary neurotrophic factor (CNTF), and leukemia inhibitory factor (LIF) are

three cytokines that can prevent neuronal death in experimental models of natural neuronal death and neurodegenerative disorders (Hagg and Varon, 1993; Barger et al.,
1995; Middleton et al., 2000). Several of these neurotrophic factors and cytokines use
a survival pathway involving the transcription factor NF-κB (Mattson and Camandola,
2000). Activation of NF-κB can protect cultured neurons against death induced by
trophic factor withdrawal and exposure to excitotoxic, oxidative, and metabolic
insults. In vivo studies that examined mice lacking the p50 subunit of NF-κB, or in
which NF-κB was inhibited by "decoy DNA," have shown that NF-κB also protects
neurons in the intact brain (Yu et al., 2000). Gene targets that mediate the survival-
promoting action of NF-κB may include manganese superoxide dismutase, Bcl-2, and
inhibitor of apoptosis proteins.

In addition to external signals that promote neuronal survival, several stress-responsive intracellular signaling pathways have been identified that can protect neurons
against apoptosis. One interesting example is a "preconditioning" mechanism in which
metabolic stress resulting from reduced food intake or high levels of activity in neuronal circuits can induce the expression of neurotrophic factors and heat-shock proteins (Yu and Mattson, 1999; Lee et al., 2000). Neurotrophic factors, in turn, act in an
autocrine or paracrine manner to activate cell surface receptor-mediated kinase signaling pathways that ultimately induce expression of genes encoding survival-promoting
proteins such as antioxidant enzymes. Heat-shock proteins can prevent apoptosis by
acting as "chaperones" for many different proteins, thereby maintaining protein stability; they may also interact directly with caspases, inhibiting their activation. Intracellular messengers that have the potential to kill neurons may also protect them from
apoptosis. For example, calcium is a prominent transducer of stress responses that can
activate transcription through the cyclic AMP response element binding protein
(CREB); this pathway can promote neuron survival in experimental models of developmental cell death (Hu et al., 1999). Calcium may also activate a rapid neuroprotective
signaling pathway in which the Ca^{2+}-activated actin-severing protein gelsolin
induces actin depolymerization resulting in suppression of Ca^{2+} influx through membrane *N*-methyl-D-aspartate (NMDA) receptors and voltage-dependent Ca^{2+} channels
(Furukawa et al., 1997); this may occur through intermediary filament actin-binding
proteins that interact with NMDA receptor and Ca^{2+} channel proteins. Increased Ca^{2+}
levels or activation of membrane receptors [such as the receptor for secreted amyloid
precursor protein α (sAPPα)] can stimulate cyclic guanosine 5'-monophosphate (GMP)
production via a nitric oxide (NO)-mediated pathway, and cyclic GMP can induce
activation of K^+ channels and the transcription factor NF-κB and thereby increase
resistance of neurons to excitotoxic apoptosis (Furukawa et al., 1996).

SYNAPTIC APOPTOSIS

Synaptic terminals are the major sites of intercellular communication between neurons, and are also sites where signaling pathways that initiate or prevent apoptosis are
highly concentrated. For example, receptors for glutamate are postsynaptic regions of
dendrites, and receptors for neurotrophic factors are in both pre- and postsynaptic terminals. The biochemical machinery involved in apoptosis can be activated in synaptic
terminals, wherein it can alter synaptic function and promote localized degeneration of
synapses and neurites (Mattson and Duan, 1999) (Fig. 3). Par-4 production, mitochon-

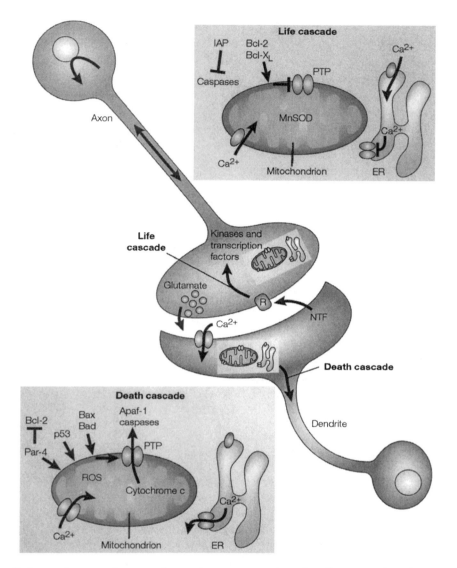

Fig. 3. Synaptic neurodegenerative and neuroprotective signaling cascades. Age- and disease-related stressors promote excessive activation of apoptotic (death) biochemical cascades in synaptic terminals and neurites. For example, overactivation of glutamate receptors under conditions of reduced energy availability or increased oxidative stress results in Ca^{2+} influx into postsynaptic regions of dendrites. Ca^{2+} entering the cytoplasm through plasma membrane channels and ER channels induces apoptotic cascades (lower left) that involve Par-4, pro-apoptotic Bcl-2 family members (Bax and Bad), and/or p53. These factors act on mitochondria to induce Ca^{2+} influx, oxidative stress, opening of permeability transition pores (PTP), and release of cytochrome c. This results in caspase activation and execution of the cell-death process. Anti-apoptotic (life) signaling pathways are also concentrated in synaptic compartments (upper right). For example, activation of receptors R for neurotrophic factors (NTF) in axon terminals stimulates kinase cascades and transcription factors and increased production of survival-promoting proteins such as Bcl-2, Bcl-XL, and MnSOD (which act at the level of mitochondria) and inhibitor of apoptosis proteins (IAPs; which inhibit caspases). Modified from Mattson (2000).

drial alterations, caspase activation, and release into the cytosol of factors that may cause nuclear apoptosis can be induced in synaptosome preparations by insults that induce apoptosis in intact neurons (Mattson et al., 1998). It was recently shown that AMPA receptor subunits are selectively degraded in hippocampal neurons after exposure to an apoptotic dose of glutamate, resulting in decreased Ca^{2+} influx, thereby preventing excitotoxic necrosis (Glazner et al., 2000). The latter findings, and the presence of many different caspase substrates in synapses (Chan and Mattson, 1999), suggest that caspase-mediated cleavage of synaptic proteins may control the process of neuronal apoptosis.

Interestingly, apoptotic pathways may also function in synaptic plasticity, particularly under conditions of stress and injury. Caspases can be activated in a reversible manner after trophic factor withdrawal or activation of glutamate receptors (Glazner et al., 2000), and such caspase activation induces a selective degradation of subunits of a specific type of glutamate receptor called the AMPA receptor; this results in suppression of cell excitability and Ca^{2+} influx (Lu and Mattson, 2000). The latter mechanism may allow neurons to "withdraw" from participation in neuronal circuits, thereby permitting them to recover from potentially lethal conditions. In addition, TNF-α and NF-κB activation can modify long-term depression and potentiation of synaptic transmission in the hippocampus (Albensi and Mattson, 2000) providing further evidence that anti-apoptotic signaling can modulate synaptic plasticity. Mitochondria are present in quite high concentrations in synaptic terminals, and mitochondrial membrane permeability in synaptic terminals has been associated with impaired synaptic plasticity in the hippocampus (Albensi et al., 2000), suggesting a role for apoptotic mitochondrial alterations in synaptic function.

APOPTOSIS, EXCITOTOXICITY, AND NEURODEGENERATIVE DISORDERS

The evidence implicating apoptosis and excitotoxicity in each of the disorders described below is based on analyses of postmortem tissue from patients, and studies of experimental animal and cell-culture models have strongly implicated neuronal apoptosis. Studies of the pathogenic mechanisms of genetic mutations that cause early-onset autosomal dominant forms of AD, HD, and ALS, that implicate apoptosis in age-related neurological disorders have been particularly valuable.

Alzheimer's Disease

Progressive impairment of cognition and emotional disturbances characterize AD; these symptoms result from degeneration of synapses and death of neurons in limbic structures such as hippocampus and amygdala, and associated regions of cerebral cortex (Fig. 2). The damaged neurons exhibit aggregates of hyperphosphorylated tau protein and evidence of excessive Ca^{2+}-mediated proteolysis and oxidative stress (Yankner, 1996; Mattson, 1997). A defining feature of AD is accumulation of amyloid plaques formed by aggregates of amyloid β-peptide (Aβ) a 40–42 amino acid fragment generated by proteolytic processing of the APP. DNA damage and caspase activation, and alterations in expression of apoptosis-related genes such as Bcl-2 family members, Par-4, and DNA damage response genes have been documented in neurons associated with amyloid deposits in the brains of AD patients (Su et al., 1994; Guo et al., 1998;

Masliah et al., 1998). Expression profile analysis of thousands of genes in brain tissue samples from AD patients and age-matched control patients revealed a marked decrease in expression of an anti-apoptotic gene called *NCKAP1* (Suzuki et al., 2000).

Cell-culture studies have shown that Aβ can induce apoptosis directly (Loo et al., 1993; Mark et al., 1995), and Aβ can greatly increase neuronal vulnerability to death induced by conditions, such as increased oxidative stress and reduced energy availability, that are known to occur in the brain during aging (Mattson, 1997). The ability of Aβ to increase the vulnerability of neurons to excitotoxicity (Mattson et al., 1992) is particularly striking. The mechanism whereby Aβ sensitizes neurons to death involves membrane lipid peroxidation, which impairs the function of membrane ion-motive ATPases and glucose and glutamate transporters resulting in membrane depolarization, ATP depletion, excessive Ca^{2+} influx, and mitochondrial dysfunction. Accordingly, antioxidants that suppress lipid peroxidation and drugs that stabilize cellular Ca^{2+} homeostasis can protect neurons against Aβ-induced apoptosis (Mattson, 1998). In addition, levels of sAPPα, which can protect neurons against excitotoxicity and apoptosis (Mattson, 1997), may be decreased in AD, and neurotrophic factors and cytokines known to prevent neuronal apoptosis can protect neurons against Aβ-induced death. Further evidence for a role for excitotoxicity in AD comes from studies showing that excessive activation of glutamate receptors can elicit changes in the cytoskeleton of neurons similar to those seen in neurofibrillary tangles in AD (Mattson, 1990; Stein-Behrens et al., 1994).

Familial forms of AD can result from mutations in three different genes, namely, those encoding APP, presenilin-1, and presenilin-2. APP mutations may cause AD by altering proteolytic processing of APP such that levels of Aβ are increased and levels of sAPPα are decreased (Mattson, 1997) (Fig. 4). Presenilin-1 mutations render neurons vulnerable to death induced by a variety of insults including trophic factor withdrawal and exposure to Aβ, glutamate, and energy deprivation (Guo et al., 1997, 1998, 1999a). The primary effect of presenilin-1 mutations may be to perturb Ca^{2+} homeostasis in the endoplasmic reticulum such that more Ca^{2+} is released when neurons are exposed to potentially damaging oxidative and metabolic insults (Guo et al., 1999b; Mattson et al., 2000a). Mutant presenilin-1 acts at an early step prior to Par-4 production, mitochondrial dysfunction, and caspase activation. Agents that suppress Ca^{2+} release, including dantrolene and xestospongin, can counteract the endangering effects of the mutations (Mattson et al., 2000a), indicating that enhanced Ca^{2+} release is central to the pathogenic action of mutant presenilin-1.

Motor System Disorders

Patients with PD exhibit profound motor dysfunction as the result of degeneration of dopaminergic neurons in their substantia nigra. The cause(s) of PD is unknown, but likely involves increased oxidative stress and mitochondrial dysfunction in dopaminergic neurons (Jenner and Olanow, 1998). Both environmental and genetic factors may sensitize dopaminergic neurons to age-related increases in oxidative stress and energy deficits (Jenner and Olanow, 1998; Polymeropoulos, 1998). The fact that the toxin 1-methyl-4-phenyl-1,2,3,6-tetrahydropyridine (MPTP) can induce PD-like neuropathology and motor symptoms in mice, monkeys, and humans demonstrates the potential for neurotoxins to cause sporadic PD. Analyses of brain tissue from PD patients implicates

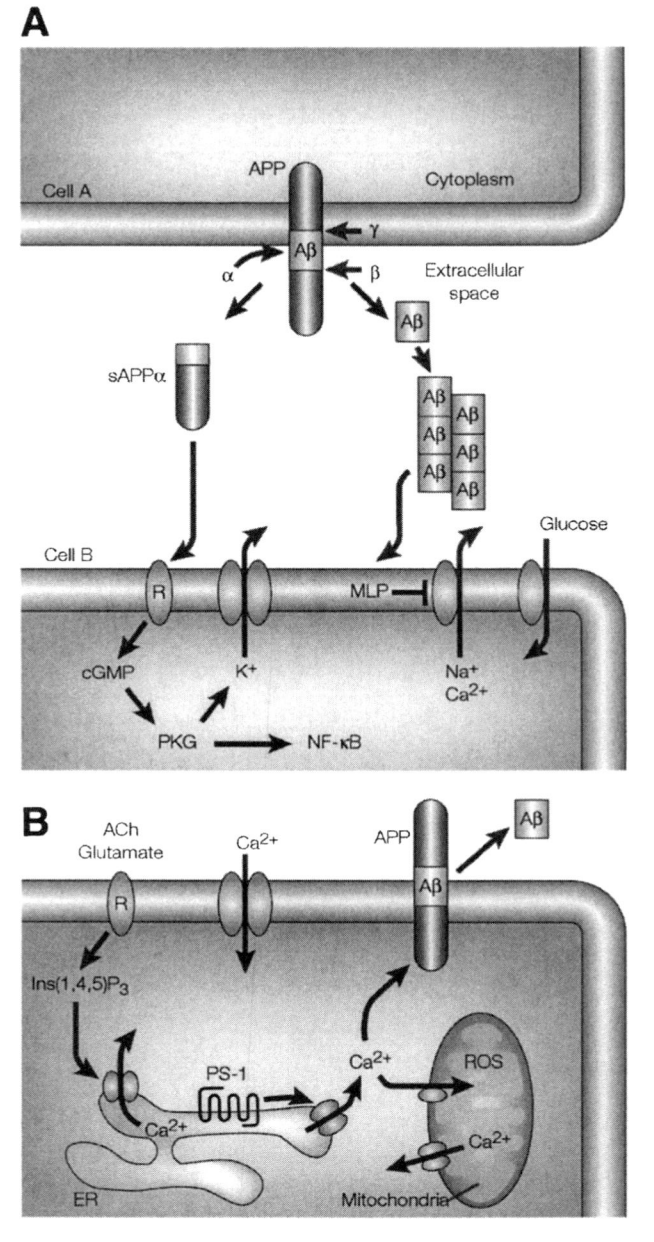

Fig. 4. Mechanisms whereby mutations in APP and presenilin-1 cause neuronal degeneration in familial AD. **(A)** The amyloid precursor protein (APP) can be proteolytically processed in two major ways. Cleavage of APP within the amyloid β-peptide (Aβ) sequence by α-secretase (a) releases a secreted form of APP (sAPPα) from the cell surface sAPPα activates a receptor (R) linked to cyclic GMP production and activation of cyclic GMP-dependent protein kinase (PKG). PKG can then promote opening of K$^+$ channels, resulting in membrane hyperpolarization, and can also activate the transcription factor NF-κB; these effects of sAPPα are believed to mediate its neuron-survival-promoting properties. A second pathway of APP processing involves cleavages at the N- and C-termini of Aβ by enzymes called β-secretase (β) and γ-secretase (γ), respectively. The latter pathway releases Aβ from cells that, under appropriate

apoptosis-related DNA damage and gene activation in the death of dopaminergic neurons. Par-4 levels are selectively increased in substantia nigra dopaminergic neurons prior to their death, and suppression of Par-4 expression protects dopaminergic neurons against death (Duan et al., 1999a). Dopaminergic neurons can be spared by treatment of animals with caspase inhibitors, drugs that suppress macromolecular synthesis, and neurotrophic factors such as glial cell-derived neurotrophic factor supporting a role for apoptosis in PD (Gash et al., 1996; Klevenyi et al., 1999). The protein α-synuclein is a major component of the PD brain lesions called Lewy bodies, and mutations in α-synuclein are responsible for a small percentage of PD cases; expression of mutant α-synuclein in cultured cells promotes apoptosis (El-Agnaf et al., 1998).

HD results from mutations in the *huntingtin* gene that are characterized by expansions of a trinucleotide (CAG) sequence producing a huntingtin protein containing polyglutamine repeats (Brandt et al., 1996). The huntingtin mutations cause degeneration of neurons in the striatum resulting in uncontrolled body movements. Studies of HD patients, and of rodents given the mitochondrial toxin 3-nitropropionic acid (3NP), suggest that impaired mitochondrial function and excitotoxic death are central to the disease. 3NP induces selective damage to striatal neurons that is associated with Par-4 production, mitochondrial dysfunction, and caspase activation; blockade of Par-4 expression or caspase activation protects striatal neurons against 3NP-induced death (Duan et al., 2000). The ability of activation of the anti-apoptotic transcription factor NF-κB to protect striatal neurons against 3NP-induced death provides further support for apoptosis as a major death pathway in HD (Yu et al., 2000). Caspase-8 is redistributed to an insoluble fraction in striatal tissue from HD patients, and expression of mutant huntingtin in cultured cells induces caspase-8-dependent apoptosis (Sanchez et al., 1999). Expression of mutant huntingtin in the brains of adult rats using viral vectors results in the formation of intraneuronal inclusions and cell death (Senut et al., 2000). However, the formation of nuclear inclusions containing huntingtin may not be required for apoptosis; in fact, such inclusions may be part of a cytoprotective response (Kim et al., 1999). Moreover, wild-type—but not mutant—huntingtin can protect cells by suppressing cell death before mitochondrial dysfunction (Rigamoni et al., 2000). Lymphoblasts from HD patients exhibit increased sensitivity to stress-induced apoptosis associated with mitochondrial dysfunction and increased caspase-3 activation (Sawa et al., 1999) suggesting an adverse effect of mutant huntingtin that is not limited to neurons.

(continued) conditions (high concentration and oxidizing environment), begin to self-aggregate. Under these conditions, Aβ induces membrane lipid peroxidation (MLP), resulting in impairment of the function of membrane ion-motive ATPases (Na^+ and Ca^{2+} pumps) and glucose transporters. Neurons are thus rendered vulnerable to apoptosis. **(B)** Presenilin-1 (PS-1) is an integral membrane protein located primarily in the endoplasmic reticulum (ER). Mutations in PS-1 perturb ER Ca^{2+} homeostasis in a manner that results in increased release of Ca^{2+} through IP_3 receptors and ryanodine receptors (RyR). The enhanced Ca^{2+} release triggers further Ca^{2+} influx through Ca^{2+} release channels in the plasma membrane, and this altered Ca^{2+} homeostasis renders neurons vulnerable to apoptosis and excitotoxicity, and alters APP processing in a manner that increases Aβ production. Modified from Mattson (2000).

Degeneration of spinal-cord motor neurons in ALS results in progressive paralysis. This selective degeneration of motor neurons involves increased oxidative stress, overactivation of glutamate receptors, and cellular Ca^{2+} overload (Cookson and Shaw, 1999). Production of autoantibodies against voltage-dependent Ca^{2+} channels may play a role in the neurodegenerative process (Smith et al., 1996). Although most cases of ALS are sporadic, mutations in the antioxidant enzyme Cu/Zn-superoxide dismutase (Cu/Zn-SOD) are responsible for some inherited cases of ALS. Expression of Cu/Zn-SOD genes containing these mutations in transgenic mice results in spinal-cord pathology remarkably similar to that of ALS patients. The mutations do not decrease antioxidant activity of the enzyme, but result in gain of an adverse pro-apoptotic activity, which may involve increased peroxidase activity. Mutant Cu/Zn-SOD causes increased oxidative damage to membranes and disturbances in mitochondrial function that may render neurons vulnerable to excitotoxic apoptosis (Kruman et al., 1999). DNA damage is evident in spinal-cord motor neurons of ALS patients and is associated with increased mitochondrial localization of Bax and decreased association of Bcl-2. Levels of Bax, but not Bcl-2, are increased in spinal-cord motor neurons of ALS patients, and a similar pattern of Bcl-2 family member expression is observed in Cu/Zn-SOD mutant mice. Apoptosis involvement in ALS is further suggested by the ability of overexpression of Bcl-2 and administration of caspase inhibitors to delay motor neuron degeneration and death in Cu/Zn-SOD mutant mice (Lee et al., 2000; Martin et al., 2000).

Ischemic Stroke

Ischemic brain damage resulting from occlusion of a cerebral blood vessel is characterized by an infarct with a necrotic core in which all cells die rapidly and a surrounding ischemic penumbra in which neurons die over days to weeks (Dirnagl et al., 1999). Metabolic compromise, overactivation of glutamate receptors, Ca^{2+} overload, and increased oxyradical production occur in neurons subjected to ischemia. In addition, complex cytokine cascades involving microglial cells and the cerebrovasculature may play important roles in promoting or preventing neuronal death after stroke. Cells in the ischemic penumbra exhibit DNA damage and activation of the DNA damage-responsive proteins PARP and Ku80. In rodent stroke models, neurons in the ischemic penumbra exhibit morphological and molecular changes consistent with apoptosis including caspase activation, expression of pro-apoptotic genes, and release of cytochrome c.

Membrane phospholipid hydrolysis may trigger neuronal apoptosis in stroke. Cleavage of membrane sphingomyelin by acidic sphingomyelinase (ASMase) generates the lipid mediator ceramide. Cerebral ischemia in mice induces large increases in ASMase activity and ceramide levels, and production of inflammatory cytokines (Yu et al., 2000). Mice lacking ASMase, or mice that are given a drug that inhibits production of ceramide, exhibit decreased cytokine production, decreased brain damage, and improved behavioral outcome after a stroke (Yu et al., 2000). In addition, mice lacking phospholipase-A2 exhibit decreased brain damage after focal cerebral ischemia, suggesting an important role for one or more lipid mediators generated by this enzyme in ischemic neuronal injury (Bonventre et al., 1997).

Studies in which caspase genes are deleted in mice, or in which drugs that inhibit caspases are given to mice, have provided strong evidence that caspases mediate much of the neuronal death that occurs in the penumbral region of an ischemic infarct (Endres et al., 1998; Schielke et al., 1998). In addition, delivery of neurotrophic factors known to prevent neuronal apoptosis can prevent neuronal death after stroke; particularly effective are bFGF, NGF, and sAPPα (Mattson and Furukawa, 1998). A pivotal role for mitochondrial alterations in stroke-induced neuron death is suggested by studies showing that lack of mitochondrial Mn-SOD exacerbates (Murukami et al., 1998), whereas overexpression of Mn-SOD decreases (Keller et al., 1998), focal ischemic brain injury. Moreover, treatment of rats with cyclosporine A decreases ischemic infarct size (Matsumoto et al., 1999). Finally, although stroke and AD are quite different disorders, they may share common pathways of neuronal death because presenilin-1 mutations increase vulnerability of cortical neurons to ischemia-induced cell death (Mattson et al., 2000b).

Traumatic Brain and Spinal-Cord Injury

The leading cause of death and disability in persons under the age of 40 is traumatic injury to the brain and spinal cord. Trauma initiates biochemical and molecular events involving many of the same neurodegenerative cascades and neuroprotective signaling mechanisms that occur in the chronic neurodegenerative diseases described above. Studies of brains of patients that died from traumatic brain injury (TBI) have documented apoptosis-related changes in neurons including the presence of DNA strand breaks, caspase activation, and increased Bax and p53 expression (Clark et al., 1999). In mice, sensory-motor and cognitive deficits after TBI are strongly correlated with numbers of neurons exhibiting apoptotic nuclear damage (Fox et al., 1998). Increased levels of the death effector proteins p53 (Napieralski, 1999) and Fas (Beer et al., 2000) occur in neurons after TBI. In addition, caspase-3 activity increases markedly in cerebral cortex of rats in response to TBI, and intraventricular administration of the caspase-3 inhibitor z-DEVD-fmk prior to injury reduced cell death and improved symptoms, indicating a central role for caspases in this brain-injury model. NGF infusion beginning 24 h after TBI results in improved learning and memory and decreased death of neurons in comparison with control rats (Yakovlev et al., 1997). Finally, cyclosporine A protects against synaptic dysfunction and cell death in rodent models of TBI, consistent with a key role for mitochondrial membrane permeability in the neurodegenerative process (Albensi et al., 2000).

In cases of spinal-cord injury, apoptosis is suggested by evidence for nuclear DNA fragmentation and caspase activation in spinal cords of 14 of 15 people that had died 3 h to 2 mo after traumatic spinal cord injury (SCI), with apoptosis of oligodendrocytes in the injury center and adjacent white matter tracts being particularly prominent (Emery et al., 1998). Experimental SCI in rodents results in neuronal apoptosis, which can be prevented by glutamate receptor antagonists (Wada et al., 1999); caspase activation occurs in neurons at the injury site within hours, and in oligodendrocytes adjacent to and distant from the injury site over a period of days, after SCI in rats (Springer et al., 1999). SCI-induced apoptosis of oligodendrocytes may involve a progressive inflammation-like process (Crowe et al., 1997), and such white-matter damage may be responsible for the bulk of the deficits observed in SCI patients.

IMPLICATIONS FOR TREATMENT AND PREVENTION

Rapid advances in our understanding of the molecular and cellular underpinnings of neuronal apoptosis have uncovered new therapeutic drug targets. The cell-death process might be blocked at several different levels. For example, disorder-specific apoptotic triggers might be held in check including Aβ production in AD and glutamate receptor activation in stroke. Drugs that target premitochondrial steps might include free-radical scavengers, agents that block calcium influx, or inhibitors of Par-4. At the level of mitochondria, agents such as cyclosporine A and creatine, which suppress mitochondrial oxyradical production and prevent ATP depletion, have proven very effective in animal models (Matsumoto et al., 1999; Albensi et al., 2000; Sullivan et al., 2000). Caspase inhibitors are now being studied intensively to determine their efficacy in animal models of both acute and chronic neurodegenerative disorders. Another general approach for suppressing excitotoxicity and neuronal apoptosis is to administer neurotrophic factors or stimulate production of endogenous neurotrophic factors (Mattson and Lindvall, 1997). A final example is the use of estrogen and other hormones, which have been shown to possess neuroprotective properties (Mattson and Keller, 1998).

Despite the rapid advances in understanding genetic and molecular biological aspects of neurodegenerative disorders, there are as yet no effective treatments for any of the disorders described above. However, the available data suggest that several of the most prevalent age-related neurodegenerative disorders may be preventable. One such preventive strategy is dietary restriction, which involves a reduced calorie intake with maintenance of micronutrient nutrition. Dietary restriction can extend the lifespan of all mammalian species examined, and reduces development of various age-related diseases. Neurons of rodents maintained on dietary restriction are more resistant to apoptosis and exhibit improved symptoms in experimental models of AD, PD, HD, and stroke (Bruce-Keller et al., 1999; Duan and Mattson, 1999; Yu and Mattson, 1999; Zhu et al., 2000). Moreover, epidemiological data indicate that a reduced calorie intake is associated with reduced risks for AD and PD (Logroscino et al., 1996; Mayeux et al., 1999). Dietary restriction may improve the survival and plasticity of neurons by a mechanism involving a preconditioning response, in which the mild metabolic stress associated with reduced energy availability induces neurons to increase their production of stress proteins and neurotrophic factors (Duan and Mattson, 1999; Lee et al., 1999; Yu et al., 1999; Lee et al., 2000). Instillation of new approaches for increasing neuronal resistance to degeneration through such dietary manipulations, as well as through changes in behavior (Ohlsson and Johansson, 1995), will provide an important complement to drugs designed to delay neurodegeneration in patients that are already symptomatic.

CONCLUSIONS

The symptoms of the most prominent of human neurological disorders including AD, PD, HD, stroke, and ALS are the result of neuronal death. Such cell death commonly involves two, often overlapping, biochemical cascades called apoptosis and excitotoxicity. Apoptosis involves oxidative stress, perturbed calcium homeostasis, mitochondrial dysfunction, and activation of cysteine proteases called caspases. Excitotoxicity involves overactivation of ionotropic glutamate receptors resulting in excessive calcium influx, and often occurs under conditions of reduced energy availability and oxidative stress. These death cascades are counteracted by cell-survival signals that inhibit oxyradical production, and stabilize calcium homeostasis and mitochondrial function. Recent studies have identified specific genetic and environmental factors responsible for neuronal apoptosis and excitotoxicity in neurodegenerative diseases. With the discoveries of such disease-initiating factors and specific components of neuronal death and life cascades has come the development of preventative and therapeutic approaches for the neurodegenerative disorders.

REFERENCES

Albensi, B. C. and Mattson, M. P. (2000) Evidence for the involvement of TNF and NF-kappaB in hippocampal synaptic plasticity. *Synapse* **35,** 151–159.

Albensi, B. C., Sullivan, P. G., Thompson, M. B., Scheff, S. W., and Mattson, M. P. (2000) Cyclosporin ameliorates traumatic brain-injury-induced alterations of hippocampal synaptic plasticity. *Exp. Neurol.* **162,** 385–389.

Ankarcrona, M., Dypbukt, J. M., Bonfoco, E., Zhivotovsky, B., Orrenius, S., Lipton, S. A., and Nicotera, P. (1995) Glutamate-induced neuronal death: a succession of necrosis or apoptosis depending on mitochondrial function. *Neuron* **15,** 961–973.

Barger, S. W., Horster, D., Furukawa, K., Goodman, Y., Krieglstein, J., and Mattson, M. P. (1995) Tumor necrosis factors alpha and beta protect neurons against amyloid beta-peptide toxicity: evidence for involvement of a kappa B-binding factor and attenuation of peroxide and Ca2+ accumulation. *Proc. Natl. Acad. Sci. USA* **92,** 9328–9332.

Beal, M. F. (1995) Aging, energy, and oxidative stress in neurodegenerative diseases. *Ann. Neurol.* **38,** 357–366.

Beer, R., Franz, G., Schopf, M., Reindl, M., Zelger, B., Schmutzhard, E., et al. (2000) Expression of Fas and Fas ligand after experimental traumatic brain injury in the rat. *J. Cereb. Blood Flow. Metab.* **20,** 669–677.

Bonventre, J. V., Huang, Z., Taheri, M. R., O'Leary, E., Li, E., Moskowitz, M. A., and Sapirstein, A. (1997) Reduced fertility and postischaemic brain injury in mice deficient in cytosolic phospholipase A2. *Nature* **390,** 622–625.

Brandt, J., Bylsma, F. W., Gross, R., Stine, O. C., Ranen, N., and Ross, C. A. (1996) Trinucleotide repeat length and clinical progression in Huntington's disease. *Neurology* **46,** 527–531.

Bruce-Keller, A. J., Umberger, G., McFall, R., and Mattson, M. P. (1999) Food restriction reduces brain damage and improves behavioral outcome following excitotoxic and metabolic insults [see comments]. *Ann. Neurol.* **45,** 8–15.

Camandola, S. and Mattson, M. P. (2000) Pro-apoptotic action of PAR-4 involves inhibition of NF-kappa B activity and suppression of BCL-2 expression. *J. Neurosci. Res.* **61,** 134–139.

Chan, S. L. and Mattson, M. P. (1999) Caspase and calpain substrates: roles in synaptic plasticity and cell death. *J. Neurosci. Res.* **58,** 167–190.

Choi, D. W. (1992) Excitotoxic cell death. *J. Neurobiol.* **23,** 1261–1276.

Clark, R. S., Kochanek, P. M., Chen, M., Watkins, S. C., Marion, D. W., Chen, J., et al. (1999) Increases in Bcl-2 and cleavage of caspase-1 and caspase-3 in human brain after head injury. *Faseb J.* **13,** 813–821.

Cookson, M. R. and Shaw, P. J. (1999) Oxidative stress and motor neurone disease. *Brain Pathol.* **9,** 165–186.

Crowe, M. J., Bresnahan, J. C., Shuman, S. L., Masters, J. N., and Beattie, M. S. (1997) Apoptosis and delayed degeneration after spinal cord injury in rats and monkeys. [published erratum appears in *Nat. Med.* **3,** 240]. *Nat. Med.* **3,** 73–76.

Dirnagl, U., Iadecola, C., and Moskowitz, M. A. (1999) Pathobiology of ischaemic stroke: an integrated view. *Trends Neurosci.* **22,** 391–397.

Duan, W. and Mattson, M. P. (2000) Participation of Par-4 in the degeneration of striatal neurons induced by metabolic compromise with 3-nitropropionic acid. *Exp. Neurol.* **165,** 1–11.

Duan, W. and Mattson, M. P. (1999) Dietary restriction and 2-deoxyglucose administration improve behavioral outcome and reduce degeneration of dopaminergic neurons in models of Parkinson's disease. *J. Neurosci. Res.* **57,** 195–206.

Duan, W., Rangnekar, V. M., and Mattson, M. P. (1999a) Prostate apoptosis response-4 production in synaptic compartments following apoptotic and excitotoxic insults: evidence for a pivotal role in mitochondrial dysfunction and neuronal degeneration. *J. Neurochem.* **72,** 2312–2322.

Duan, W., Zhang, Z., Gash, D. M., and Mattson, M. P. (1999b) Participation of prostate apoptosis response-4 in degeneration of dopaminergic neurons in models of Parkinson's disease. *Ann. Neurol.* **46,** 587–597.

El-Agnaf, O. M., Jakes, R., Curran, M. D., Middleton, D., Ingenito, R., Bianchi, E., et al. (1998) Aggregates from mutant and wild-type alpha-synuclein proteins and NAC peptide induce apoptotic cell death in human neuroblastoma cells by formation of beta-sheet and amyloid-like filaments. *FEBS Lett.* **440,** 71–75.

Emery, E., Aldana, P., Bunge, M. B., Puckett, W., Srinivasan, A., Keane, R. W., et al. (1998) Apoptosis after traumatic human spinal cord injury. *J. Neurosurg.* **89,** 911–920.

Endres, M., Namura, S., Shimizu-Sasamata, M., Waeber, C., Zhang, L., Gomez-Isla, T., et al. (1998) Attenuation of delayed neuronal death after mild focal ischemia in mice by inhibition of the caspase family. *J. Cereb. Blood Flow Metab.* **18,** 238–247.

Fox, G. B., Fan, L., Levasseur, R. A., and Faden, A. I. (1998) Sustained sensory/motor and cognitive deficits with neuronal apoptosis following controlled cortical impact brain injury in the mouse. *J. Neurotrauma.* **15,** 599–614.

Fu, W., Begley, J. G., Killen, M. W., and Mattson, M. P. (1999) Anti-apoptotic role of telomerase in pheochromocytoma cells. *J. Biol. Chem.* **274,** 7264–7271.

Fu, W., Killen, M., Culmsee, C., Dhar, S., Pandita, T. K., and Mattson, M. P. (2000) The catalytic subunit of telomerase is expressed in developing brain neurons and serves a cell survival-promoting function [In Process Citation]. *J. Mol. Neurosci.* **14,** 3–15.

Furukawa, K., Barger, S. W., Blalock, E. M., and Mattson, M. P. (1996) Activation of K+ channels and suppression of neuronal activity by secreted beta-amyloid-precursor protein. *Nature* **379,** 74–78.

Furukawa, K., Fu, W., Li, Y., Witke, W., Kwiatkowski, D. J., and Mattson, M. P. (1997) The actin-severing protein gelsolin modulates calcium channel and NMDA receptor activities and vulnerability to excitotoxicity in hippocampal neurons. *J. Neurosci.* **17,** 8178–8186.

Gash, D. M., Zhang, Z., Ovadia, A., Cass, W. A., Yi, A., Simmerman, L., et al. (1996) Functional recovery in parkinsonian monkeys treated with GDNF. *Nature* **380,** 252–255.

Glazner, G. W., Chan, S. L., Lu, C., and Mattson, M. P. (2000) Caspase-mediated degradation of AMPA receptor subunits: a mechanism for preventing excitotoxic necrosis and ensuring apoptosis. *J. Neurosci.* **20,** 3641–3649.

Green, D. R. and Reed, J. C. (1998) Mitochondria and apoptosis. *Science* **281,** 1309–1312.

Guo, Q., Fu, W., Sopher, B. L., Miller, M. W., Ware, C. B., Martin, G. M., and Mattson, M. P. (1999a) Increased vulnerability of hippocampal neurons to excitotoxic necrosis in presenilin-1 mutant knock-in mice. *Nat. Med.* **5,** 101–106.

Guo, Q., Fu, W., Xie, J., Luo, H., Sells, S. F., Geddes, J. W., et al. (1998) Par-4 is a mediator of neuronal degeneration associated with the pathogenesis of Alzheimer disease [see comments]. *Nat. Med.* **4,** 957–962.

Guo, Q., Sebastian, L., Sopher, B. L., Miller, M. W., Glazner, G. W., Ware, C. B., et al. (1999b) Neurotrophic factors [activity-dependent neurotrophic factor (ADNF) and basic fibroblast growth factor (bFGF)] interrupt excitotoxic neurodegenerative cascades promoted by a PS1 mutation. *Proc. Natl. Acad. Sci. USA* **96,** 4125–4130.

Guo, Q., Sopher, B. L., Furukawa, K., Pham, D. G., Robinson, N., Martin, G. M., and Mattson, M. P. (1997) Alzheimer's presenilin mutation sensitizes neural cells to apoptosis induced by trophic factor withdrawal and amyloid beta-peptide: involvement of calcium and oxyradicals. *J. Neurosci.* **17,** 4212–4222.

Hagg, T. and Varon, S. (1993) Ciliary neurotrophic factor prevents degeneration of adult rat substantia nigra dopaminergic neurons in vivo. *Proc. Natl. Acad. Sci. USA* **90,** 6315–6319.

Hu, S. C., Chrivia, J., and Ghosh, A. (1999) Regulation of CBP-mediated transcription by neuronal calcium signaling. *Neuron* **22,** 799–808.

Jenner, P. and Olanow, C. W. (1998) Understanding cell death in Parkinson's disease. *Ann. Neurol.* **44,** S72–84.

Keller, J. N., Kindy, M. S., Holtsberg, F. W., St Clair, D. K., Yen, H. C., Germeyer, A., et al. (1998) Mitochondrial manganese superoxide dismutase prevents neural apoptosis and reduces ischemic brain injury: suppression of peroxynitrite production, lipid peroxidation, and mitochondrial dysfunction. *J. Neurosci.* **18,** 687–697.

Kim, M., Lee, H. S., LaForet, G., McIntyre, C., Martin, E. J., Chang, P., et al. (1999) Mutant huntingtin expression in clonal striatal cells: dissociation of inclusion formation and neuronal survival by caspase inhibition. *J. Neurosci.* **19,** 964–973.

Klevenyi, P., Andreassen, O., Ferrante, R. J., Schleicher, J. R., Jr., Friedlander, R. M., and Beal, M. F. (1999) Transgenic mice expressing a dominant negative mutant interleukin-1beta converting enzyme show resistance to MPTP neurotoxicity. *Neuroreport* **10,** 635–638.

Kroemer, G., Dallaporta, B., and Resche-Rigon, M. (1998) The mitochondrial death/life regulator in apoptosis and necrosis. *Annu. Rev. Physiol.* **60,** 619–642.

Kruman, II, Pedersen, W. A., Springer, J. E., and Mattson, M. P. (1999) ALS-linked Cu/Zn-SOD mutation increases vulnerability of motor neurons to excitotoxicity by a mechanism involving increased oxidative stress and perturbed calcium homeostasis. *Exp. Neurol.* **160,** 28–39.

Lee, J., Bruce-Keller, A. J., Kruman, Y., Chan, S., and Mattson, M. P. (1999) 2-Deoxy-D-glucose protects hippocampal neurons against excitotoxic and oxidative injury: involvement of stress proteins. *J. Neurosci. Res.* **57,** 48–61.

Lee, J., Duan, W., Long, J. M., Ingram, D. K., and Mattson, M. P. (2000) Dietary restriction increases survival of newly-generated neural cells and induces BDNF expression in the dentate gyrus of rats. *J. Mol. Neurosci.* **15,** 99–108.

Li, M., Ona, V. O., Guegan, C., Chen, M., Jackson-Lewis, V., Andrews, L. J., et al. (2000) Functional role of caspase-1 and caspase-3 in an ALS transgenic mouse model [see comments]. *Science* **288,** 335–339.

Liu, J. P. (1999) Studies of the molecular mechanisms in the regulation of telomerase activity. *Faseb J.* **13,** 2091–2104.

Logroscino, G., Marder, K., Cote, L., Tang, M. X., Shea, S., and Mayeux, R. (1996) Dietary lipids and antioxidants in Parkinson's disease: a population- based, case-control study. *Ann. Neurol.* **39,** 89–94.

Loo, D. T., Copani, A., Pike, C. J., Whittemore, E. R., Walencewicz, A. J., and Cotman, C. W. (1993) Apoptosis is induced by beta-amyloid in cultured central nervous system neurons. *Proc. Natl. Acad. Sci. USA* **90,** 7951–7955.

Lu, C. and Mattson, M. P. (2000) Caspase-mediated degradation of AMPA receptors subunits: a mechanism for preventing excitotoxic necrosis and ensuring apoptosis. *J. Neurosci.* **20,** 3641–3649..

Mark, R. Murphy, J., Hensley, K., Butterfield, D. A., and Mattson, M. P. (1995) Amyloid beta-peptide impairs ion-motive ATPase activities: evidence for a role in loss of neuronal Ca2+ homeostasis and cell death. *J. Neurosci.* **15,** 6239–6249.

Martin, L. J., Price, A. C., Kaiser, A., Shaikh, A. Y., and Liu, Z. (2000) Mechanisms for neuronal degeneration in amyotrophic lateral sclerosis and in models of motor neuron death (Review). *Int. J. Mol. Med.* **5,** 3–13.

Martinou, J. C., Dubois-Dauphin, M., Staple, J. K., Rodriguez, I., Frankowski, H., Missotten, M., et al. (1994) Overexpression of BCL-2 in transgenic mice protects neurons from naturally occurring cell death and experimental ischemia. *Neuron* **13,** 1017–1030.

Masliah, E., Mallory, M., Alford, M., Tanaka, S., and Hansen, L. A. (1998) Caspase dependent DNA fragmentation might be associated with excitotoxicity in Alzheimer disease. *J. Neuropathol. Exp. Neurol.* **57,** 1041–1052.

Matsumoto, S., Friberg, H., Ferrand-Drake, M., and Wieloch, T. (1999) Blockade of the mitochondrial permeability transition pore diminishes infarct size in the rat after transient middle cerebral artery occlusion. *J. Cereb. Blood Flow Metab.* **19,** 736–741.

Mattson, M. P. (1990) Antigenic changes similar to those seen in neurofibrillary tangles are elicited by glutamate and calcium influx in cultured hippocampal neurons. *Neuron* **4,** 105–117.

Mattson, M. P. (1997) Cellular actions of beta-amyloid precursor protein and its soluble and fibrillogenic derivatives. *Physiol. Rev.* **77,** 1081–1132.

Mattson, M. P. (1998) Modification of ion homeostasis by lipid peroxidation: roles in neuronal degeneration and adaptive plasticity. *Trends Neurosci.* **21,** 53–57.

Mattson, M. P. (2000) Apoptosis in neurodegenerative disorders. *Nature Rev. Mol. Cell Biol.* **1,** 120–129.

Mattson, M. P. and Camandola, S. (2000) NF-kappaB in neuronal plasticity and neurodegenerative disorders. *J. Clin. Invest.* **107,** 247–254 .

Mattson, M. P., Cheng, B., Davis, D., Bryant, K., Lieberburg, I., and Rydel, R. E. (1992) β-amyloid peptides destabilize calcium homeostasis and render human cortical neurons vulnerable to excitotoxicity. *J. Neurosci.* **12,** 376–389.

Mattson, M. P. and Duan, W. (1999) "Apoptotic" biochemical cascades in synaptic compartments: roles in adaptive plasticity and neurodegenerative disorders. *J. Neurosci. Res.* **58,** 152–166.

Mattson, M. P. and Furukawa, K. (1998) Signaling events regulating the neurodevelopmental triad. Glutamate and secreted forms of beta-amyloid precursor protein as examples. *Perspect. Dev. Neurobiol.* **5,** 337–352.

Mattson, M. P. and Keller, J. N. (1998) Neuroprotective actions of estrogen: a preventative and therapeutic approach for Alzheimer's disease and stroke. *Curr. Res. Alzheimer's Dis.* **3,** 236–242.

Mattson, M. P., Keller, J. N., and Begley, J. G. (1998) Evidence for synaptic apoptosis. *Exp. Neurol.* **153,** 35–48.

Mattson, M. P., LaFerla, F. M., Chan, S. L., Leissring, M. A., Shepel, P. N., and Geiger, J. D. (2000a) Calcium signaling in the ER: its role in neuronal plasticity and neurodegenerative disorders. *Trends Neurosci.* **23,** 222–229.

Mattson, M. P. and Lindvall, O. (1997) Neurotrophic factor and cytokine signaling in the aging brain, in *The Aging Brain* (Mattson, M. P. and Geddes, J. W., eds.), JAI Press, Greenwich, CT, pp. 299–345.

Mattson, M. P., Pedersen, W. A., Duan, W., Culmsee, C., and Camandola, S. (1999) Cellular and molecular mechanisms underlying perturbed energy metabolism and neuronal degeneration in Alzheimer's and Parkinson's diseases. *Ann. NY Acad. Sci.* **893,** 154–175.

Mattson, M. P., Zhu, H., Yu, J., and Kindy, M. S. (2000b) Presenilin-1 mutation increases neuronal vulnerability to focal ischemia in vivo and to hypoxia and glucose deprivation in cell culture: involvement of perturbed calcium homeostasis. *J. Neurosci.* **20,** 1358–1364.

Mayeux, R., Costa, R., Bell, K., Merchant, C., Tung, M. X., and Jacobs, D. (1999) Reduced risk of Alzheimer's disease among individuals with low calorie intake. *Neurology* **59,** S296–297.

Middleton, G., Hamanoue, M., Enokido, Y., Wyatt, S., Pennica, D., Jaffray, E., et al. (2000) Cytokine-induced nuclear factor kappa B activation promotes the survival of developing neurons [see comments]. *J. Cell Biol.* **148,** 325–332.

Murakami, K., Kondo, T., Kawase, M., Li, Y., Sato, S., Chen, S. F., and Chan, P. H. (1998) Mitochondrial susceptibility to oxidative stress exacerbates cerebral infarction that follows permanent focal cerebral ischemia in mutant mice with manganese superoxide dismutase deficiency. *J. Neurosci.* **18,** 205–213.

Napieralski, J. A., Raghupathi, R., and McIntosh, T. K. (1999) The tumor-suppressor gene, p53, is induced in injured brain regions following experimental traumatic brain injury. *Brain Res. Mol. Brain Res.* **71,** 78–86.

Ohlsson, A. L. and Johansson, B. B. (1995) Environment influences functional outcome of cerebral infarction in rats. *Stroke* **26,** 644–649.

Oppenheim, R. W. (1991) Cell death during development of the nervous system. *Annu. Rev. Neurosci.* **14,** 453–501.

Pedersen, W. A., Luo, H., Kruman, I., Kasarskis, E., and Mattson, M. P. (2000) The prostate apoptosis response-4 protein participates in motor neuron degeneration in amyotrophic lateral sclerosis. *Faseb J.* **14,** 913–924.

Pellegrini, M. and Strasser, A. (1999) A portrait of the Bcl-2 protein family: life, death, and the whole picture. *J. Clin. Immunol.* **19,** 365–377.

Polymeropoulos, M. H. (1998) Autosomal dominant Parkinson's disease and alpha-synuclein. *Ann. Neurol.* **44,** S63–64.

Rigamonti, D., Bauer, J. H., De-Fraja, C., Conti, L., Sipione, S., Sciorati, C., et al. (2000) Wild-type huntingtin protects from apoptosis upstream of caspase-3. *J. Neurosci.* **20,** 3705–3713.

Sanchez, I., Xu, C. J., Juo, P., Kakizaka, A., Blenis, J., and Yuan, J. (1999) Caspase-8 is required for cell death induced by expanded polyglutamine repeats [see comments]. *Neuron* **22,** 623–633.

Sastry, P. S. and Rao, K. S. (2000) Apoptosis and the nervous system. *J. Neurochem.* **74,** 1–20.

Sawa, A., Wiegand, G. W., Cooper, J., Margolis, R. L., Sharp, A. H., Lawler, J. F., Jr., et al. (1999) Increased apoptosis of Huntington disease lymphoblasts associated with repeat length-dependent mitochondrial depolarization. *Nat. Med.* **5,** 1194–1198.

Schielke, G. P., Yang, G. Y., Shivers, B. D., and Betz, A. L. (1998) Reduced ischemic brain injury in interleukin-1 beta converting enzyme-deficient mice. *J. Cereb. Blood Flow Metab.* **18,** 180–185.

Senut, M. C., Suhr, S. T., Kaspar, B., and Gage, F. H. (2000) Intraneuronal aggregate formation and cell death after viral expression of expanded polyglutamine tracts in the adult rat brain. *J. Neurosci.* **20,** 219–229.

Sinson, G., Perri, B. R., Trojanowski, J. Q., Flamm, E. S., and McIntosh, T. K. (1997) Improvement of cognitive deficits and decreased cholinergic neuronal cell loss and apoptotic cell death following neurotrophin infusion after experimental traumatic brain injury. *J. Neurosurg.* **86,** 511–518.

Smith, R. G., Siklos, L., Alexianu, M. E., Engelhardt, J. I., Mosier, D. R., Colom, L., et al. (1996) Autoimmunity and ALS. *Neurology* **47,** S40–S45; discussion S45–S46.

Springer, J. E., Azbill, R. D., and Knapp, P. E. (1999) Activation of the caspase-3 apoptotic cascade in traumatic spinal cord injury. *Nat. Med.* **5,** 943–946.

Stein-Behrens, B., Mattson, M. P., Chang, I., Yeh, M., and Sapolsky, R. M. (1994) Stress exacerbates neuron loss and cytoskeletal pathology in the hippocampus. *J. Neurosci.* **14,** 5373–5380.

Su, J. H., Anderson, A. J., Cummings, B. J., and Cotman, C. W. (1994) Immunohistochemical evidence for apoptosis in Alzheimer's disease. *Neuroreport* **5,** 2529–2533.

Sullivan, P. G., Geiger, J. D., Mattson, M. P., and Scheff, S. W. (2000) Dietary supplement creatine protects against traumatic brain injury. *Ann. Neurol.* **48,** 723–729.

Suzuki, T., Nishiyama, K., Yamamoto, A., Inazawa, J., Iwaki, T., Yamada, T., et al. (2000) Molecular cloning of a novel apoptosis-related gene, human Nap1 (NCKAP1), and its possible relation to Alzheimer disease. *Genomics* **63,** 246–254.

Tamatani, M., Ogawa, S., Nunez, G., and Tohyama, M. (1998) Growth factors prevent changes in Bcl-2 and Bax expression and neuronal apoptosis induced by nitric oxide. *Cell Death Differ.* **5**, 911–919.

Wada, S., Yone, K., Ishidou, Y., Nagamine, T., Nakahara, S., Niiyama, T., and Sakou, T. (1999) Apoptosis following spinal cord injury in rats and preventative effect of *N*-methyl-D-aspartate receptor antagonist. *J. Neurosurg.* **91**, 98–104.

White, F. A., Keller-Peck, C. R., Knudson, C. M., Korsmeyer, S. J., and Snider, W. D. (1998) Widespread elimination of naturally occurring neuronal death in Bax-deficient mice. *J. Neurosci.* **18**, 1428–1439.

Wong, P. C., Rothstein, J. D., and Price, D. L. (1998) The genetic and molecular mechanisms of motor neuron disease. *Curr. Opin. Neurobiol.* **8**, 791–799.

Wyllie, A. H. (1997) Apoptosis and carcinogenesis. *Eur. J. Cell Biol.* **73**, 189–197.

Yakovlev, A. G., Knoblach, S. M., Fan, L., Fox, G. B., Goodnight, R., and Faden, A. I. (1997) Activation of CPP32-like caspases contributes to neuronal apoptosis and neurological dysfunction after traumatic brain injury. *J. Neurosci.* **17**, 7415–7424.

Yankner, B. A. (1996) Mechanisms of neuronal degeneration in Alzheimer's disease. *Neuron* **16**, 921–932.

Yu, Z. F. and Mattson, M. P. (1999) Dietary restriction and 2-deoxyglucose administration reduce focal ischemic brain damage and improve behavioral outcome: evidence for a preconditioning mechanism. *J. Neurosci. Res.* **57**, 830–839.

Yu, Z., Nikolova-Krakashian, M., Zhou, D., Cheng, G., Schuchman, E. H., and Mattson, M. P. (2000) Pivotal role for acidic sphingomyelinase in cerebral ischemia-induced ceramide and cytokine production, and neuronal death. *J. Mol. Neurosci.* **15**, 85–97 .

Yu, Z., Zhou, D., Bruce-Keller, A. J., Kindy, M. S., and Mattson, M. P. (1999) Lack of the p50 subunit of nuclear factor-kappaB increases the vulnerability of hippocampal neurons to excitotoxic injury. *J. Neurosci.* **19**, 8856–8865.

Yu, Z., Zhou, D., Cheng, G., and Mattson, M. P. (2000) Neuroprotective role for the p50 subunit of NF-kappaB in an experimental model of Huntington's disease. *J. Mol. Neurosci.* **15**, 31–44 .

Zhu, H., Fu, W., and Mattson, M. P. (2000) The catalytic subunit of telomerase protects neurons against amyloid beta-peptide-induced apoptosis. *J. Neurochem.* **75**, 117–124.

2

Oxidative Alterations
in Neurodegenerative Diseases

William R. Markesbery, Thomas J. Montine, and Mark A. Lovell

INTRODUCTION

Age-related neurodegenerative diseases represent a major medical problem for modern society. As longevity of our population increases, these disorders could reach epidemic proportions. The key to understanding age-related neurodegenerative disorders is to determine why neurons degenerate and die in specific brain regions in different disorders. Major research is underway to understand the etiology and pathogenesis of these disorders to facilitate rational development of effective therapies. Numerous partially overlapping hypotheses about the pathogenesis of neurodegenerative diseases include genetic defects, altered membrane metabolism, trace element neurotoxicity, excitotoxicity, reduced energy metabolism, and free-radical-mediated damage. Accumulating evidence indicates that increased free-radical-mediated damage to cellular function contributes to the aging process and age-related neurodegenerative disorders. Indeed, increased free-radical-mediated damage relates closely to the reduced energy metabolism, trace element toxicity, and excitotoxicity hypotheses in neurodegeneration.

Free-radical-mediated damage occurs when free radicals and their products are in excess of antioxidant defense mechanisms, a condition often referred to as oxidative stress. Considerable recent data indicate that oxidative stress may play a role in Alzheimer's disease (AD), Parkinson's disease (PD), amyotrophic lateral sclerosis (ALS), Huntington's disease (HD), and Pick's disease. These diseases share late-life onset and clinical symptoms that relate to region-specific neuron loss in the central nervous system (CNS). Although free-radical damage to neurons may not be the primary event initiating these diseases, it appears that free-radical damage is involved in the pathogenetic cascade of these disorders. The brain is especially vulnerable to free radical damage because of its abundant lipid content, high oxygen consumption rate, and endogenous neurochemical reactions of dopamine oxidation and glutamate excitotoxicity.

A free radical is defined as any atom or molecule with an unpaired electron in its outer shell. Multiple radicals exist, but the most common are formed from the reduction of molecular oxygen to water, and are typically referred to as reactive oxygen species (ROS):

From: *Pathogenesis of Neurodegenerative Disorders* Edited by: M. P. Mattson © Humana Press Inc., Totowa, NJ

$$\begin{array}{ccccccccc} e^- & & e^- & & e^- & & e^- \\ O_2 & \rightarrow & O_2^- & \rightarrow & H_2O_2 & \rightarrow & {}^\bullet OH & \rightarrow & H_2O \\ | & & | & 2H^+ & | & & | & H^+ & | \end{array}$$

| Molecular | Superoxide | Hydrogen | Hydroxyl | Water |
| oxygen | anion | peroxide | radical | |

Single-electron reduction of molecular oxygen forms superoxide anion (O_2^-), a free radical that participates in cellular signaling but does not appear to be reactive with DNA or lipids. A second single-electron reduction converts O_2^- to hydrogen peroxide (H_2O_2), a reaction catalyzed by several forms of superoxide dismutase (SOD). Hydrogen peroxide does not have an unpaired electron and is not a free radical; however, it is an effective oxidant for many biological molecules. The hydroxyl radical (${}^\bullet OH$) is formed from O_2^- and H_2O_2 by the Haber–Weiss reaction:

$$O_2^- + H_2O_2 \rightarrow O_2 + OH^- + {}^\bullet OH$$

${}^\bullet OH$ also can be formed by the Fenton reaction, in which H_2O_2 in the presence of the ferrous ion (Fe^{2+}) forms ${}^\bullet OH$:

$$Fe^{2+} + H_2O_2 \rightarrow Fe^{3+} + OH^- + {}^\bullet OH$$

${}^\bullet OH$ is the most reactive oxygen radical and capable of oxidizing lipids, carbohydrates, proteins, and DNA. Thus, the toxicity of O_2^- and H_2O_2 is primarily due to their conversion to ${}^\bullet OH$.

Nitric oxide ($NO^\bullet$) contains an unpaired electron and is a free radical, which has several physiologic functions including vasodilation. It is synthesized by the enzymatic oxidation of L-arginine to form citrulline through the action of calcium-activated, calmodulin-dependent nitric oxide synthase (NOS). Nitric oxide is produced in excitotoxicity, inflammation, and ischemia-reperfusion injury, and can react with O_2^- to produce peroxynitrite ($ONOO^-$):

$$NO^\bullet + O_2^- \rightarrow ONOO^-$$

Peroxynitrite is a powerful oxidant capable of damaging lipids, proteins, and DNA. It also can form ${}^\bullet OH$ and the nitrogen dioxide radical ($NO_2^\bullet$) as follows:

$$ONOO^- + H^+ \rightarrow {}^\bullet OH + NO_2^\bullet$$

Antioxidants are defined as substances that, when present at low concentrations compared with those of an oxidative substrate, significantly delay or inhibit oxidation of that substrate (Halliwell and Gutteridge, 1989). To defend against free radicals and maintain homeostasis, organisms have developed extensive antioxidant systems and repair enzymes to remove and repair oxidized molecules. Antioxidants have multiple mechanisms of action including preventing initiation of oxidation by radical scavenging, binding or removing catalyzing metal ions, limiting the propagation of the oxidative reaction, and decomposing peroxide (Halliwell and Gutteridge, 1989). Some antioxidants are shown in Table 1. Important enzymatic antioxidants present in the brain include copper–zinc (Cu/Zn)-SOD, manganese (Mn)-SOD, glutathione peroxidase (GSH-Px), and glutathione reductase (GSSG-R). The brain also contains a small amount of catalase (CAT). Numerous other nonenzymatic antioxidants and metal chelators are present in the brain.

Table 1
Antioxidants

Enzymatic	Nonenzymatic
Copper–zinc superoxide dismutase	Ascorbic acid
Manganese superoxide dismutase	Ceruloplasmin
Catalase	Uric acid
Glutathione peroxidase	Bilirubin
Glutathione reductase	Melatonin
Alpha and gamma tocopherol	Glutathione
	Methionine

Antioxidant-defense mechanisms can be upregulated in response to increased free-radical production (Cohen and Werner, 1994). Although upregulation of antioxidant defenses may confer protection, they are not completely effective in preventing oxidative damage, especially with aging of the organism.

SOURCES OF FREE RADICALS

Numerous sources of free radicals are present in the brain but the most common is from oxidative phosphorylation of adenosine 5'-diphosphate (ADP) to adenosine triphosphate (ATP) via the electron transport chain in the inner membranes of mitochondria. ATP is generated through the reduction of molecular oxygen to water by the sequential addition of four electrons and four H^+. Leakage of electrons along the electron transport chain causes O_2^- to form with the potential of forming OH^- via the Fenton reaction. Neurons are highly dependent on oxidative phosphorylation to generate ATP, and because the brain consumes larger amounts of oxygen than other organ, it is more vulnerable to oxidative stress. More active neurons or specific neuron compartments that contain mitochondria, such as synapses, may be particularly vulnerable to oxidative stress through this mechanism.

Excitotoxicity refers to the process by which glutamate and aspartate cause heightened neuronal excitability leading to toxicity and death through a mechanism that includes free-radical formation. Glutamate is the major excitatory neurotransmitter in the brain and glutamate-receptor-mediated excitotoxicity contributes to neuron damage in numerous pathological entities. This process is characterized by excessive influx of calcium into neurons resulting from activation of glutamate receptors, especially the N-methyl-D-aspartate (NMDA) receptor. The increase in calcium causes activation of phospholipase A2, which leads to release of arachidonic acid. The latter generates O_2^- via its metabolism by cyclooxygenases and lipooxygenases to form eicosanoids. In addition, the increase in intracellular calcium activates proteases that catalyze conversion of xanthine dehydrogenase to xanthine oxidase, which in turn catabolizes purine bases to form O_2^-. Thus, diminished energy metabolism can increase intraneuronal calcium, which leads to excitotoxicity, and these converging mechanisms are capable of generating ROS. Both of these mechanisms are thought to occur in neurodegenerative diseases, especially AD (Beal, 1995).

Another source of oxidative stress in the brain is through the enzymatic oxidative deamination of catecholamines by monoamine oxidase (MAO) to yield H_2O_2 (Fig. 1).

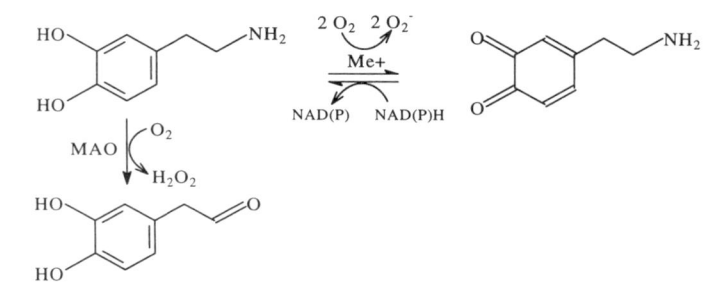

Fig. 1. Dopamine oxidation. Dopamine may be oxidized through both enzyme-dependent and enzyme-independent mechanisms. Enzymatic oxidation of dopamine is catalyzed by monoamine oxidase (MAO) to generate the aldehyde that then is oxidized further to dihydroxyphenylacetic acid. Oxidative deamination by MAO is accompanied by the production of hydrogen peroxide (H_2O_2). Enzyme-independent (autoxidation) of dopamine is catalyzed by paramagnetic metal ions (Me^+) such as iron, copper, or manganese. Autoxidation occurs via sequential one electron oxidation of the catechol nucleus to generate the o-quinone and superoxide anion (O_2^-).

In addition, catecholamines undergo trace-metal-catalyzed autoxidation to generate O_2^- (Cohen and Werner, 1994; Picklo et al., 1999). H_2O_2, generated in catecholaminergic neurons, can be converted by the iron-mediated Fenton chemistry to produce toxic ·OH, which enhances oxidative neuronal damage and possibly contributes to neurodegeneration in PD and AD (Fu et al., 1998).

Microglial cells, the resident macrophages of brain, are capable of generating free radicals when stimulated. Activated microglia, which are markedly increased in the brain in AD (Carpenter et al., 1993), release O_2^- and H_2O_2 in vivo (Colton et al., 1994). Astrocytes and microglia stimulated with cytokines express NOS and generate NO-derived species including $ONOO^-$ (Goodwin et al., 1995; Ii et al., 1996).

Hensley et al. (1994) demonstrated that aggregated amyloid (Aβ)-peptides are capable of generating free radicals and inducing oxidative events. Dyrks et al. (1992) showed that an in vitro iron-catalyzed oxidation system caused transformation of nonaggregated Aβ-peptides into aggregated forms. Aβ-peptides cause H_2O_2 accumulation in cultured hippocampal neurons (Mattson et al., 1995) and antioxidants are capable of preventing experimental Aβ-peptide-induced neuron death in cultured cells (Goodman and Mattson, 1994; Mattson, 1997). Aggregated Aβ-peptides are capable of generating NO· in cultured neural and microglial cells (Goodwin et al., 1995; Keller et al., 1998), which can produce $ONOO^-$. Thus, it is possible that the increase in aggregated Aβ-peptides in the brain in AD may increase free radical production that could play a role in its neurotoxicity.

TRACE ELEMENTS INVOLVED IN OXIDATION

Iron

An important trace element in oxidative reactions is iron (Fe) because it acts as a catalyst of free-radical generation through the Fenton reaction with formation of ·OH (*see* above). Iron, an essential element, is bound to proteins such as hemoglobin and myoglobin or as a nonheme protein-bound complex such as transferrin, ferritin, and hemosiderin. Following absorption from the gastrointestinal tract, Fe is bound to transferrin, which delivers Fe to tissue where it is stored as ferritin. Brain cells have a high-affinity receptor for transferrin.

An early instrumental neutron activation analysis (INAA) study showed that following growth and development, Fe levels remain relatively stable from age 20 to 80 yr in normal brains, after which there is a small decline (Markesbery et al., 1984). In an INAA study of bulk brain specimens from numerous different regions, we found an elevation of Fe in AD compared with age-matched control subjects (reviewed in Markesbery and Ehmann, 1994). Significant elevations of Fe in AD gray matter were found compared with white matter. In a more recent study of seven brain regions of 58 AD and 21 control subjects, we found significant elevations of Fe in the frontal, temporal, and parietal neocortex, hippocampus, and amygdala, but not in the cerebellum (Cornett et al., 1998). Laser microprobe analyses show a significant elevation of Fe in neurofibrillary tangle (NFT)-bearing neurons in the hippocampus in AD (Good et al., 1992b). Using micro particle-induced X-ray emission (micro-PIXE) analysis, we found a significant elevation of Fe in the cores and rims of senile plaques (SP) in the amygdala of AD subjects (Lovell et al., 1998a). Two other studies (Candy et al., 1986; Edwardson et al., 1991) found elevated Fe in NFT-bearing neurons and SP in AD using a microprobe system.

M. Smith et al. (1997a) showed that redox-active Fe is associated with SP and NFT in AD and catalyzes an H_2O_2-dependent oxidation. Redox-active Fe bound to these pathological lesions of AD suggests the potential for generation of free radicals at the expense of cellular reductants. Iron regulatory protein 2 is associated with NFT, SP, and neuropil threads in AD and co-localizes with redox-active Fe, suggesting impaired Fe homeostasis in AD (Smith et al., 1998a).

Ferritin is present in SP in AD (Grundke-Iqbal et al., 1990) and ferritin from AD patients contains more Fe than brains of age-matched controls (Fleming and Joshi, 1987). An increase in heavy-chain isoferritin (H) to light-chain isoferritin (L) ratio is present in the frontal lobe of AD, but not in PD and the H/L isoferritin ratio is higher in caudate/putamen in PD than AD, indicating regional Fe alterations in both disorders (Connor et al., 1995). Transferrin is present around SP in AD (Connor et al., 1992), but transferrin-receptor density is significantly reduced in the hippocampus and neocortex in AD (Kalaria et al., 1992). The C2 allele of transferrin is significantly elevated in the blood of late-onset AD compared with age-matched controls and is twice as high in AD patients homozygous for apolipoprotein ε4 alleles compared with AD patients with one or no copies of the ε4 allele (Namekata et al., 1997). The Fe-binding protein, P97 or melanotransferrin, is elevated in the serum, cerebrospinal fluid (CSF), and brains of AD patients (Jefferies et al., 1996; Kennard et al., 1996).

Iron interaction with Aβ-peptides is of considerable interest. Iron promotes the aggregation of Aβ-peptides in vitro (Mantyh et al., 1993), and may be capable of modulating amyloid precursor protein (APP) processing (Bodovitz et al., 1995). Low Fe decreases soluble APP production and elevated levels increase soluble APP production. High levels of Fe inhibit the maturation of APP production of downstream catabolites. Iron modulation of APP may be at the level of α-secretase cleavage. Fe and lipid peroxidation increase the vulnerability of neurons to Aβ-peptide toxicity (Goodman and Mattson, 1994), further supporting a role for Fe in the pathogenesis of AD.

Altered Fe homeostasis may play a role in dopaminergic neuron loss in PD. Elevated Fe levels were observed in neurons of the substantia nigra and in Lewy bodies using microprobe techniques (Hirsch et al., 1991; Good et al.; 1992a; Jellinger et al., 1992). Aluminum, known to increase lipid peroxidation caused by Fe salts (Gutteridge et al., 1985), is increased in the substantia nigra in PD (Hirsch et al., 1991; Good et al., 1992a). There is an increase in lipid peroxidation in the substantia nigra in PD as noted below. Dexter et al. (1990) showed that ferritin was decreased in the substantia nigra in PD, whereas Riederer et al. (1989) found it increased. Faucheux et al. (1995) observed an increase in lactoferrin receptors in substantia nigra neurons in PD that could be related to the accumulation of Fe within nigral neurons. If free Fe is increased in the substantia nigra in PD, it could enhance free-radical production through catechol autooxidation and Fenton chemistry, and could possibly be important in the pathogenesis of neuron loss.

Copper

Copper is an essential element that plays an important role in many enzymes and modulates numerous regulatory responses in cells. Copper is extremely efficient in generating free radicals owing to its ability to engage in redox reactions. The brain has a high Cu content compared with other organs and its highest level is in gray matter. Copper is bound to numerous enzymes and proteins including Cu/Zn-SOD, cytochrome oxidase, neurocuprein, and ceruloplasmin. Bulk brain studies of Cu show no significant differences in AD and control subjects or a decrease in Cu in AD (Plantin et al., 1987; Tandon et al., 1994; Deibel et al., 1996). Our micro-PIXE study demonstrated a significant increase in Cu in SP in the amygdala in AD (Lovell et al., 1998a). Multhaup et al. (1996) showed that the APP of AD reduces bound Cu^{2+} to Cu^+, which leads to disulfide bond formation in the APP. The reduction of Cu involves an electron-transfer reaction that could enhance the formation of $^{\bullet}OH$. The increase in Cu in SP may relate to the finding that soluble Aβ binds one Cu ion, but the aggregated state binds three Cu ions (Atwood et al., 1998).

Zinc

Zinc is an essential element important in numerous brain enzymes and proteins. It is redox inert and not directly involved in free-radical generation. The brain contains three Zn pools: (1) a membrane-bound metallothionein protein, (2) a pool in synaptic vesicles, and (3) a pool of free or loosely bound Zn in cytoplasm (Frederickson et al., 1989). Zinc is maintained within a relatively narrow range in brain and excess levels are neurotoxic (Cuajungco and Lees, 1997). Our INAA study demonstrated that brain Zn remains relatively constant in the brain throughout adult life (Markesbery et al., 1984). Two INAA studies demonstrated elevated Zn levels in frontal, temporal, and parietal lobes, hippocampus, and amygdala in AD (Deibel et al., 1996; Cornett et al., 1998). A micro-PIXE study showed increased Zn in the rims and cores of SP in AD and in AD neuropil compared with control neuropil (Lovell et al., 1998a). The relationship between Zn and Aβ is of considerable interest. Bush et al. (1994a) demonstrated Aβ specifically and saturably binds to Zn. In vitro concentrations of Zn above 300 nM rapidly destabilized a human $Aβ_{1-40}$ solution and induced aggregation of Aβ fibrils.

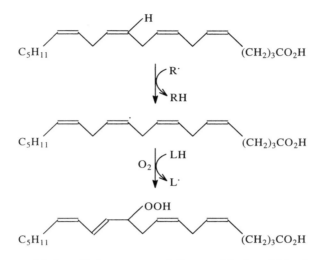

Fig. 2. Lipid peroxidation. Polyunsaturated fatty acids (arachidonic acid shown as an example) begins with hydrogen atom abstraction by a radical (R) to generate a lipid radical that then reacts with O_2 to generate a lipid hydroperoxyl radical (not shown). Propagation of lipid peroxidation occurs when the lipid hydroperoxyl radical abstracts a hydrogen atom from another lipid molecule (LH) to generate a lipid hydroperoxide and another lipid radical (L).

Zinc did not have this effect on rat $A\beta_{1-40}$. The Zn-containing transcription factor NF-κB is one of the regulators of APP synthesis. Zinc binding inhibits the cleavage of APP by α-secretase and inhibits α-secretase cleavage of $A\beta$ (Bush et al., 1994b). Thus, it is possible that elevated Zn may lead to increased levels of transcription factors or influence APP processing (Bush et al., 1994b; Atwood et al., 1999). Hensley et al. (1994) demonstrated aggregation of $A\beta$ has the potential of generating free radicals that can alter membranes and oxidative-sensitive enzymes. This suggests a mechanism by which elevated Zn concentration could contribute to oxidative stress through the accumulation of aggregated $A\beta$.

Overall, changes in Fe, Cu, and Zn could provide a microenvironment in the brain in which excess generation of free radicals could lead to increased lipid, protein, and DNA oxidation and, in conjunction with multiple other factors, contribute to the pathophysiological cascade of neuron injury in neurodegenerative diseases.

LIPID PEROXIDATION

Lipid peroxidation is one of the major outcomes of free-radical-mediated injury to tissue. Peroxidation of fatty acyl groups, mostly in membrane phospholipids, has three phases: initiation, propagation, and termination. Initiation occurs when a hydrogen atom is abstracted from a fatty acyl chain, leaving a carbon-based radical (Fig. 2). Hydrogen atoms can be abstracted by carbon-, nitrogen-, oxygen-, or sulfur-based radicals. Among the oxygen-based radicals, ˙OH is the most reactive at hydrogen atom abstraction. Allylic hydrogens are most labile to abstraction because their carbon–hydrogen bond is made more acidic by the adjacent carbon–carbon double bond. Therefore, polyunsaturated fatty acids are the most vulnerable to lipid peroxidation. The second phase, propagation of lipid peroxidation, begins with reaction of the carbon-based radical on the fatty acyl chain with molecular oxygen to form a hydroperoxyl radical (Fig. 2). These are extremely reactive species that abstract a second hydrogen atom from nearby fatty acyl chains to generate a lipid hydroperoxide and a new carbon-based radical, thus propagating peroxidation. Finally, termination of lipid peroxidation occurs when two radical species react with each other to form a nonradical product. Thus, lipid peroxidation is a self-propagating process that will proceed until the substrate is consumed or termination occurs. Cellular antioxidant systems may intercede by either preventing initiation of lipid peroxidation (e.g., SOD, CAT, or Fe chelators) or limiting propagation (e.g., ascorbate, α-tocopherol, and reduced glutathione).

There are two broad outcomes to lipid peroxidation, viz., structural damage to membranes and generation of bioactive secondary products. Membrane damage derives from the generation of fragmented fatty acyl chains, lipid–lipid crosslinks, and lipid–protein crosslinks (Farber, 1995). In addition, lipid hydroperoxyl radicals can undergo endocyclization to produce novel fatty acid esters that may disrupt membranes. Two classes of cyclized fatty acids are the isoprostanes and neuroprostanes, derived *in situ* from free-radical-mediated peroxidation of arachidonyl or docosahexadonyl esters, respectively (Morrow and Roberts, 1997; Roberts et al., 1998). In total, these processes combine to produce changes in the biophysical properties of membranes that can have profound effects on the activity of membrane-bound proteins.

Fragmentation of lipid hydroperoxides, in addition to producing abnormal fatty acid esters, also liberates a number of diffusible products, some of which are potent electrophiles (Esterbauer et al., 1991; Porter et al., 1995). The most abundant diffusible products of lipid peroxidation are chemically reactive aldehydes such as malondialdehyde, acrolein, 4-hydroxy-2-nonenal (HNE) from ω-6 fatty acyl groups, 4-hydroxy-2-hexenal (HHE) from ω-3 fatty acyl groups, and alkanes (Esterbauer et al., 1991). Alternatively, hydrolysis of abnormal fatty acyl groups generated by lipid peroxidation can liberate abnormal products from damaged lipids. For example, free isoprostanes and neuroprostanes are easily detectable in plasma and CSF (Morrow and Roberts, 1997; Montine et al., 1998b, 1999a, b; Roberts et al., 1998).

Some lipid peroxidation products are thought to contribute to the deleterious effects of lipid peroxidation in tissue. Reactive aldehydes from lipid peroxidation react with a number of cellular nucleophiles, including protein, nucleic acids, and some lipids (Esterbauer et al., 1991). Indeed, many of the cytotoxic effects of lipid peroxidation can be reproduced directly by electrophilic lipid peroxidation products such as HNE (Farber, 1995). These include depletion of glutathione, dysfunction of structural proteins, reduction in enzyme activities, and induction of cell death. Chemically stable products of lipid peroxidation also may contribute to the pathogenesis of lipid peroxidation through receptor-mediated signaling. For example, peroxidation and fragmentation of polyunsaturated fatty acyl groups in phosphatidylcholines can generate platelet-activating-factor analogs that stimulate cellular receptors (McIntyre et al., 1999). Also, at least one isomer of the isoprostanes is a potent vasoconstrictor, likely through a receptor-mediated mechanism (Morrow and Roberts, 1997).

In addition to being potential mediators of tissue damage, products of lipid peroxidation are commonly used to quantify the extent of lipid peroxidation. When considering the quantification of lipid peroxidation, it is necessary to define whether the assay is being applied in vitro or in vivo. Assays such as those for thiobarbituric-reactive substances (TBARS) or chromatography for specific secondary products are accurate measures of lipid peroxidation in vitro when metabolism of the lipid peroxidation products does not occur. However, in more complicated model systems with metabolic activity and in vivo, extensive metabolism of electrophilic lipid peroxidation products compromises the accuracy of these assays (Gutteridge and Halliwell, 1990; Moore and Roberts, 1998). One solution to the problem of accurately quantifying lipid peroxidation in vivo is to measure one class of isoprostanes, the F_2-isoprostanes. F_2-isoprostanes are chemically stable products of free-radical-mediated damage to arachidonyl esters that are not extensively metabolized *in situ* (Morrow and Roberts, 1997).

Lipid Peroxidation in Neurodegenerative Diseases

Alzheimer's Disease

There is compelling evidence that the magnitude of lipid peroxidation in the brains of AD patients examined postmortem exceeds that in age-matched control individuals. Seminal experiments demonstrated significantly increased TBARS in diseased regions of AD brain obtained postmortem compared with age-matched control individuals (Lovell et al., 1995). Other studies measured free HNE and acrolein in AD brain tissue and showed that both are elevated in diseased regions of AD brain compared with controls (Markesbery and Lovell, 1998; Lovell et al., 2000a).

F_2-isoprostane levels are elevated in the frontal lobe and hippocampus of AD patients compared with controls with short postmortem intervals (Pratico et al., 1998; Montine et al., 1999a). In addition, F_2-isoprostanes are elevated in the cerebral cortex of aged homozygous apolipoprotein E (apoE) gene deficient mice (Montine et al., 1999c; Pratico et al., 1999). A class of free-radical-generated products analogous to the F_2-isoprostanes, but generated from docosahexenoic rather than arachidonic acid, has been described and termed F_4-neuroprostanes (Roberts et al., 1998). Because docosahexenoic acid is more labile to peroxidation than arachidonic acid and docosahexenoic acid is relatively enriched in brain, it was proposed that F_4-neuro-prostanes might be more sensitive markers of brain oxidative damage than F_2-iso-prostanes. Indeed, F_4-neuroprostanes are significantly more abundant than F_2-isoprostanes in cerebral cortex of aged homozygous apolipoprotein E gene deficient mice (Montine et al., 1999c). One group reported that F_4-neuroprostanes (called F_4-isoprostanes in their publication) are elevated in temporal and occipital lobes, but not parietal lobe of AD patients compared with controls, and that F_4-neuroprostane levels are higher than F_2-isoprostanes in these regions (Nourooz-Zadeh et al., 1999). However, interpretation of data from this study is limited by excessively long postmortem intervals (47 h average in AD patients) (Nourooz-Zadeh et al., 1999).

In contrast to quantification, several groups have studied the localization of lipid peroxidation products in AD brain. These studies used immunochemical detection of protein covalently modified by lipid peroxidation products or displaying protein carbo-nyls. There is broad agreement among these studies. Consistent with the quantitative studies described above, hippocampus and cerebral cortex from AD patients display protein modifications that are not detectable or are barely detectable in the corresponding brain regions from age-matched control individuals (Sayre et al., 1997; Montine et al., 1997a, b, 1998a; Calingasan et al., 1999; Smith et al., 1998b). Also, in AD patients, proteins modified by lipid peroxidation products are present in diseased regions of brain but not in uninvolved regions. In diseased regions of AD brain, neuronal cytoplasm and NFT are the major focus of protein modification. Importantly, none of these studies observed modified proteins in or adjacent to neuritic plaques. This stands in sharp contrast to genetically modified mice expressing mutant human APP, where increased HNE-protein adduct immunoreactivity and advanced glycation end product immunoreactivity (*vida infra*) are localized adjacent to or within amyloid deposits (Smith et al., 1998c). One group observed that the tissue distribution of HNE-protein adducts varies with apoE genotype. In these studies, one chemical form of HNE-protein adducts, the 2-pentylpyrrole adduct, co-localized with NFT and was significantly associated with homozygosity for the ε4 allele of apoE (Montine et al., 1997 a, b). Michael adducts, the most abundant chemical form of HNE-protein adducts, were observed in pyramidal neuron and astrocyte cytoplasm of AD patients with an ε3 allele of apoE, but only in pyramidal neuron cytoplasm of AD patients homozygous for ε4 allele of apoE (Montine et al., 1998a). These authors suggested that the tissue distribution, but not the apparent quantity, of HNE-protein adducts in AD is influenced by apoE genotype, perhaps related to the essential role of apolipoprotein E in CNS lipid trafficking. Recent findings suggest that apoE4 promotes, whereas apoE3 and apoE2 suppress, oxidative damage to neurons. The three apoE isoforms differ in that E2 contains two cysteine residues, E3 contains two cysteines, and E4 lacks the cysteines. Apolipoprotein E2, and to a lesser extent E3, can protect neurons against oxidative insults that induce lipid peroxidation, whereas E4 is ineffective (Pedersen et al., 2000). The neuroprotective effect of E2 is correlated with an increased HNE-binding capacity of the protein. Thus, apolipoprotein E genotype appears to affect risk for AD by modifying the antioxidant capacity (specifically, the ability to bind HNE) of the protein.

More recently, CSF has been investigated as a source of CNS tissue, for the assessment of lipid peroxidation, in AD brain. One study measured free HNE in CSF obtained from the lateral ventricles postmortem and showed that its concentration is significantly elevated in AD patients compared with age-matched controls (Lovell et al., 1997). Two studies determined the concentration of F_2-isoprostanes, and in one case F_4-neuroprostanes, in CSF obtained from the lateral ventricles postmortem and showed significant elevations in AD patients compared with age-matched controls (Montine et al., 1999a, b). Importantly, CSF F_2-isoprostane concentrations in AD patients are significantly correlated with decreasing brain weight, degree of cerebral cortical atrophy, and increasing Braak stage, but not with apoE genotype or the tissue density of neuritic plaques or NFT (Montine et al., 1999b). It is noteworthy that all of the postmortem CSF studies in AD patients used material from individuals with very short (average 2–3 h) postmortem intervals.

The aforementioned studies of brain tissue and CSF were performed using material collected postmortem. AD patients undergoing postmortem examination typically have advanced disease and an average duration of dementia of 8–12 yr. Therefore, a serious limitation to analysis of tissue obtained postmortem is that the increased brain lipid peroxidation in AD patients might be a consequence of late-stage disease. Obviously, a late-stage consequence of AD would be a less attractive therapeutic target than a process contributing to disease progression at an earlier stage. A recent study of probable AD patients early in the course of dementia showed that F_2-isoprostanes are significantly elevated in CSF obtained from the lumbar cistern compared with age-matched hospitalized patients without neurological disease (Montine et al., 1999a). The average duration of dementia in these probable AD patients was less than 2 yr. In combination with the postmortem studies, this recent study with probable AD patients demonstrates that brain lipid peroxidation is elevated both early and late in the course of AD, and provides a rationale for slowing the progression of AD by suppressing brain lipid peroxidation (Sano et al., 1997).

Data obtained from studies of experimental animal and cell-culture models of AD suggest that Aβ-peptide and Fe may be important initiators of lipid peroxidation and neuronal degeneration (Mattson, 1998). Exposure of cultured neurons and synapses to Aβ-peptide results in membrane lipid peroxidation and HNE production. HNE renders neurons vulnerable to apoptosis and excitotoxicity by covalently modifying, and impairing the function of, membrane ion-motive ATPases, glucose transporters, and glutamate transporters (Mark et al., 1995, 1997a, b; Keller et al., 1997). Conversely, HNE enhances calcium influx through membrane glutamate receptor channels and voltage-dependent channels. Aβ-peptide, lipid peroxidation, and HNE can alter neurotransmitter and neurotrophic factor signal transduction pathways involving GTP-binding proteins and transcription factors, which may contribute to cognitive dysfunction in AD (Kelly et al., 1996; Blanc et al., 1997). Recent findings suggest that lipid peroxidation can be suppressed, and AD prevented or delayed, by dietary and behavioral manipulations that either directly suppress lipid peroxidation or activate neurotrophic factor- and stress-responsive signaling pathways (Bruce-Keller et al., 1999; Zhu et al., 1999).

Parkinson's Disease

Idiopathic PD is the second most common neurodegenerative disease. Similar to AD, there is compelling evidence from human postmortem tissue that increased oxidative stress occurs in the midbrain of patients with PD compared with age-matched controls (Coyle and Puttfarcken, 1993; Cohen and Werner, 1994; Jenner and Olanow, 1998). However, the mode of oxidative stress to brain may differ somewhat from AD. For example, depletion of nigral reduced glutathione is proposed to be a relatively early and specific event in PD (Jenner, 1994). Lipid peroxidation appears to be a component of nigral degeneration in PD when examined postmortem. Compared with controls, PD patients have elevated TBARS and lipid hydroperoxides in midbrain, immunochemically detectable HNE-protein adducts in midbrain, and elevated free HNE in CSF (Dexter et al., 1989, 1994; Yoritaka et al., 1996; Shelley, 1998). Interestingly, HNE-protein adducts were present in several midbrain nuclei, and not just in the substantia nigra of PD patients (Yoritaka et al., 1996), a pattern similar to 8-hydroxyguanosine immunoreactivity in midbrain from PD patients (Zhang et al., 1999). Despite the postmortem evidence that associates PD with increased midbrain lipid peroxidation, pharmacologic studies have questioned the significance of lipid peroxidation earlier in the course of PD (Parkinson Study Group, 1989).

In humans and some mammals including mice, exposure to 1-methyl-4-phenyltetrahydropyridine (MPTP) produces selective degeneration of dopaminergic neurons in the CNS. MPTP-induced nigral degeneration in animals has been used widely as a model of PD (Langston, 1994). MPTP-induced dopaminergic degeneration is thought to involve mitochondrial dysfunction and oxidative damage. Indeed, mice lacking both alleles of the Cu/Zn-SOD1 or the glutathione peroxidase gene (*GPX1*) are significantly more vulnerable to MPTP-induced dopaminergic neurodegeneration than littermate controls (Klivenyi et al., 2000; Zhang et al., 2000). Moreover, the brainstem of mice systemically exposed to MPTP show an acute eightfold increase in HNE concentrations, a 50% reduction in reduced glutathione levels, and a sixfold increase in the concentration of HNE-glutathione adducts within 24 h of exposure (Shelley, 1998).

In summary, these data suggest that some lipid-peroxidation products may participate in the pathogenesis of MPTP-induced dopaminergic neurodegeneration and are elevated in the substantia nigra in late stage PD. The role of nigral lipid peroxidation in earlier stages of PD is not clear.

Amyotrophic Lateral Sclerosis

Similar to AD and PD, ALS has both familial and sporadic forms. Research into the role of oxidative damage in ALS has been fueled by the discovery that mutations in the gene for Cu/Zn-SOD1 are the cause of a subset of familial ALS (Rosen et al., 1993). Indeed, some lines of mice expressing SOD1 with mutations linked to ALS develop a disease phenotype that closely mimics familial ALS and is thought to derive in part from increased free-radical-mediated damage (Dal Canto and Gurney, 1994; Gurney et al., 1994; Tu et al., 1997). Data on postmortem human tissue from patients with ALS suggest that the role of lipid peroxidation in this disease may be complex. Postmortem examination of patients with sporadic ALS showed increased protein carbonyl formation and other signatures of oxidative damage, but not elevated TBARS, in motor cortex and immunoreactivity for malondialdehyde protein adducts in spinal cord (Ferrante et al., 1997). Familial ALS patients, both with and without mutations in SOD1, do not have motor-cortex changes but have spinal-cord changes similar to sporadic ALS patients (Ferrante et al., 1997). Others showed that lumbar spinal cord obtained postmortem from sporadic ALS patients has immunohistochemically detectable HNE-protein adducts in the anterior horn (Pedersen et al., 1998). These investigators also showed by immunoprecipitation that one of the modified proteins is an excitatory amino acid transporter (EAAT2), a molecule whose dysfunction has been implicated in sporadic ALS (Browne and Beal, 1994).

Two studies examined CSF from living ALS patients for evidence of increased lipid peroxidation. One study measured free HNE in CSF obtained from the lumbar cistern of sporadic ALS patients at initial diagnosis and before therapy (Smith et al., 1998). HNE was measured using high-performance liquid chromatography (HPLC) with fluorescence detection following derivatization of CSF, an analytically less rigorous approach than chromatography followed by mass spectrometry, the method used to quantify free HNE in CSF from PD patients. Nevertheless, this study demonstrated a significant elevation in free HNE levels in CSF from sporadic ALS patients compared to patients with several other neurodegenerative diseases, but not when compared to patients with Guillain-Barré syndrome or chronic inflammatory demyelinating polyneuropathy. CSF levels of free HHE were not elevated in the same CSF samples (Smith et al., 1998). HHE is generated by the same chemistry as HNE except ω-3 fatty acids are the substrates for HHE and ω-6 fatty acids are the substrates for HNE. In a separate, smaller study, F_2-isoprostanes in CSF from sporadic ALS patients were not significantly different from age-matched controls (Montine et al., 1999a). The average disease duration in this group of ALS patients was approx 2 yr and many of the patients had already initiated therapy. In summary, data from postmortem studies consistently associate ALS with increased lipid peroxidation in spinal cord and suggest a mechanism whereby lipid-peroxidation products could contribute to disease progression. Moreover, a closely related animal model of familial ALS indicates that free-radical-mediated damage may contribute to disease progression. Increased oxidative stress and increased vulnerability to excitotoxicity have been documented in studies of motor neurons from Cu/Zn-SOD mutant ALS mice (Kruman et al., 1999). In addition, Fe and HNE have been shown to impair glucose and glutamate transport in cultured motor neurons, suggesting a role for lipid peroxidation in the deficit in glutamate transport and energy metabolism documented in studies of ALS patients (Pedersen et al., 1999). Finally, some but not all data from CSF support a role for increased lipid peroxidation early in the course of sporadic ALS.

Huntington's Disease

HD is an autosomal dominant disease in which striatal degeneration is proposed to derive significant contributions from impaired energy metabolism, excitotoxicity, and increased oxidative stress (Beal, 1995). Postmortem tissue studies of striatum from HD patients have shown increased indices of a number of oxidative damage markers, including TBARS (Beal, 1995). Moreover, transgenic mouse models of HD have increased levels of free-radical production (Browne et al., 1999). A possible role for lipid peroxidation in HD was recently highlighted by a study showing elevated F_2-isoprostanes in CSF obtained from the lumbar cistern of HD patients compared to patients with multiple system atrophy-parkinsonian type (striatonigral degeneration) or hospitalized controls without neurological disease (Montine et al., 1999b).

PROTEIN OXIDATION

Free radicals can attack amino acids and lead to damage and inactivation of enzymes or receptors as well as cause protein–protein crosslinking. Oxidative modifications of protein residues in neurons are mediated by a variety of systems including lipid peroxidation, as noted above. Oxidation of amino acid residues in protein can have deleterious effects on functional properties of cell homeostatic balance. Oxidative damage to enzymes critical to brain function can cause neuron degeneration and result in accelerated brain aging (Stadtman, 1992). Stadtman described a two-step process in which the first step is oxidation of enzyme amino acids by free radicals yielding carbonyl derivatives. The second step involves further degradation of the enzyme by proteases to amino acids and peptides.

Protein Oxidation in Neurodegenerative Disease

Alzheimer's Disease

Protein carbonyl formation, a measure of oxidative damage to proteins can be quantified by derivatization with 2,4-dinitrophenylhydrazide (DNP). An early study demonstrated that there is increased brain protein carbonyls in aging and AD (Smith et al., 1991). Subsequently, other studies demonstrated an increase in protein carbonyls in the brain in AD (Hensley et al., 1995; Gabbita et al., 1999), and in Pick's disease and dementia with Lewy bodies (Aksenova et al., 1999). Smith et al. (1996), using immu-nocytochemical techniques with *in situ* DNP labeling linked to an antibody system against DNP, demonstrated the presence of protein carbonyls in NFT and glia, but not in NFT-free neurons in AD. These changes were not found in control brains and are similar to what has been observed for protein adducts from lipid-peroxidation products.

The enzymes glutamine synthetase (GS) and creatine kinase (CK) are especially sensitive to oxidative modification (Stadtman, 1992). Two studies reported a signifi-cant decline in glia-specific GS activity in the hippocampus and neocortex in AD com-pared with age-matched controls (Smith et al., 1991; Hensley et al., 1995). Decreased levels of GS could result in diminished glutamate turnover causing prolonged NMDA receptor activation and neuron injury in brain areas susceptible to glutamate toxicity. In addition, because glutamate is converted to glutamine, loss of the enzyme could alter nitrogen balance, pH, and glutathione synthesis in astrocytes.

Creatine kinase BB is a member of the CK gene family and the predominant cytoso-lic CK isoform in the brain. Creatine kinases are a family of enzymes that catalyze a reversible transfer of a phosphoryl group between ATP and creatine. Cells respond to external challenges by mobilizing the creatine phosphate (CrP)/CK system (Struzynska et al., 1997) and increasing CK BB expression (Aksenov et al., 1998). Activation of the CrP/CK system and changes in CK expression may be an early indicator of oxidative and bioenergetic stress. Hensley et al. (1995) demonstrated that decreased CK BB activity in brain correlates well with the presence of NFT in severely affected regions. Aksenov et al. (1997) and Aksenova et al. (1999) demonstrated that CK BB is decreased in several neurodegenerative diseases including AD, Pick's disease, and dementia with Lewy bodies, but is decreased most in AD. These studies demonstrated that the CK BB decline is not a result of gene expression, which suggests that posttranslational oxida-tive modification of the enzyme contributes to the loss of CK BB activity (Aksenov et al., 1997, 2000a). A study using two-dimensional fingerprinting of oxidatively modified protein demonstrated decreased specific activity of CK BB that correlated with an increase in protein carbonyl content in the enzyme (Aksenov et al., 2000a). This sug-gests that introduction of carbonyl groups into amino acid residues in the active site of CK BB by oxidation or reaction with reactive aldehydes participates in the inactivation of the enzyme. A subsequent two-dimensional fingerprinting study demonstrated sig-nificantly elevated protein carbonyls in β-actin and CK BB in AD, but insignificant elevations of β-tubulin and undetectable protein carbonyl immunoreactivity in tau isoforms and glial fibrillary acidic proteins (Aksenov et al., 2000b). David et al. (1998) described a correlation between reduced CK activity and decreased ATP binding to CK BB in the brain in AD. The ATP-binding domain of CK BB contains arginine, histi-dine, and lysine, amino acid residues that may be converted into carbonyl derivatives by free radicals and modified by lipid peroxidation products. Oxidative inactivation of CK BB may involve the modification of these amino acid residues in the ATP-binding site of CK BB.

Methionine, cysteine, and tryptophan are also amino acid vulnerable to oxidation. The principal product of methionine oxidation is methionine sulfoxide (MetSO). Sev-eral studies have demonstrated that proteins lose their biological activity when specific methionine residues are oxidized to MetSO (Brot and Weissbach, 1983; Swaim and Pizzo, 1988; Vogt, 1995). Levine et al. (1996) suggested that methionine residues act as a last line of antioxidant defense in proteins. High concentrations of methionine in proteins allow effective scavenging of oxidants to form MetSO before an attack on residues critical to structure or function. The enzyme responsible for reduction of MetSO back to methionine is methionine sulfoxide reductase (MsrA). Most biological systems contain disulfide reductases and MetSO reductases that can convert the oxidized forms of cysteine and methionine residues to their unmodified forms. According to Berlett and Stadtman (1997), methionine and cysteine oxidative modification products are the only residues that can be repaired within protein. Moskovitz et al. (1998) found high levels of MetSO in yeast MsrA mutants when exposed to oxidative stress compared with wild-type strain, which indicates MsrA possesses an antioxidant function. This suggests that MsrA could have an important role in providing cells with a defense system against oxidative stress. A recent study showed MsrA activity in the brain of AD patients was diminished in all regions studied and reached statistical significance in the superior and middle temporal gyri, inferior parietal lobule, and hippocampus (Gabbita et al., 1999). Messenger RNA analysis suggested that the loss in enzyme activity may be the result of a posttranslational modification of MsrA or defective translation resulting in inferior processing of the mRNA. This study suggests that a decline in MsrA activity could reduce the antioxidant defenses and increase the oxidation of critical proteins in neurons in the brain in AD.

Peroxynitrite causes nitration of tyrosine residues that yield nitrotyrosine, which is used as an indicator of $ONOO^-$ activity. Investigators using immunohistochemistry found nitrotyrosine present in NFT in the hippocampus in AD (Good et al., 1996; Smith et al., 1997b). Hensley et al. (1998) demonstrated significant elevations of protein-bound 3-nitrotyrosine and 3,3'-dityrosine in the hippocampus, neocortex, and ventricular CSF of AD subjects compared with normal control subjects. These studies indicate that nitric oxide and its redox congeners, especially $ONOO^-$, are likely involved in protein oxidation in AD.

Parkinson's Disease

Alam et al. (1997b) described a generalized increase in protein carbonyls in the brain in PD, but not in brain of patients with incidental Lewy body disease (putative presymptomatic PD), suggesting that oxidative protein damage occurs late in PD or that L-Dopa treatment contributes to protein oxidation. Good et al. (1998) demonstrated the presence of nitrotyrosine immunoreactivity in Lewy bodies and in amorphous deposits in intact and degenerating neurons of the substantia nigra in autopsied PD patients, indicating that oxidative modification of proteins has occurred in the target cells of this disorder. A relatively selective inhibitor of the neuronal isoform of nitric oxide synthase, 7-nitroindazole, protects against MPTP-induced dopamine depletion in mice (Schulz et al., 1995) and against dopamine depletion and loss of tyrosine hydroxylase-positive neurons in the substantia nigra of MPTP-treated baboons (Hantraye et al., 1996). Knockout mice deficient in the neuronal isoform of NOS are resistant to MPTP neurotoxicity (Przedborski et al., 1996). These studies suggest that $NO^•$ may play a role in PD and an experimental model of PD.

GLYCATION

Analogous to protein adduction by lipid-peroxidation products, nonenzyme-cata-lyzed posttranslational modification of proteins by reducing sugars is associated also with some neurodegenerative diseases. This process, termed glycation, is initiated by reversible Schiff-base formation between a protein-bound amino group and an aldose that undergoes Amadori rearrangement to regenerate carbonyl activity. Subsequent irreversible rearrangements, fragmentations, dehydrations, and condensations yield a complex mixture of protein-bound products termed advanced glycation endproducts (AGEs). The chemical structures of AGEs have been partially characterized; the best studied are pentosidine, pyrraline, and *N*-(carboxymethyl)lysine (CML). Both the reversible adduct formation and subsequent evolution to AGEs are accelerated by oxy-gen in a process called glycoxidation (Smith et al., 1995).

AGEs have been studied most extensively from the perspective of diabetes mellitus and its complications where their accumulation is correlated with the degree of hyper-glycemia. However, accumulation of AGEs in tissue is also associated with advancing age and with some disorders that are not characterized by hyperglycemia but are asso-ciated with oxidative stress, e.g., uremia and AD. Moreover, molecules other than reducing sugars can lead to AGE formation under conditions of increased oxidative stress. For example, other carbohydrates and even ascorbate can lead to AGE forma-tion (Dunn et al., 1990; Dyer et al., 1991; Grandhee and Monnier, 1991).

Like lipid-peroxidation products, the biological consequences of AGEs may be viewed as deriving either from biochemical reactions or from receptor-mediated pro-cesses. AGEs can damage structural proteins and enzymes, thus rendering them dys-functional. For example, enzymatic activity of human Cu/Zn-SOD is inactivated by glycation (Arai et al., 1987). AGEs can also be redox active, generate oxidative stress themselves, and induce neuronal apoptosis (Yan et al., 1994; Kikuchi et al., 1999). In addition to these deleterious biochemical reactions, AGE-modified proteins may bind to a receptor (RAGE) that exists on several cell types including neurons and glia. One role of RAGE is thought to be incorporation of AGE-modified proteins for degrada-tion. However, binding of ligands, including Aβ-peptide, to RAGE on neurons in cul-ture stimulates production of reduced oxygen species, and binding to RAGE on microglia in culture leads to cellular activation (Du Yan et al., 1997).

Glycoxidation in Neurodegenerative Diseases

Alzheimer's Disease

AGE formation in neurodegenerative diseases has been investigated most exten-sively in AD. These studies have focused on immunohistochemical localization of AGEs in brain. The results are not as consistent as those for localizing lipid-peroxidation products. Antibodies to pentosidine or pyrraline protein adducts are immunoreactive with neuritic and diffuse SP as well as intracellular and extracellular NFT in the hippocampus of AD patients (Smith et al., 1994). This study also observed immunoreactivity with both antibodies to vessel walls. The same pattern of immuno-reactivity is present in hippocampal sections from control individuals as well as patients with diabetes; however, the extent of staining is much less than in AD because SP and NFT densities were much lower (Smith et al., 1994; Sasaki et al., 1998). Using differ-ent antibodies to AGEs, others observed a similar pattern of immunoreactivity in AD brain (Dickson et al., 1996; Munch et al., 1998; Sasaki et al., 1998). One group noted that intracellular NFT are more strongly immunoreactive than extracellular NFT and that granulovacuolar degeneration is immunoreactive for AGEs (Sasaki et al., 1998). Others showed that Hirano bodies are immunoreactive for AGEs (Munch et al., 1998). Diffuse neuron cytoplasmic immunoreactivity was not reported in these studies. Other groups using antibodies directed at CML demonstrated diffuse pyramidal neuron cytoplasmic immunoreactivity that apparently increases with age and is present in hippocampal sections from AD patients (Kimura et al., 1996; Takedo et al., 1996). One of these studies noted that while CML immunoreactivity in hippocampus from AD patients is also present extracellularly, CML immunoreactivity is not co-localized with SP (Takedo et al., 1996).

Other Neurodegenerative Diseases

Investigation of AGEs in neurodegenerative diseases other than AD has been limited. Pentosidine and pyrraline immunoreactivity has been reported in the substantia nigra and locus ceruleus of patients with PD, and in the frontal and temporal cortex, but not hippocampus, of patients with dementia with Lewy bodies (Castellani et al., 1996). Staining was localized to the periphery of some Lewy bodies. Another study detected AGEs on intraneuronal hyaline inclusions of spinal cord from patients with ALS linked to mutations in SOD1 and in mice that express mutant human SOD1 (Shibata et al., 1999). AGE immunoreactivity has been demonstrated in NFT from patients with progressive supranuclear palsy, Guamanian parkinsonism–dementia complex, and Guamanian ALS (Sasaki et al., 1998). Finally, AGE immunoreactivity has been observed with Pick bodies and granulovacuolar degeneration in temporal lobe sections of patients with Pick's disease (Sasaki et al., 1998).

DNA OXIDATION

Accumulated damage to DNA in nondividing mammalian cells may play a role in aging and age-associated diseases. Oxidation of DNA causes strand breaks, sister chromatid exchange, DNA–DNA and DNA–protein crosslinking and base modification. Defects in DNA synthesis can produce mutant genes that lead to altered proteins. These damaged proteins can alter functions and result in functional defects and eventually cell death.

Multiple oxidants of DNA exist including $^{\bullet}OH$, $ONOO^-$, and singlet oxygen. Other species, O_2^- and H_2O_2 are linked to DNA damage by the Fe-mediated Fenton reaction. Secondary effects of oxidation can lead to DNA damage through aldehyde breakdown products of lipid peroxidation that form aldehyde–DNA adducts (Douki and Ames, 1994; Ames et al., 1995). Multiple oxidative DNA adducts have been identified but the most thoroughly studied is the adduct involving the C-8 hydroxylation of guanine, 8-hydroxy-2-deoxyguanosine (8-OHdG), which has become the most popular method for monitoring DNA oxidation in vivo (Beckman and Ames, 1997).

Cells respond to DNA alterations by repairing the damage and restoring the physical and functional state of the genotype. Three modes of DNA repair are: (1) base excision repair, (2) nucleotide excision repair, and (3) mismatch repair (Friedberg and Wood, 1996). Base excision repair is carried out by a group of enzymes termed DNA glycosylases that recognize the damaged base and cleaves its glycosylic bond. Nucleotide excision repair removes damaged oligonucleotide fragments via large enzyme complexes. Defective DNA repair may be of considerable importance in aging and age-associated neurodegenerative disorders.

DNA Oxidation in Neurodegenerative Disease

Alzheimer's Disease

The study of oxidative DNA damage in the brain is in its infancy and most of the reports are from aging, AD, or PD studies. Mullaart et al. (1990) described a twofold increase in DNA strand breaks in the brain in AD. Mecocci et al. (1993) reported an increase in 8-OHdG in nuclear and mitochondrial brain fractions in aging. Using HPLC, they demonstrated a 10-fold increase in brain mitochondrial DNA oxidation compared with brain nuclear DNA oxidation and a 15-fold increase in brain DNA oxidation in subjects older than 70 yr of age. These same investigators demonstrated a threefold increase in mitochondrial DNA oxidation in the parietal lobe in AD (Mecocci et al., 1994) and a small but significant increase in oxidative damage to brain nuclear DNA.

More recent studies of DNA oxidation have used gas chromatography with mass spectroscopy (GC-MS), a highly sensitive method for identifying oxidative adducts from DNA bases. Lyras et al. (1997), using GC-MS, found various bases increased or decreased in total brain DNA in different brain regions in AD. The most consistent elevations were in 8-OHdG, 8-hydroxyadenine and 5-hydroxycytosine in the parietal lobe in AD. Using GC-MS with stable isotope-labeled oxidized base analogs for standards, we studied nuclear DNA from four brain regions in AD patients and prospectively evaluated control subjects, all with short postmortem intervals (Gabbita et al., 1998). This study demonstrated statistically significant elevations of 8-OHdG, 5-hydroxyuracil, and 8-hydroxyadenine in frontal, parietal, and temporal lobes, and 5-hydroxycytosine in the parietal and temporal lobes in AD. The increases in mean 8-OHdG were the largest elevations of all the base adducts analyzed, indicating that guanine is the most vulnerable base to oxidation. The pattern of damage to multiple bases in the brain suggests that this is due to •OH attack on DNA. Wade et al. (1998) used immunocytochemistry to show that 8-OHdG was increased in mitochondria in neurons in AD compared with controls. The same investigators showed that the 5 kb deletion, the most common alteration in human mitochondria, was prominent in large hippocampal pyramidal neurons in AD (Hirai et al., 1998). This deletion was present in neurons following immunostaining for 8-OHdG. These authors concluded that oxidative modification of mitochondrial DNA is an early event in the neuropathologic changes in AD and accumulation of deleted mitochondrial DNA may potentiate oxidative damage in vulnerable neurons.

Recently, we evaluated the levels of 8-OHdG in intact DNA and free 8-OHdG, which represents the repair product, in ventricular CSF from AD and control subjects (Lovell et al., 1999). A significant elevation of 8-OHdG in intact DNA in AD compared with age-matched control subjects was found. In contrast, levels of free 8-OHdG were significantly decreased in AD samples. This indicates that there may be a double insult of increased oxidative damage and a deficiency of repair mechanisms responsible for removal of oxidized bases in AD. Indeed, fibroblasts and lymphocytes from familial AD patients have a deficiency of DNA repair after exposure to fluorescent light (Parshad et al., 1996). Hermon et al. (1998) described increased protein levels of two excision-repair cross-complementing genes for nucleotide excision repair in the brain of AD patients, suggesting ongoing oxidative DNA damage. We found statistically significant decreases in 8-oxyguanine glycosylase activity, which is responsible for the excision of 8-oxyguanine, in the nuclear fraction in the hippocampus, superior and middle temporal gyri, and inferior parietal lobule in AD (Lovell et al., 2000b). DNA helicase activity was elevated in all nuclear DNA samples and statistically significantly elevated in the hippocampus and cerebellum. This study demonstrates that the base excision repair capability for 8-oxyguanine is decreased in AD. The increase in DNA helicase activity in some brain regions may interfere with base-excision repair mechanisms.

One of the consequences of DNA strand breaks is the activation of poly (ADP-ribose) polymerase (PARP), a Zn-finger DNA binding protein (Lautier et al. 1990). Over-activation of PARP in response to oxidative damage causes depletion of intracellular NAD^+ resulting in depletion of energy stores and cell death (Zhang et al., 1994; Eliasson et al., 1997). Love et al. (1999) demonstrated enhanced PARP activity in the brain in AD using immunohistochemical staining of frontal and temporal lobes. Su et al. (1996) showed that bcl-2 expression is increased in neurons with DNA damage in AD and that bcl-2 has an antioxidant effect. Deng et al. (1999) demonstrated that bcl-2 in PC12 cells inhibited peroxynitrite-induced cell death and enhanced DNA damage recovery, which suggest neuronal upregulation of bcl-2 may facilitate DNA repair following oxidative stress.

Parkinson's Disease

Sanchez-Ramos et al. (1994) described increased 8-OHdG in the substantia nigra in PD but also observed elevations in the basal ganglia and cerebral cortex. Alam et al. (1997a) described an increase in 8-hydroxyguanine in total cellular DNA in the substantia nigra, frontal pole, and putamen, plus a decrease in fapy guanine in PD. They suggested that the rise in 8-hydroxyguanine could be due to change in the 8-hydroxyguanine/fapy guanine ratio rather than an increase in total oxidative guanine damage. Zhang et al. (1999), using immunohistochemical methods, demonstrated increased 8-OHdG in the cytoplasm of substantia nigra neuron in PD and determined that the oxidative damage targeted RNA and mitochondrial DNA, similar to what has been described by others in the hippocampus of AD patients (Wade, 1998).

Further studies to define the role of oxidative DNA damage and DNA repair in the brain in aging and neurodegenerative diseases could yield important information about the pathogenesis of neuron degeneration.

ANTIOXIDANTS IN NEURODEGENERATIVE DISEASES

As described above, a broad spectrum of enzymatic and nonenzymatic antioxidants provides important protection against oxidative stress (Fig. 1). Most of the research on antioxidants in neurodegenerative diseases has been in AD. Studies of the activity and expression of brain antioxidants in AD have yielded inconsistent results. For example, two reports showed no significant difference in Cu/Zn- or Mn-SOD activity in AD and control brains (Kato et al., 1991; Gsell et al., 1995), whereas two studies found a reduction of SOD activity in several brain regions in AD (Richardson et al., 1993; Marcus et al., 1998). CAT activity was elevated in the amygdala in one report (Gsell et al., 1995), but reduced in several brain regions in another (Marcus et al., 1998). Two studies demonstrated no difference in GSH-Px activity in AD and controls (Kish et al., 1986; Marcus et al., 1998), whereas one study showed a elevation of GSH-Px (Lovell et al., 1995). One possible explanation for the variability in brain antioxidants in AD is that most of these studies used brains with prolonged postmortem intervals, which can affect enzyme activity. It has been well demonstrated that the overall efficiency of antioxidant defenses is attenuated during aging and, thus, closely age-matched controls are critical in such studies.

In a study using short postmortem interval AD and age-matched, longitudinally evaluated normal control subjects, we found significant elevations of GSH-Px activity in the hippocampus, GSSG-R activity in hippocampus and amygdala, and CAT activity in hippocampus and temporal neocortex in AD (Lovell et al., 1995). These changes correlated with elevations of lipid peroxidation in the same regions. We did not find a significant alteration in Cu/Zn- or Mn-SOD activity in AD. In a separate series of AD and control subjects, we evaluated mRNA expression of oxidative stress handling genes and found significant elevations of GSH-Px, GSSG-R, and CAT mRNA in hippocampus and inferior parietal lobule in AD (Aksenov et al., 1998). It is possible that the elevations observed in these studies represent a compensatory rise in antioxidant activity in response to increased free-radical generation. It should be emphasized that none of our studies defined a deficiency of antioxidant enzyme activity or mRNA expression, which suggests that the increased oxidative stress in AD is not related to failure of these defense mechanisms.

As noted above, MsrA may play a role in antioxidant defenses in proteins by its ability to reverse MetSO back to methionine, which can scavenge oxidants. Our study showed a marked decrease in MsrA in multiple brain regions in AD, which suggests an important defense mechanism from protein oxidation may be impaired (Gabbita et al., 1999).

Thioredoxin (Trx), a ubiquitous protein containing redox-active disulfide/dithiol within its conserved active site (Holmgren, 1985), functions in the reduction of protein disulfides and as an ROS scavenger (Mitsui et al., 1992). Thioredoxin reductase (TR), a selenium-containing enzyme, functions to change oxidized Trx to reduced Trx. Together, Trx and TR operate as a potent NADPH-dependent protein disulfide reductase system that can repair oxidized proteins or maintain levels of reduced Trx, which can interact directly with ROS. We found decreased Trx protein levels in five AD brain regions compared with normal control subjects, with statistically significantly decreases in amygdala and hippocampus (Lovell et al., 2000c). Thioredoxin reductase activity was increased in five AD brain regions and reached statistical significance in amygdala and cerebellum, which suggests a compensatory rise in response to oxidative stress. Neuron culture studies showed that Trx was protective against Aβ-peptide-induced neuron degeneration, which is mediated in part through free-radical mechanisms.

Uric acid acts as a scavenger of $ONOO^-$ in vitro and inhibits nitration of tyrosine in cultured neurons challenged with oxidative stress (Whiteman and Halliwell, 1996; Mattson et al., 1997). Hensley et al. (1998) described decreased uric acid in multiple brain regions and in ventricular CSF in short-postmortem-interval AD patients compared with age-matched control subjects, suggesting that diminished defenses against $ONOO^-$ may be important in AD.

Defenses against the secondary products of oxidation may also play a meaningful role in the brain in neurodegenerative disorders. Glutathione transferases (GST) are a multigene family of enzymes that catalyze the nucleophilic conjugation of glutathione with many electrophilic compounds. These enzymes protect against toxic compounds including aldehydes such as HNE. Exogenous GST protects cultured hippocampal neurons against HNE toxicity (Xie et al., 1998). In a study using short postmortem interval AD and age-matched control brains, we found GST activity and protein levels significantly depleted in multiple brain regions and in ventricular CSF in AD (Lovell et al., 1998b). This indicates that a protective enzyme against aldehydes formed from lipid peroxidation is significantly decreased in AD and may play a role in neuron degeneration in this disorder.

Although the study of antioxidant defenses in the brain is in the early descriptive stage, there are indications that several defense mechanisms in the brain are diminished, at least in late-stage AD subjects compared with age-matched controls. Perhaps, as we learn more about the antioxidant defenses in the brain, it will enhance our understanding of oxidative stress in brain, more clearly define the mechanisms of neurodegeneration, and determine new therapeutic targets.

CONCLUSION

This review demonstrates that, although a number of environmental or genetic factors may cause or trigger neuronal degeneration in neurodegenerative diseases, free radical-mediated damage is at least a part of the pathogenetic cascade of events. Oxidative damage to lipids, proteins, or nucleic acids is found in AD, PD, ALS, HD, and Pick's disease. In addition, there is a decline in some antioxidant systems in AD and ALS.

The presence of free-radical-mediated damage is supported by clinical findings that suggest protective mechanisms against oxidative stress may play a role in slowing, or perhaps, eventually preventing disorders such as AD. It also appears that upregulating the natural defense mechanisms developed by cells against the constant challenge of oxidative stress could be beneficial.

Further research will be aimed at defining specifically how oxidative events are involved in the pathogenetic cascade leading to neuron death, and developing methods to decrease oxidative stress and enhance antioxidant defense mechanisms with the ultimate goal of preventing neurodegenerative disorders.

ACKNOWLEDGMENTS

This work was supported by National Institute of Health grants AG0 5119 (WRM), AG0 5144 (WRM), AG0 0774 (TJM), AG 16835 (TJM), and grants from the Abercrombie Foundation (WRM), Kleberg Foundation (WRM), the American Foundation for Aging Research (TJM), and the Alzheimer's Disease and Related Disorders Association (TJM). The authors thank Jane Meara, Lisa von Wiegen, and Paula Thomason for technical and editorial assistance.

REFERENCES

Aksenov, M. Y., Aksenova, M. V., Butterfield, D. A., and Markesbery, W. R. (2000a) Oxidative modification of creatine kinase BB in Alzheimer's disease brain. *J. Neurochem.* **74,** 2520–2527.

Aksenov, M. Y., Aksenova, M. V., Butterfield, D. A., Geddes, J. W., and Markesbery, W. R. (2000b) Protein oxidation in the Alzheimer's disease brain: analysis of protein carbonyls by immunocytochemistry and two-dimensional Western blotting. *Neuroscience*, in press.

Aksenov, M. Y., Aksenova, M. V., Payne, R. M., Smith, C. D., Markesbery, W. R., and Carney, J. M. (1997) The expression of creatine kinase isoenzymes in neocortex of patients with neurodegenerative disorders: Alzheimer's and Pick's disease. *Exp. Neurol.* **146,** 458–465.

Aksenov, M. Y., Tucker, H. M., Nair, P., Aksenova, M. V., Butterfield, D. A., Estus, S., and Markesbery, W. R. (1998) The expression of key oxidative stress-handling genes in different brain regions in Alzheimer disease. *J. Mol. Neurosci.* **11,** 151–164.

Aksenova, M. V., Aksenov, M. Y., Payne, R. M., Trojanowski, J. Q., Schmidt, M. L., Carney, J. M., et al. (1999) Oxidation of cytosolic proteins and expression of creatine kinase BB in frontal lobe in different neurodegenerative disorders. *Dement. Geriatr. Cogn. Disord.* **10,** 158–165.

Alam, Z. I., Jenner, A., Daniel, S. E., Lees, A. J., Cairns, N., Marsden, C. D., et al. (1997a) Oxidative DNA damage in the parkinsonian brain: an apparent selective increase in 8-hydroxyguanine levels in substantia nigra. *J. Neurochem.* **69,** 1196–1203.

Alam, Z. I., Daniel, S. E., Lees, A. J., Marsden, D. C., Jenner, P., and Halliwell, B. (1997b) A generalized increase in protein carbonyls in the brain in Parkinson's disease but not incidental Lewy body disease. *J. Neurochem.* **69,** 1326–1329.

Ames, B. N., Shigenaga, M. K., and Hagen, T. M. (1995) Mitochondrial decay in aging. *Biochim. Biophys. Acta* **1271,** 165–170.

Arai, K., Maguchi, S., Fujii, S., Ishibashi, H., Oikawa, K., and Taniguchi, N. (1987) Glycation and inactivation of human Cu-Zn-superoxide dismutase. Identification of the in vivo glycated sites. *J. Biol. Chem.* **262,** 16,969–16,972.

Atwood, C. S., Huang, X., Moir, R. D., Tanzi, R. E., and Bush, A. I. (1999) Role of free radicals and metal ions in the pathogenesis of Alzheimer's disease, in Interrelations Between Free Radicals and Metal Ions in Life Processes (Sigel, A., Sigel, H., eds.), Marcel Dekker, New York, pp. 309–363.

Atwood, C. S., Moir, R. D., Huang, X., Scarpa, R. C., Bacarra, N. M., Romano, D. M., et al. (1998) Dramatic aggregation of Alzheimer Aβ by Cu (II) is induced by conditions representing physiological acidosis. *J. Biol. Chem.* **273,** 12,817–12,826.

Beal, M. F. (1995) Aging, energy and oxidative stress in neurodegenerative diseases. *Ann. Neurol.* **38,** 357–366.

Beckman, K. B. and Ames, B. N. (1997) Oxidative decay of DNA. *J. Biol. Chem.* **272,** 19,633–19,636.

Berlett, B. S. and Stadtman, E. R. (1997) Protein oxidation in aging, disease, and oxidative stress. *J. Biol. Chem.* **272,** 20,313–20,316.

Blanc, E. M., Kelly, J. F., Mark, R. J., and Mattson, M. P. (1997) 4-Hydroxynonenal, an aldehydic product of lipid peroxidation, impairs signal transduction associated with muscarinic acetylcholine and metabotropic glutamate receptors: possible action on $G\alpha_{(q/11)}$. *J. Neurochem.* **69,** 570–580.

Bodovitz, S., Falduto, M. T., Frail, D. E., and Klein, W. L. (1995) Iron levels modulate α-secretase cleavage of amyloid precursor protein. *J. Neurochem.* **64,** 307–315.

Brot, N. and Weissbach, H. (1983) Biochemistry and physiological role of methionine sulfoxide residues in proteins. *Arch. Biochem. Biophys.* **223,** 271–281.

Browne, S. E. and Beal, M. F. (1994) Oxidative damage and mitochondrial dysfunction in neurodegenerative diseases. *Biochem. Soc. Trans.* **22,** 1002–1006.

Browne, S. E., Ferrante, R. J., and Beal, M. F. (1999) Oxidative stress in Huntington's disease. *Brain Pathol.* **9,** 147–163.

Bruce-Keller, A. J., Li, Y. J., Lovell, M. A., Kraemer, P. J., Gary, D. S., Brown, R. R., et al. (1998) 4-Hydroxynonenal, a product of lipid peroxidation, damages cholinergic neurons and impairs visuospatial memory in rats. *J. Neuropathol. Exp. Neurol.* **57,** 257–267.

Bruce-Keller, A. J., Umberger, G., McFall, R., and Mattson, M. P. (1999) Food restriction reduces brain damage and improves behavioral outcome following excitotoxic and metabolic insults. *Ann. Neurol.* **45,** 8–15.

Bush, A. I., Pettingell, W. H., Multhaup, G., Paradis, M., Vonsattel, J. P., Gusella, J. F., et al. (1994a) Rapid induction of Alzheimer Aβ-amyloid formation by zinc. *Science* **265,** 1464–1467.

Bush, A. I., Pettingell, W. H., Paradis, M. D., and Tanzi, R. E. (1994b) Modulation of Aβ-adhesiveness and secretase site cleavage by zinc. *J. Biol. Chem.* **269,** 12,152–12,158.

Calingasan, N. Y., Uchida, K., and Gibson, G. E. (1999) Protein-bound acrolein: a novel marker of oxidative stress in Alzheimer's disease. *J. Neurochem.* **72,** 751–756.

Candy, J. M., Oakley, A. E., Klinowski, J., Carpenter, T. A., Perry, R. H., Atack, J. R., et al. (1986) Aluminosilicates and senile plaque formation in Alzheimer's disease. *Lancet* **1,** 354–357.

Carpenter, A. F., Carpenter, P. W., and Markesbery, W. R. (1993) Morphometric analysis of microglia in Alzheimer's disease. *J. Neuropathol. Exp. Neurol.* **52,** 601–608.

Castellani, R., Smith, M. A., Richey, P. L., and Perry, G. (1996) Glycoxidation and oxidative stress in Parkinson's disease and diffuse Lewy body disease. *Brain Res.* **737,** 195–200.

Cohen, G. and Werner, P. (1994) Free radicals, oxidative stress and neurodegeneration, in *Neurodegenerative Diseases* (Calne, D. B., ed.), W. B. Saunders, Philadelphia, PA, pp. 139–161.

Colton, C. A., Snell, J., Chernyshev, O., and Gilbert, D. L. (1994) Induction of superoxide anion and nitric oxide production in cultured microglia. *Ann. NY Acad. Sci.* **738,** 54–63.

Connor, J. R., Snyder, B. S., Arosio, P., Loeffler, D. A., and LeWitt, P. (1995) A quantitative analysis of isoferritins in select regions of aged, parkinsonian, and Alzheimer's diseased brains. *J. Neurochem.* **65,** 717–724.

Connor, J. R., Snyder, B. S., Beard, J. L., Fine, R. E., and Mufson, E. J. (1992) Regional distribution of iron and iron-regulatory proteins in the brain in aging and Alzheimer's disease. *J. Neurosci. Res.* **31,** 327–335.

Cornett, C. R., Markesbery, W. R., and Ehmann, W. D. (1998) Imbalances of trace elements related to oxidative damage in Alzheimer's disease brain. *Neurotoxicology* **19,** 339–346.

Coyle, J. T. and Puttfarcken, P. (1993) Oxidative stress, glutamate, and neurodegenerative disorders. *Science* **262,** 689–695.

Cuajungco, M. P. and Lees, G. J. (1997) Zinc metabolism in the brain: relevance to human neurodegenerative disorders. *Neurobiol. Dis.* **4,** 137–169.

Dal Canto, M. C. and Gurney, M. E. (1994) Development of central nervous system pathology in a murine transgenic model of human amyotrophic lateral sclerosis. *Am. J. Pathol.* **145,** 1271–1279.

David, S., Shoemaker, M., and Haley, B. E. (1998) Abnormal properties of creatine kinase in Alzheimer's disease brain: correlation of reduced enzyme activity and active site photolabeling with aberrant cytosol-membrane partitioning. *Brain Res. Mol. Brain Res.* **54,** 276–287.

Deibel, M. A., Ehmann, W. D., and Markesbery, W. R. (1996) Copper, iron, and zinc imbalances in severely degenerated brain regions in Alzheimer's disease: possible relation to oxidative stress. *J. Neurol. Sci.* **143,** 137–142.

Deng, G., Su, J. H., Ivins, K. J., Van Houten, B., and Cotman, C. W. (1999) Bcl-2 facilitates recovery from DNA damage after oxidative stress. *Exp. Neurol.* **159,** 309–318.

Dexter, D. T., Carayon, A., Vidailhet, M., Ruberg, M., Agid, F., Agid, Y., et al. (1990) Decreased ferritin levels in brain in Parkinson's disease. *J. Neurochem.* **55,** 16–20.

Dexter, D. T., Carter, C. J., Wells, F. R., Javoy-Agid, F., Agid, Y., Lees, A., et al. (1989) Basal lipid peroxidation in substantia nigra is increased in Parkinson's disease. *J. Neurochem.* **52,** 381–389.

Dexter, D. T., Holley, A. E., Flitter, W. D., Slater, T. F., Wells, F. R., Daniel, S. E., et al. (1994) Increased levels of lipid hydroperoxides in the parkinsonian substantia nigra: an HPLC and ESR study. *Mov. Disord.* **9,** 92–97.

Dickson, D. W., Sinicropi, S., Yen, S. H., Ko, L. W., Mattiace, L. A., Bucala, R., and Vlassara, H. (1996) Glycation and microglial reaction in lesions of Alzheimer's disease. *Neurobiol. Aging* **17,** 733–743.

Douki, T. and Ames, B. N. (1994) An HPLC-EC assay for 1, N2-propano adducts of 2'-deoxyguanosine with 4-hydroxynonenal and other, αβ-unsaturated aldehydes. *Chem. Res. Toxicol.* **7,** 511–518.

Du Yan, S., Zhu, H., Fu, J., Roher, A., Tourtellotte, W. W., Rajavashisth,T., et al. (1997) Amyloid-β peptide-receptor for advanced glycation endproduct interaction elicits neuronal expression of macrophage-colony stimulating factor: a pro-inflammatory pathway in Alzheimer disease. *Proc. Natl. Acad. Sci. USA* **94,** 5296–5301.

Dunn, J. A., Ahmed, M. U., Murtiashaw, M. H., Richardson, J. M., Walla, M. D., Thorpe, S. R., and Baynes, J. W. (1990) Reaction of ascorbate with lysine and protein under autoxidizing conditions: formation of Nε-(carboxymethyl)lysine by reaction between lysine and products of autoxidation of ascorbate. *Biochemistry* **29,** 10,964–10,970.

Dyer, D. G., Blackledge, J. A., Thorpe, S. R., and Baynes, J. W. (1991) Formation of pentosidine during nonenzymatic browning of proteins by glucose. Identification of glucose and other carbohydrates as possible precursors of pentosidine in vivo. *J. Biol. Chem.* **266,** 11,654–11,660.

Dyrks, T., Dyrks, E., Hartmann, R., Masters, C., and Beyreuther, K. (1992) Amyloidogenicity of βA4 and βA4-bearing amyloid protein precursor fragments by metal-catalyzed oxidation. *J. Biol. Chem.* **267,** 18,210–18,217.

Edwardson, J. A., Ferrer, I. N., McArthur, F. K., McKeith, I. G., McLaughlin, I., Morris, C. M., et al. (1991) Alzheimer's disease and the aluminum hypothesis, in *Aluminum in Chemistry, Biology and Medicine* vol. 1. (Nicolini, M., Zatta, P. F., Corain, B., eds.), Raven Press, New York, NY, pp. 85–86.

Eliasson, M. J., Sampei, K., Mandir, A. S., Hurn, P. D., Traystman, R. J., Bao, J., et al. (1997) Poly (ADP-ribose) polymerase gene disruption renders mice resistant to cerebral ischemia. *Nat. Med.* **3,** 1089–1095.

Esterbauer, H., Schaur, R. J., and Zollner, H. (1991) Chemistry and biochemistry of 4-hydroxynonenal, malondialdehyde and related aldehydes. *Free Radic. Biol. Med.* **11,** 81–128.

Farber, J. L. (1995) Mechanisms of cell injury, in *Pathology of Environmental and Occupational Disease* (Craighead, J. E., ed.), Mosby, St. Louis, MO, pp. 287–302.

Faucheux, B. A., Nillesse, N., Damier, P., Spik, G., Mouatt-Prigent, A., Pierce, A., et al. (1995) Expression of lactoferrin receptors is increased in the mesencephalon of patients with Parkinson disease. *Proc. Natl. Acad. Sci. USA* **92,** 9603–9607.

Ferrante, R. J., Browne, S. E., Shinobu, L. A., Bowling, A. C., Baik, M. J., MacGarvey, U., et al. (1997) Evidence of increased oxidative damage in both sporadic and familial amyotrophic lateral sclerosis. *J. Neurochem.* **69,** 2064–2074.

Fleming, J. and Joshi, J. G. (1987) Ferritin: isolation of aluminum-ferritin complex from brain. *Proc. Natl. Acad. Sci. USA* **84,** 7866–7870.

Frederickson, C. J., Hernandez, M. D., and McGinty, J. F. (1989) Translocation of zinc may contribute to seizure-induced death of neurons. *Brain Res.* **480,** 317–321.

Friedberg, E. C. and Wood, R. C. (1996) DNA excision repair pathways, in *DNA Replication in Eukaryotic Cells* (DePamphilis, M. L., ed.), Cold Spring Harbor Laboratory Press, Cold Spring Harbor, NY, pp. 249–269.

Fu, W., Luo, H., Parthsarathy, S., and Mattson, M. P. (1998) Catecholamines potentiate amyloid β-peptide neurotoxicity: involvement of oxidative stress, mitochondrial dysfunction, and perturbed calcium homeostasis. *Neurobiol. Dis.* **5,** 229–243.

Gabbita, S. P., Lovell, M. A., and Markesbery, W. R. (1998) Increased nuclear DNA oxidation in the brain in Alzheimer's disease. *J. Neurochem.* **71,** 2034–2040.

Gabbita, S. P., Aksenov, M. Y., Lovell, M. A., and Markesbery, W. R. (1999) Decrease in peptide methionine sulfoxide reductase in Alzheimer's disease brain. *J. Neurochem.* **73,** 1660–1666.

Good, P. F., Hsu, A., Werner, P., Perl, D. P., and Olanow, C. W. (1998) Protein nitration in Parkinson's disease. *J. Neuropathol. Exp. Neurol.* **57,** 338–342.

Good, P. F., Olanow, C. W., and Perl, D. P. (1992a) Neuromelanin-containing neurons of the substantia nigra accumulate iron and aluminum in Parkinson's disease: a LAMMA study. *Brain Res.* **593,** 343–346.

Good, P. F., Perl, D. P., Bierer, L. M., and Schmeidler, J. (1992b) Selective accumulation of aluminum and iron in the neurofibrillary tangles of Alzheimer's disease: a laser microprobe (LAMMA) study. *Ann. Neurol.* **31,** 286–292.

Good, P. F., Werner, P., Hsu, A., Olanow, C. W., and Perl, D. P. (1996) Evidence of neuronal oxidative damage in Alzheimer's disease. *Am. J. Pathol.* **149,** 21–28.

Goodman, Y. and Mattson, M. P. (1994) Secreted forms of β-amyloid precursor protein protect hippocampal neurons against amyloid β-peptide-induced oxidative injury. *Exp. Neurol.* **128,** 1–12.

Goodwin, J. L., Uemura, E., and Cunnick, J. E. (1995) Microglial release of nitric oxide by the synergistic action of β-amyloid and IFN-gamma. *Brain Res.* **692,** 207–214.

Grandhee, S. K. and Monnier, V. M. (1991) Mechanism of formation of the Maillard protein cross-link pentosidine. Glucose, fructose, and ascorbate as pentosidine precursors. *J. Biol. Chem.* **266,** 11,649–11,653.

Grundke-Iqbal, I., Fleming, J., Tung, Y. C., Lassmann, H., Iqbal, K., and Joshi, J. G. (1990) Ferritin is a component of the neuritic (senile) plaque in Alzheimer dementia. *Acta Neuropathol. (Berl.)* **81,** 105–110.

Gsell, W., Conrad, R., Hickethier, M., Sofic E., Frolich, L., Wichart, I., et al. (1995) Decreased catalase activity but unchanged superoxide dismutase activity in brains of patients with dementia of Alzheimer type. *J. Neurochem.* **64,** 1216–1223.

Gurney, M. E., Pu, H., Chiu, A. Y., Dal Canto, M. C., Polchow, C. Y., Alexander, D. D., et al. (1994) Motor neuron degeneration in mice that express a human Cu,Zn superoxide dismutase mutation. *Science* **264,** 1772–1775.

Gutteridge, J. M. and Halliwell, B. (1990) The measurement and mechanism of lipid peroxidation in biological systems. *Trends Biochem. Sci.* **15,** 129–135.

Gutteridge, J. M., Quinlan, G. J., Clark, I., and Halliwell, B. (1985) Aluminum salts accelerate peroxidation of membrane lipids stimulated by iron salts. *Biochem. Biophys. Acta* **835,** 441–447.

Halliwell, B. and Gutteridge, J. M. C. (1989) *Free Radicals in Biology and Medicine.* Oxford University Press, New York, NY.

Hantraye, P., Brouillet, E., Ferrante, R., Palfi, S., Dolan, R., Matthews, R. T., and Beal, M. F. (1996) Inhibition of neuronal nitric oxide synthase prevents MPTP-induced parkinsonism in baboons. *Nat. Med.* **2,** 1017–1021.

Hensley, K., Carney, J. M., Mattson, M. P., Aksenova, M., Harris, M., Wu, J. F., et al. (1994) A model for β-amyloid aggregation and neurotoxicity based on free radical generation by the peptide: relevance to Alzheimer disease. *Proc. Natl. Acad. Sci. USA* **91,** 3270–3274.

Hensley, K., Hall, N., Subramaniam, R., Cole, P., Harris, M., Aksenov, M., et al. (1995) Brain regional correspondence between Alzheimer's disease histopathology and biomarkers of protein oxidation. *J. Neurochem.* **65,** 2146–2156.

Hensley, K., Maidt, M. L., Yu, Z., Sang, H., Markesbery, W. R., and Floyd, R. A. (1998) Electrochemical analysis of protein nitrotyrosine and dityrosine in the Alzheimer brain indicates region-specific accumulation. *J. Neurosci.* **18,** 8126–8132.

Hermon, M., Cairns, N., Egly, J. M., Fery, A., Labudova, O., and Lubec, G. (1998) Expression of DNA excision-repair-cross-complementing protein p80 and p89 in brain of patients with Down syndrome and Alzheimer's disease. *Neurosci. Lett.* **251,** 45–48.

Hirai, K., M., Smith, M. A., Wade, R., and Perry, G. (1998) Vulnerable neurons in Alzheimer disease accumulate mitochondrial DNA with the common 5KB deletion. *J. Neuropathol. Exp. Neurol.* **57,** 511.

Hirsch, E. C., Brandel, J. P., Galle, P., Javoy-Agid, F., and Agid, Y. (1991) Iron and aluminum increase in the substantia nigra of patients with Parkinson's disease: an x-ray microanalysis. *J. Neurochem.* **56,** 446–451.

Holmgren, A., (1985) Thioredoxin. *Annu. Rev. Biochem.* **54,** 237–271.

Ii, M., Sunamoto, M., Ohnishi, K., and Ichimori, Y. (1996) β-amyloid-protein-dependent nitric oxide production from microglial cells and neurotoxicity. *Brain Res.* **720,** 93–100.

Jefferies, W. A., Food, M. R., Gabathuler, R., Rothenberger, S., Yamada, T., Yasuhara, O., and McGeer, P. L. (1996) Reactive microglia specifically associated with amyloid plaques in Alzheimer disease brain tissue express melanotransferrin. *Brain Res.* **712,** 122–126.

Jellinger, K., Kienzl, E., Rumpelmair, G., Riederer, P., Stachelberger, H., Ben-Shachar, and Youdim, M. B. (1992) Iron-melanin complex in substantia nigra of parkinsonian brains: an x-ray microanalysis. *J. Neurochem.* **59,** 1168–1171.

Jenner, P. (1994) Oxidative damage in neurodegenerative disease. *Lancet* **344,** 796–798.

Jenner, P. and Olanow, C. W. (1998) Understanding cell death in Parkinson's disease. *Ann. Neurol.* **44,** S72–S84.

Kalaria, R. N., Sromek, S. M., Grahovac, I., and Harik, S. I. (1992) Transferrin receptors of rat and human brain and cerebral microvessels and their status in Alzheimer's disease. *Brain Res.* **585,** 87–93.

Kato, K., Kurobe, N., Suzuki, R., Morishita, R., Asano, T., Sato, T., and Inagaki, T. (1991) Concentrations of several proteins characteristic of nervous tissue in cerebral cortex of patients with Alzheimer's disease. *J. Mol. Neurosci.* **3,** 95–99.

Keller, J. N., Pang, Z., Geddes, J. W., Begley, J. G., Germeyer, A., Waeg, G., and Mattson, M. P. (1997) Impairment of glucose and glutamate transport and induction of mitochondrial oxidative stress and dysfunction in synaptosomes by amyloid β-peptide: role of the lipid peroxidation product 4-hydroxynonenal. *J. Neurochem.* **69,** 273–284.

Keller, J. N., Kindy, M. S., Holtsberg, F. W, St. Clair, D. K., Yen, H. C., Germeyer, A., et al. (1998) Mitochondrial maganese superoxide dismutase present neural apoptosis and reduces ischemic brain injury: suppression of peroxynitrite production, lipid peroxidation, and mitochondrial dysfunction. *J. Neurosci.* **18,** 687–697.

Kelly, J., Furukawa, K., Barger, S. W., Mark, R. J., Rengen, M. R., Blanc, E. M., et al. (1996) Amyloid β-peptide disrupts carbachol-induced muscarinic cholinergic signal transduction in cortical neurons. *Proc. Natl. Acad. Sci. USA* **93,** 6753–6758.

Kennard, M. L., Feldman, H., Yamada, T., and Jefferies, W. A. (1996) Serum levels of the iron binding protein p97 are elevated in Alzheimer's disease. *Nat. Med.* **2,** 1230–1235.

Kikuchi, S., Shinpo, K., Moriwaka, F., Makita, Z., Miyata, T., and Tashiro, K. (1999) Neurotoxicity of methyglyoxal and 3-deoxyglucosone on cultured cortical neurons: synergism between glycation and oxidative stress, possibly involved in neurodegenerative diseases. *J. Neurosci. Res.* **57,** 280–289.

Kimura, T., Takamatsu, J., Ikeda, K., Kondo, A., Miyakawa, T., and Horiuchi, S. (1996) Accumulation of advanced glycation end products of the Maillard reaction with age in human hippocampal neurons. *Neurosci. Lett.* **208,** 53–56.

Kish, S. J., Morito, C. L., and Hornykiewicz, O. (1986) Brain glutathione peroxidase in neurodegenerative disorders. *Neurochem. Pathol.* **4,** 23–28.

Klivenyi, P., Andreassen, O. A., Ferrante, R. J., Dedeoglu, A., Mueller, G., Lancelot, E., et al. (2000) Mice deficient in cellular glutathione peroxidase show increased vulnerability to malonate, 3-nitropropionic acid, and 1-methyl-4-phenyl-1,2,5,6-tetrahydropyridine. *J. Neurosci.* **20,** 1–7.

Kruman, I., Pedersen, W. A., and Mattson, M. P. (1999) ALS-linked Cu/Zn-SOD mutation increases vulnerability of motor neurons to excitotoxicity by a mechanism involving increased oxidative stress and perturbed calcium homeostasis. *Exp. Neurol.* **160,** 28–39.

Langston, J. W. (1994) Organic neurotoxicants, in *Neurodegenerative Diseases* (Calne, D. B., ed.), W. B. Saunders, Philadelphia, PA, pp. 225–240.

Lautier, D., Poirier, D., Boudreau, A., Alaoui Jamali, M. A., Castonguay, A., and Poirier, G. (1990) Stimulation of poly (ADP-ribose) synthesis by free radicals in C3H10T1/2 cells: relationship with NAD metabolism and DNA breakage. *Biochem. Cell Biol.* **68,** 602–608.

Levine, R. L., Mosoni, L., Berlett, B. S., and Stadtman, E. R. (1996) Methionine residues as endogenous antioxidants in proteins. *Proc. Natl. Acad. Sci. USA* **93,** 15,036–15,040.

Love, S., Barber, R., and Wilcock, G. K. (1999) Increased poly (ADP-ribosyl)ation of nuclear proteins in Alzheimer's disease. *Brain* **122,** 247–253.

Lovell, M. A., Ehmann, W. D., Butler, S. M., and Markesbery, W. R. (1995) Elevated thiobarbituric acid-reactive substances and antioxidant enzyme activity in the brain in Alzheimer's disease. *Neurology* **45,** 1594–1601.

Lovell, M., Ehmann, W., Mattson, M., and Markesbery, W. (1997) Elevated 4-hydroxynonenal in ventricular fluid in Alzheimer's disease. *Neurobiol. Aging* **18,** 457–461.

Lovell, M. A., Robertson, J. D., Teesdale, W. J., Campbell, J. L., and Markesbery, W. R. (1998a) Copper, iron and zinc in Alzheimer's disease senile plaques. *J. Neurol. Sci.* **158,** 47–52.

Lovell, M. A., Xie, C., and Markesbery, W. R. (1998b) Decreased glutathione activity transferase in brain and ventricular fluid in Alzheimer's disease. *Neurology* **51,** 1562–1566.

Lovell, M. A., Gabbita, S. P., and Markesbery, W. R. (1999) Increased DNA oxidation and decreased levels of repair products in Alzheimer's disease ventricular CSF. *J. Neurochem.* **72,** 771–776.

Lovell, M. A., Xie, C. S., and Markesbery, W. R. (2000a) Acrolein, a product of lipid peroxidation is increased in the brain in Alzheimer's disease and is neurotoxic to primary neuronal culture. *Neurobiol. Aging*, in press.

Lovell, M. A., Xie, C., and Markesbery, W. R. (2000b) Decreased base excision repair and increased helicase activity in Alzheimer's disease brain. *Brain Res.* **855,** 116–123.

Lovell, M. A., Xie, C., Gabbita, S. P., and Markesbery, W. R. (2000c) Decreased thioredoxin and increased thioredoxin reductase levels in Alzheimer's disease brain. *Free Radic. Biol. Med.* **28,** 418–427.

Lyras, L., Cairns, N. J., Jenner, A., Jenner, P., and Halliwell, B. (1997) An assessment of oxidative damage to proteins, lipids and DNA in brains from patients with Alzheimer's disease. *J. Neurochem.* **68,** 2061–2069.

Mantyh, P. W., Ghilardi, J. R., Rogers, S., DeMaster, E., Allen, C. J., Stimson, E. R., and Maggio, J. E. (1993) Aluminum, iron, and zinc ions promote aggregation of physiological concentrations of β-amyloid peptide. *J. Neurochem.* **61,** 1171–1174.

Marcus, D. L., Thomas, C., Rodriguez, C., Simberkoff, K., Tsai, J. S., Strafaci, J. A., and Freedman, M. L. (1998) Increased peroxidation and reduced antioxidant enzyme activity in Alzheimer's disease. *Exp. Neurol.* **150,** 40–44.

Mark, R. J., Hensley, K., Butterfield, D. A., and Mattson, M. P. (1995) Amyloid β-peptide impairs ion-motive ATPase activities: evidence for a role in loss of neuronal Ca^{2+} homeostasis and cell death. *J. Neurosci.* **15,** 6239–6249.

Mark, R. J., Lovell, M. A., Markesbery, W. R., Uchida, K., and Mattson, M. P. (1997a) A role for 4-hydroxynonenal in disruption of ion homeostasis and neuronal death induced by amyloid β-peptide. *J. Neurochem.* **68,** 255–264.

Mark, R. J., Pang, Z., Geddes, J. W., and Mattson, M. P. (1997b) Amyloid β-peptide impairs glucose uptake in hippocampal and cortical neurons: involvement of membrane lipid peroxidation. *J. Neurosci.* **17,** 1046–1054.

Markesbery, W. R., Ehmann, W. D., Alauddin, M., and Hossain, T. I. (1984) Brain trace element concentrations in aging. *Neurobiol. Aging* **5,** 19–28.

Markesbery, W. R. and Ehmann, W. D. (1994) Brain trace elements, in *Alzheimer's Disease* (Terry, R. D., et al., eds.), Raven Press, New York, NY, pp. 353–368.

Markesbery, W. R. and Lovell, M. A. (1998) Four-hydroxynonenal, a product of lipid peroxidation, is increased in the brain in Alzheimer's disease. *Neurobiol. Aging* **19,** 33–36.

Mattson, M. P. (1997) Cellular actions of β-amyloid precursor protein and its soluble and fibrillogenic derivatives. *Physiol. Rev.* **77,** 1081–1132.

Mattson, M. P. (1998) Modification of ion homeostasis by lipid peroxidation: roles in neuronal degeneration and adaptive plasticity. *Trends Neurosci.* **21,** 53–57.

Mattson, M. P., Lovell, M. A., Furukawa, K., and Markesbery, W. R. (1995) Neurotrophic factors attenuate glutamate-induced accumulation of peroxides, elevation of intracellular Ca^{2+} concentration, and neurotoxicity and increase antioxidant enzyme activities in hippocampal neurons. *J. Neurochem.* **65,** 1740–1751.

Mattson, M. P., Goodman, Y., Luo, H., Fu, W., and Furukawa, K. (1997) Activation of NF-κB protects hippocampal neurons against oxidative stress-induced apoptosis: evidence for induction of manganese superoxide dismutase and suppression of peroxynitrite production and protein tyrosine nitration. *J. Neurosci. Res.* **49,** 681–697.

McIntyre, T. M., Zimmerman, G. A., and Prescott, S. M. (1999) Biologically active oxidized phospholipids. *J. Biol. Chem.* **274,** 25,189–25,192.

Mecocci, P., MacGarvey, U., Kaufman, A. E., Koontz, D., Shoffner, J. M., Wallace, D. C., and Beal, M. F. (1993) Oxidative damage to mitochondrial DNA shows marked age-dependent increases in human brain. *Ann. Neurol.* **34,** 609–616.

Mecocci, P., MacGarvey, U., and Beal, M. F. (1994) Oxidative damage to mitochondrial DNA is increased in Alzheimer's disease. *Ann. Neurol.* **36,** 747–751.

Mitsui, A., Hirakawa, T., and Yodi, S. (1992) Reactive oxygen-reducing and protein-refolding activities of adult T-cell leukemia-derived factor/human thioredoxin. *Biochem. Biophys. Res. Commun.* **186,** 1220–1226.

Montine, K. S., Kim, P. J., Olson, S. J., Markesbery, W. R., and Montine, T. J. (1997a) 4-Hydroxy-2-nonenal pyrrole adducts in human neurodegenerative disease. *J. Neuropathol. Exp. Neurol.* **56,** 866–871.

Montine, K. S., Olson, S. J., Amarnath, V., Whetsell, W. O., Graham, D. G., and Montine, T. J. (1997b) Immunohistochemical detection of 4-hydroxy-2-nonenal adducts in Alzheimer's disease is associated with inheritance of apoE4. *Am. J. Pathol.* **150,** 437–443.

Montine, K., Reich, E., Neely, M. D., Sidell, K. R., Olson, S. J., Markesbery, W. R., and Montine, T. (1998a) Distribution of reducible 4-hydroxynonenal adduct immunoreactivity in Alzheimer's disease is associated with apoE genotype. *J. Neuropathol. Exp. Neurol.* **57,** 415–425.

Montine, T. J., Markesbery, W. R., Morrow, J. D., and Roberts, L. J. (1998b) Cerebrospinal fluid F_2-isoprostane levels are increased in Alzheimer's disease. *Ann. Neurol.* **44,** 410–413.

Montine, T. J., Beal, M. F., Cudkowicz, M. E., O'Donnell, H., Margolin, R. A., McFarland, L., et al. (1999a) Increased CSF F_2-isoprostane concentration in probable AD. *Neurology* **52,** 562–565.

Montine, T. J., Beal, M. F., Robertson, D., Cudkowicz, M. E., Biaggioni, I., O'Donnell, H., et al. (1999b) Cerebrospinal fluid F2-isoprostanes are elevated in Huntington's disease. *Neurology* **52,** 1104–1105

Montine, T., Montine, K., Olson, S. J., Graham, D. G., Roberts, L. J., Morrow, J. D., et al. (1999c) Increased cerebral cortical lipid peroxidation and abnormal phospholipids in aged homozygous apoE-deficient C57BL/6J mice. *Exp. Neurol.* **158,** 234–241.

Moore, K. and Roberts, L. J. (1998) Measurement of lipid peroxidation. *Free Radic. Res.* **28,** 659–671.

Morrow, J. D. and Roberts, L. J. (1997) The isoprostanes: unique bioactive products of lipid peroxidation. *Prog. Lipid Res.* **36,** 1–21.

Moskovitz, J., Flescher, E., Berlett, B. S., Azare, J., Poston, J. M., and Stadtman, E. R. (1998) Overexpression of peptide-methionine sulfoxide reductase in Saccharomyces cerevisiae and human T cells provides them with high resistance to oxidative stress. *Proc. Natl. Acad. Sci. USA* **95,** 14,071–14,075.

Mullaart, E., Boerrigter, M. E., Ravid, R., Swabb, D. F., and Vijg, J. (1990) Increased levels of DNA breaks in cerebral cortex of Alzheimer's disease patients. *Neurobiol. Aging* **11,** 169–173.

Multhaup, G., Schlicksupp, A., Hesse, L., Beher, D., Ruppert, T., Masters, C. L., and Beyreuther, K. (1996) The amyloid precursor protein of Alzheimer's disease in the reduction of copper (II) to copper. *Science* **271,** 1406–1409.

Munch, G., Cunningham, A. M., Riederer, P., and Braak, E. (1998) Advanced glycation end products are associated with Hirano bodies in Alzheimer's disease. *Brain Res.* **796,** 307–310.

Namekata, K., Imagawa, M., Terashi, A., Ohta, S., Oyama, F., and Ihara, Y. (1997) Association of transferrin C2 allele with late-onset Alzheimer's disease. *Hum. Genet.* **101,** 126–129.

Nourooz-Zadeh, J., Liu, E. H., Yhlen, B., Anggard, E. E., and Halliwell, B. (1999) F4-isoprostanes as specific marker of docosahexaenoic acid peroxidation in Alzheimer's disease. *J. Neurochem.* **72,** 734–740.

Parkinson Study Group. (1989) Effect of deprenyl on the progression of disability in early Parkinson's disease. *N. Engl. J. Med.* **321,** 1364–1371.

Parshad, R. P., Sanford, K. K., Price, F. M., Melnick, L. K., Nee, L. E., Schapiro, M. B., et al. (1996) Fluorescent light-induced chromatid breaks distinguish Alzheimer disease cells from normal cells in tissue culture. *Proc. Natl. Acad. Sci. USA* **93,** 5146–5150.

Pedersen, W. A., Fu, W., Keller, J. N., Markesbery, W. R., Appel, S., Smith, R. G., et al. (1998) Protein modification by the lipid peroxidation product 4-hydroxynonenal in the spinal cords of amyotrophic lateral sclerosis patients. *Ann. Neurol.* **44,** 819–824.

Pedersen, W. A., Cashman, N., and Mattson, M. P. (1999) The lipid peroxidation product 4-hydroxynonenal impairs glutamate and glucose transport and choline acetyltransferase activity in NSC-19 motor neuron cells. *Exp. Neurol.* **155,** 1–10.

Pedersen, W. A., Chan, S. L., and Mattson, M. P. (2000) A mechanism for the neuroprotective effect of apolipoprotein E: isoform-specific modification by the lipid peroxidation product 4-hydroxynonenal. *J. Neurochem.* **74,** 1426–1433.

Picklo, M. J., Amarnath, V., Graham, D. G., and Montine, T. J. (1999) Endogenous catechol thioethers may be pro-oxidant or antioxidant. *Free Radic. Biol. Med.* **27,** 271–277.

Plantin, L. O., Lying-Tunell, U., and Kristensson, K. (1987) Trace elements in the human central nervous system studied with neutron activation analysis. *Biol. Trace Elem. Res.* **13,** 69–75.

Porter, N. A., Caldwell, S. E., and Mills, K. A. (1995) Mechanisms of free radical oxidation of unsaturated lipids. *Lipids* **30,** 277–290.

Pratico, D., Lee, V. M., Trojanowski, J. Q., Rokach, J., and Fitzgerald, G. A. (1998) Increased F_2-isoprostanes in Alzheimer's disease: evidence for enhanced lipid peroxidation in vivo. *FASEB J.* **12,** 1777–1783.

Pratico, D., Rokach, J., and Tangirala, R. K. (1999) Brains of aged apolipoprotein E-deficient mice have increased levels of F_2-isoprostanes, in vivo markers of lipid peroxidation. *J. Neurochem.* **73,** 736–741.

Przedborski, S., Jackson-Lewis, V., Yokoyama, R., Shibata, T., Dawson, V. L., and Dawson, T. M. (1996) Role of nitric oxide in 1-methyl-4-phenyl-1,2,3,6-tetrahydropyridine (MPTP)-induced dopaminergic neurotoxicity. *Proc. Natl. Acad. Sci. USA* **93,** 4565–4571.

Richardson, J. S. (1993) Free radicals in the genesis of Alzheimer's disease. *Ann. NY Acad. Sci.* **695,** 73–76.

Riederer, P., Sofic, E. M., Rausch, W. D., Schmidt, B., Reynolds, G. P., Jellinger, K., and Youdim, M. B. (1989) Transition metals, ferritin, glutathione, and ascorbic acid in parkinsonian brains. *J. Neurochem.* **52,** 515–520.

Roberts, L. J., Montine, T. J., Markesbery, W. R., Tapper, A. R., Hardy, P., Chemtob, S., et al. (1998) Formation of isoprostane-like compounds (neuroprostanes) in vivo from docosahexaenoic acid. *J. Biol. Chem.* **273,** 13,605–13,612.

Rosen, D. R., Siddique, T., Patterson, D., Figlewicz, D. A., Sapp, P., Hentati, A., et al. (1993) Mutations in Cu/Zn superoxide dismutase gene are associated with familial amyotrophic lateral sclerosis. *Nature* **362,** 59–61.

Sanchez-Ramos, J. R., Overvik, E., and Ames B. N. (1994) A marker of oxyradical-mediated DNA damage (8-hydroxy-2'-deoxyguanosine) is increased in nigrostriatum of Parkinson's disease brain. *Neurodegeneration* **3**, 197–204.

Sano, M., Ernesto, C., Thomas, R. G., Klauber, M. R., Schafer, K., Grundman, M., et al. (1997) A controlled trial of selegiline, α-tocopherol, or both as a treatment for Alzheimer's disease. The cooperative study. *N. Engl. J. Med.* **336**, 1216–1222.

Sasaki, N., Fukatsu, R., Tsuzuki, K., Hayashi, Y., Yoshida, T., Fujii, N., et al. (1998) Advanced glycation end products in Alzheimer's disease and other neurodegenerative diseases. *Am. J. Pathol.* **153**, 1149–1155.

Sayre, L. M., Zelasko, D. A., Harris, P. L., Perry, G., Salomon, R. G., and Smith, M. A. (1997) 4-Hydroxynonenal-derived advanced lipid peroxidation end products are increased in Alzheimer's disease. *J. Neurochem.* **68**, 2092–2097.

Schulz, J. B., Matthews, R. T., Muqit, M. M., Browne, S. E., and Beal, M. F. (1995) Inhibition of neuronal nitric oxide synthase by 7-nitroindazole protects against MPTP-induced neurotoxicity in mice. *J. Neurochem.* **64**, 936–939.

Shelley, M. L. (1998) 4-Hydroxy-2-nonenal may be involved in the pathogenesis of Parkinson's disease. *Free Radic. Biol. Med.* **25**, 169–174.

Shibata, N., Hirano, A., Kato, S., Nagai, R., Horiuchi, S., Komori, T., et al. (1999) Advanced glycation endproducts are deposited in neuronal hyaline inclusions: a study of familial amyotrophic lateral sclerosis with superoxide dismutase-1 mutation. *Acta Neuropathol. (Berl.)* **97**, 240–246.

Smith, C. D., Carney, J. M., Starke-Reed, P. E., Oliver, C. N., Stadtman, E. R., Floyd, R. A., and Markesbery, W. R. (1991) Excess brain protein oxidation and enzyme dysfunction in normal aging and Alzheimer disease. *Proc. Natl. Acad. Sci. USA* **88**, 10,540–10,543.

Smith, M. A., Taneda, S., Richey, P. L., Miyata, S, Yan, S. D., Stern, D., et al. (1994) Advanced Maillard reaction end products are associated with Alzheimer's disease pathology. *Proc. Natl. Acad. Sci. USA* **91**, 5710–5714.

Smith, M. A., Sayre, L., Monnier, V., and Perry, G. (1995) Radical AGEing in Alzheimer's disease. *Trends Neurosci.* **18**, 172–176.

Smith, M. A., Perry, G., Richey, P. L., Sayre, L. M., Anderson, V. E., Beal, M. F., and Kowall, N. (1996) Oxidative damage in Alzheimer's. *Nature* **382**, 120–121.

Smith, M. A., Harris, P. L. R., Sayre, L. M., and Perry, G. (1997a) Iron accumulation in Alzheimer disease is a source of redox-generated free radicals. *Proc. Natl. Acad. Sci. USA* **94**, 9866–9868.

Smith, M. A., Richey-Harris, P., Sayre, L. M., Beckman, J. S., and Perry, G. (1997b) Widespread peroxynitrite-mediated damage in Alzheimer's disease. *J. Neurosci.* **17**, 2653–2657.

Smith, M. A., Wehr, K., Harris, P. L. R., Siedlak, S. L., Connor, J. R., and Perry, G. (1998a) Abnormal localization of iron regulatory protein in Alzheimer's disease. *Brain Res.* **788**, 232–236.

Smith, M. A., Sayre, L. M., Anderson, V. E., Harris, P. L., Beal, M. F., Kowall, N., and Perry, G. (1998b) Cytochemical demonstration of oxidative damage in Alzheimer's disease by immunochemical enhancement of the carbonyl reaction with 2,4-dinitrophenylhydrazine. *J. Histochem. Cytochem.* **46**, 731–735.

Smith, M. A., Hirai, K., Hsiao, K., Pappolla, M. A., Harris, P. L., Siedlak, S. L., et al. (1998c) Amyloid-β deposition in Alzheimer transgenic mice is associated with oxidative stress. *J. Neurochem.* **70**, 2212–2215.

Smith, R. G., Henry, Y. K., Mattson, M. P., and Appel, S. H. (1998) Presence of 4-hydroxynonenal in cerebrospinal fluid of patients with sporadic amyotrophic lateral sclerosis. *Ann. Neurol.* **44**, 696–699.

Stadtman, E. R. (1992) Protein oxidation and aging. *Science* **257**, 1220–1224.

Struzynska, L., Dabrowska-Bouta, B., and Rafalowska, U. (1997) Acute lead toxicity and energy metabolism in rat brain synaptosomes. *Acta Neurobiol. Exp. (Warsz)* **57**, 275–281.

Su, J. H., Satou, T., Anderson, A. J., and Cotman, C. W. (1996) Up-regulation of Bcl-2 is associated with neuronal DNA damage in Alzheimer's disease. *Neuroreport* **7,** 437–440.

Swaim, M. W. and Pizzo S. V. (1988) Methionine sulfoxide and the oxidative regulation of plasma proteinase inhibitors. *J. Leukoc. Biol.* **43,** 365–379.

Takedo, A., Yasuda, T., Miyata, T., Mizuno, K., Li, M., Yoneyama, S., et al. (1996) Immuno-histochemical study of advanced glycation end products in aging and Alzheimer's disease brain. *Neurosci. Lett.* **221,** 17–20.

Tandon, L., Ni, B. F., Ding, X. X., Ehmann, W. D., Kasarskis, E. J., and Markesbery, W. R. (1994) RNAA for arsenic, cadmium, copper and molybdenum in CNS tissues from subjects with age-related neurodegenerative diseases. *J. Radioanal. Nucl. Chem.* **179,** 331–339.

Tu, P. H., Gurney, M. E., Julien, J. P., Lee, V. M., and Trojanowski, J. Q. (1997) Oxidative stress, mutant SOD1, and neurofilament pathology in transgenic mouse models of human motor neuron disease. *Lab. Invest.* **76,** 441–456.

Vogt, W. (1995) Oxidation of methionyl residues in proteins: tools, targets, and reversal. *Free Radic. Biol. Med.* **18,** 93–105.

Wade, R., Hirai, K., Perry, G., and Smith, M. A. (1998) Accumulation of 8-hydroxyguanosine in neuronal cytoplasm indicates mitochondrial damage and radical production are early features of Alzheimer disease. *J. Neuropathol. Exp. Neurol.* **57,** 511.

Whiteman, M. and Halliwell, B. (1996) Protection against peroxynitrite-dependent tyrosine nitration and α-1-antiproteinase inactivation by ascorbic acid. A comparison with other biological antioxidants. *Free Radic. Res.* **25,** 275–283.

Xie, C., Lovell, M. A., and Markesbery, W. R. (1998) Glutathione transferase protects neuronal cultures against four hydroxynonenal toxicity. *Free Radic. Biol. Med.* **25,** 979–988.

Yan, S. D., Chen, X., Schmidt, A. M., Brett, J., Godman, G., Zou, Y. S., et al. (1994) Glycated tau protein in Alzheimer disease: a mechanism for induction of oxidant stress. *Proc. Natl. Acad. Sci. USA* **91,** 7787–7791.

Yoritaka, A., Hattori, N., Uchida, K., Tanaka, M., Stadtman, E. R., and Mizuno, Y. (1996) Immunohistochemical detection of 4-hydroxynonenal protein adducts in Parkinson disease. *Proc. Natl. Acad. Sci. USA* **93,** 2696–2701.

Zhang, J. Z., Graham, D. G., Montine, T. J., and Ho, Y. S. (2000) Enhanced N-methyl-4-phenyl-1,2,3,6-tetrahydropyridine toxicity in mice deficient in Cu/Zn-superoxide dismutase or glutathione peroxidase. *J. Neuropathol. Exp. Neurol.* **59,** 53–61.

Zhang, J., Dawson, V. L., Dawson, T. M., and Snyder, S. H. (1994) Nitric oxide activation of poly (ADP-ribose) synthetase in neurotoxicity. *Science* **263,** 687–689.

Zhang, J., Perry, G., Smith, M. A., Robertson, D., Olson, S. J., Graham, D. G., and Montine, T. J. (1999) Parkinson's disease is associated with oxidative damage to cytoplasmic DNA and RNA in substantia nigra neurons. *Am. J. Pathol.* **154,** 1423–1429.

Zhu, H., Guo, Q., and Mattson, M. P. (1999) Dietary restriction protects hippocampal neurons against the death-promoting action of a presenilin-1 mutation. *Brain Res.* **842,** 224–229.

3

Parkinson's Disease

M. T. Silva and A. H. V. Schapira

INTRODUCTION

Parkinson's disease (PD) is one of the commonest neurological disorders with a prevalence of 1 in 350 giving an overall lifetime risk of approx 1 in 40 (Roman et al., 1995). Onset is typically between the fourth and seventh decade. Currently there are at least 500,000 PD cases in the United States but as the average age of Western populations increases, the number of patients affected may increase up to fourfold from the current levels (Tanner and Ben-Shlomo, 1999). PD remains a clinical diagnosis but may be supported by the presence of typical pathological features: the degeneration of pigmented brain-stem nuclei (e.g., the substantia nigra pars compacta, the locus coeruleus, and the substantia innominata) with Lewy bodies in some of the surviving neurons. Lewy bodies are intracytoplasmic inclusions containing a dense eosinophilic core of filamentous and granular material surrounded by radially oriented filaments and contain both ubiquitin and α-synuclein (Spillantini et al., 1997). Although other neurotransmitters are also affected in PD, the nigrostriatal dopaminergic tract bears the brunt of the neurodegenerative process with approx 50–70% loss of striatal dopamine at the time of clinical presentation (Hornykiewicz, 1975; Riederer and Wuketich, 1976).

There have been important recent advances in our understanding of the etiology and pathogenesis of PD. The study of familial parkinsonism has led to the identification of genes such as the α-synuclein and parkin genes. As a result of these advances, a theory of etiology involving both genetic susceptibility factors and environmental triggers is emerging. It is unlikely that a single etiological agent is capable of explaining all cases of idiopathic PD and thus it is more probable that multiple causes may result in the pathological and clinical features of PD. This chapter is an overview of the etiology and pathogenesis of PD. Genetic and environmental factors in the etiology of PD are discussed. Biochemical abnormalities implicated in the pathogenesis of nigral cell death are reviewed in the context of the clues that they may provide in understanding the mechanisms involved in the development of PD.

GENETICS OF PARKINSON'S DISEASE

Although Charcot suggested that PD was inherited (Charcot, 1867), for many years the role of genetic factors in the etiology and pathogenesis of idiopathic PD was thought to be minimal. More recent research including epidemiological surveys, twin studies,

From: *Pathogenesis of Neurodegenerative Disorders* Edited by: M. P. Mattson © Humana Press Inc., Totowa, NJ

and the study of familial PD have provided evidence that genetic factors may be much more important than previously suspected.

Epidemiological Family Studies

Over a century ago, Gowers suggested an increased prevalence of PD in the relatives of PD cases (Gowers, 1888). Several epidemiological studies have sought to evaluate the prevalence of PD among family members, but they have been fraught with methodological problems, such as ensuring the adequate examination of family members, the possibility of presymptomatic PD, the difficulty in definitively diagnosing idiopathic PD, as well as the necessity for an adequate control population. In studies where living secondary cases have been examined (Duvoisin et al., 1969; Martin et al., 1973; Alonso et al., 1986), first-degree relatives with PD were seen in 2–27% of probands. In other control studies where relatives were not examined (Butterfield et al., 1993; Marttila and Rinne, 1976; Vieregge et al., 1992a; Semchuk et al., 1993), this figure varied between 6% and 25%. This compared to control populations in which PD was seen in 0.7–8.5% of controls. These wide variations probably reflect both the different methodologies used and the populations involved. Several recent case control studies have also shown that PD is more common in relatives of PD cases compared to matched controls (Payami et al., 1994, Bonifati V et al., 1995; Vieregge and Heberlein, 1995, De Michele et al., 1996, Marder et al., 1996). Between 6% and 30% of these PD cases had an affected relative (first- or second-degree). The risk increases with the number of individuals affected in a family, suggesting a multifactorial etiology with an inheritable component in a subset of families (Lazzarini et al., 1994). The relative risk of developing PD in family members has varied widely between different studies, but in many of these incomplete family information was obtained and only one has been population based (Marder et al., 1996). Overall, the relative risk in first-degree relatives of PD cases is increased approx two- to threefold (Gasser, 1998).

Twin Studies

Demonstrating an increased concordance rate of a disease in monozygotic vs dizygotic twins provides strong evidence in favor of a genetic etiology for a disease. The first controlled twin study in PD was published in 1983 (Ward et al., 1983). Sixty-two twin pairs were identified, 43 of whom were monozygotic. Only one of the monozygotic twin pairs was diagnosed as having definite idiopathic PD. Other twin pairs had atypical parkinsonism and were not included in the analysis.

Other twin studies have since been published (Bharucha et al, 1986; Marsden, 1987; Marttila et al., 1988a; Vieregge et al., 1992b) but none of them show any significant difference in concordance rate between monozygotic and dizygotic twin pairs. These data would therefore mitigate against a significant genetic component to PD, although they do not exclude a mitochondrial DNA (mtDNA) influence. However, methodological review of the twin studies has indicated that they have insufficient power to exclude a genetic basis for PD (Johnson et al., 1990).

All of the problems applicable to pedigree analysis in PD are relevant to twin studies. Concordance rates are particularly difficult to establish in neurodegenerative disorders of late onset where there may be several years between the emergence of clinical features in co-twins. In addition, the selectivity of diagnosis and the exclusion of other disorders such as essential tremor may skew twin studies toward a negative result.

In an attempt to address some of these problems, positron emission tomography (PET) using [18]fluorodopa (^{18}F-dopa) has been used to evaluate the integrity of the nigrostriatal tract. ^{18}F-dopa uptake is significantly reduced in patients with PD and may be used to identify patients who may be presymptomatic (Brooks et al., 1990). An early study of asymptomatic co-twins with PET and ^{18}F-dopa uptake did not demonstrate a statistically significant difference in concordance rates (Burn et al., 1992). However, ^{18}F-dopa uptake was reduced in some asymptomatic co-twins and also in those who had isolated postural tremor. These results also suggest that essential tremor may need to be included as an inheritable factor that co-segregates with and may be a *forme fruste* of PD. PET would be extremely useful in PD longitudinal studies in co-twin analysis. Recently Piccini et al. (1999) have reported an ^{18}F-dopa study of 18 individuals who have a twin with PD, repeating the procedure 7 yr later. All of the twins were environmentally concordant in early life. At follow-up, the combined concordance levels for subclinical dopaminergic dysfunction and clinical PD were 75% in the 12 monozygotic twins and 22% in the 9 dizygotic twins evaluated twice, suggesting a substantial role for inheritance in sporadic PD.

The largest twin study in PD examined concordance rates in US Army veterans (Tanner et al., 1999). No increase in concordance was found overall between mono- and dizygotic twins. However, there was a significant increase in concordance when only patients and co-twins, the former with onset below age 50, were studied as a separate group. These results imply that young onset PD is more likely to be genetic.

Familial Parkinsonism

Early studies of PD suggested a high familial incidence, but there were a number of problems with these studies such as ascertainment bias and incomplete examination of family members (Allen, 1937; Mjönes, 1949). In the study by Mjönes (1949), a first-degree relative was affected in 68 out of 79 familial cases from a total of 149 patients with PD. In a high proportion of these families, the parents survived beyond the age of 60. In a disease which is age-related, this obviously skews the result toward a positive familial basis. There have been numerous reports of Parkinson kindreds (e.g., Nukada et al., 1978; Degl'Innocenti et al., 1989). There is autosomal dominant inheritance in the majority of families. Many of these and other studies have been complicated by how closely the affected individuals conform to a diagnosis of sporadic idiopathic PD. Thus, in not all cases was dopa responsiveness established and in only a few cases was pathological confirmation obtained. In many cases where pathology was available, atypical features have been found.

Autosomal Dominant PD (ADPD) and α-Synuclein Mutations

There is a large Italian/American kindred, known as the Contursi kindred, an extended family pedigree of 592 members with 60 affected individuals, who have highly penetrant autosomal dominant PD. The clinical picture of the PD phenotype in this family has been well documented (Golbe et al., 1990, 1996), exhibiting typical PD features but with infrequent tremor and an average of onset of 46 ± 13 yr with a fairly rapid progression to death and a mean survival of 10 yr. Affected family members showed pharmacological parallels to idiopathic PD in that they were dopa-responsive. Patients show typical neuropathological findings with Lewy bodies. In 1996, analysis of this kindred demonstrated linkage to chromosome 4q21-23 (Polymeropoulos et al., 1996). A year later a mutation in the α-synuclein gene was identified in all but one of the affected members (Polymeropoulos et al., 1997). This was an A→G substitution at base 209; this mutation codes for a substitution of threonine for alanine at amino acid 53. The same mutation was also found in three Greek families from Patras with autosomal dominant PD. The A53T mutation was not found in 230 patients of European descent with familial PD (Vaughan et al., 1998), in sporadic PD cases in the UK (Warner and Schapira, 1998), or in young onset PD cases (Chan et al., 1998).

A different α-synuclein mutation was identified in a German family with a G→C substitution at base 88 producing amino acid substitution of proline to alanine at amino acid 30, which is not far from the Contursi kindred mutation (Kruger et al., 1998). In both mutations, a hydrophobic residue substitutes for alanine, and both mutations occur in a coding region linking repeats of a consensus sequence of amino acids, KTKEGV (Golbe et al., 1999). The substitution in the Contursi kindred lies between repeats 3 and 4, whereas that of the German kindred occurs between repeats 2 and 3. The two protein defects might therefore be expected to produce a similar change in the chemical behavior or structure of α-synuclein.

So far five families with mutations in α-synuclein have been described and sequencing of the whole gene in 20 European and American kindreds failed to find any further mutations (Vaughan et al., 1998), making it a rare cause of PD. However, understanding of the mechanisms by which mutations in this gene can result in the clinical, pharmacological and pathological features of these familial cases of PD should provide important insights in the pathogenesis of idiopathic sporadic PD.

α-Synuclein

α-Synuclein is a small, flexible monomeric protein of 140 amino acids (14 kDa). It was initially described, in 1988, in the cholinergic synapses of an electric fish and, in 1991, in mammalian central nervous system (CNS) synaptic terminals (Maroteaux et al., 1988; Maroteaux and Scheller, 1991). It is abundantly expressed in the nervous system in which it is concentrated in presynaptic terminals. α-Synuclein mRNA has been found at low levels in all tissues except liver (Ueda et al., 1993). α-Synuclein is composed of three separate domains: the amphipathic region from residues 1 to 60, the NAC region from 61 to 89, and the acidic region from 100 to 130. It is a member of a highly conserved family of proteins (Borden, 1998). The normal function of α-synuclein is unknown. Its homolog in the zebra finch is upregulated during song learning, suggesting a role in neuronal plasticity (George et al., 1995). α-Synuclein is the precursor protein to the non-β-amyloid component of amyloid plaques in Alzheimer's disease (Ueda et al., 1993). This nonamyloid component of plaque is a 35-residue segment derived from α-synuclein.

An important finding was that Lewy bodies are predominately composed of α-synuclein together with neurofilament and ubiquitin (Spillantini et al., 1997). The amino acid change may disrupt an α-helix section of the protein and extend its β-sheet structure. The relationship of this action to neurodegeneration is uncertain, although such a change may promote self-aggregation. Ubiquitin conjugates with proteins targeted for proteolysis and, along with several other Lewy body constituents, appears to be the part of the cellular response to the abnormal neurofilament aggregation. The neurofilament is mostly intact but hyperphosphorylated, as if being prepared for another type of proteolysis. Ubiquitin staining is usually in the outer halo of the Lewy body, while α-synuclein staining is uniform and in the core of the Lewy body (Spillantini et al., 1997). α-Synuclein staining has now replaced ubiquitin staining as the standard diagnostic test for Lewy bodies. Lewy bodies stain with antibodies to both the C- and N-termini of α-synuclein, and the staining pattern aligns closely to the filaments seen in Lewy bodies. This suggests that α-synuclein itself aggregates abnormally into filamentous structures, which may precede the aggregation of neurofilament and ubiquitination steps, thus implying that aggregation of α-synuclein is a key factor in the pathogenesis of PD.

Other Autosomal Dominant PD Mutations

In 1998, two members of a German family were identified with an autosomal dominant PD missense mutation in the ubiquitin carboxy-terminal hydrolase L1 (UCH-L1) gene (Leroy et al., 1998). UCH-L1 is one of the most abundant proteins in the brain, comprising up to 2% of total brain protein and immunoreactivity for this protein has been found in Lewy bodies. UCH-L1 is involved in the ubiquitin-dependent proteolytic pathway (Leroy et al., 1998). This mutation, Ile93Met, has been shown to cause a partial loss of the catalytic activity of this thiol protease, which could lead to aberrations in the proteolytic pathway and aggregation of proteins (Leroy et al., 1998).

In 1998, a third locus for autosomal dominant PD with Lewy body pathology was described located on chromosome 2p13 (Gasser, 1998). Clinical features closely resemble those of idiopathic sporadic PD with a mean age of onset of 59 yr. The penetrance of this mutation was estimated to be 40%, suggesting it might play a role in idiopathic sporadic PD. Although the locus has been found, the gene has not yet been cloned.

Autosomal Recessive Juvenile PD (ARJPD)

ARJPD has been most commonly seen in the Japanese population and is characterized by onset before age 40 yr, symptomatic improvement following sleep, mild dystonia, and a good response to L-dopa (Yamamura et al., 1973). Resting tremor is seen less frequently than in idiopathic sporadic PD, and patients may have brisk tendon reflexes but no other pyramidal, cerebellar, or autonomic features. Pathologically there is dopaminergic cell loss in the substantia nigra pars compacta and locus coeruleus, but no Lewy bodies are seen (Takahashi et al., 1994). The gene responsible for ARJPD was mapped to 6q25.2-q27 (Matsumine et al., 1997), and, in 1998, the gene was discovered and named parkin (Kitada et al., 1998). Affected patients all carry deletions in parts of the parkin gene and in a subsequent analysis of 18 unrelated families, 34 patients were found to carry deletions of the parkin gene (Hattori et al., 1998). The function of the parkin protein is not known but it has homology at the *N*-terminus to ubiquitin and a ring finger motif at the C-terminus. Parkin possibly has ubiquitin-type properties, but it is interesting that Lewy bodies are not seen in ARJPD cases. Abbas et al. (1999) analyzed the coding regions of the parkin gene in 35 mostly European families with early-onset autosomal recessive PD, and found a wide variety of point mutations in addition to deletions in the parkin gene. They found a mutation in the parkin gene associated with a patient of 58 yr demonstrating that mutations in this gene are not always associated with early-onset PD. Recently, mutations in the parkin gene have been found in sporadic cases of PD.

Mitochondrial Inheritance

The mitochondrial genome encodes 13 proteins of the oxidative phosphorylation system in addition to two ribosomal and 22 transfer RNAs. The discovery of complex I deficiency in PD substantia nigra raised the possibility that the mutation of genes (nuclear or mitochondrial) encoding complex I subunits might be involved in determining the enzyme's defective activity. As mtDNA is inherited in a strictly maternal pattern, if there were full penetrance of such a mtDNA gene defect, mitochondrial inheritance should be identifiable in pedigrees with parkinsonism. Such maternal inheritance has been described in PD (Wooten, et al., 1997). Furthermore, 40% of patients with proven mitochondrial diseases and mtDNA mutations appear as sporadic cases. Thus maternal inheritance is not a *sine qua non* of mtDNA gene defects. However, molecular genetic investigations of mtDNA have so far been unable to identify any specific mutation that clearly co-segregates with PD.

Allelic Associations

It is possible that a given combination of gene alleles can determine an increased risk for developing PD. Attention has mainly focused on genes encoding enzymes involved in the metabolism of xenobiotics and enzymes involved in dopamine metabolism (Gasser et al., 1994).

One area of interest has been the cytochrome P450 system, which is involved in the metabolism of endogenous and exogenous molecules and is widely expressed in tissues, including brain and liver. It is possible that abnormalities of P450 may generate neurotoxins. Another area of interest was the "fast" and "slow" acetylator of debrisoquine, but a number of factors complicated interpretation of the results, including the effect that some anti-parkinsonian and other medications had on debrisoquine hydroxylation. A better approach is to determine the frequency of certain alleles of the gene family encoding these P450 enzymes. Two reports, each using Caucasian patients with PD and ethnically matched controls, demonstrated that the CYP2D6B allele was present in increased frequency among patients with PD (Armstrong et al., 1992; Smith et al., 1992). The increase was a modest 2.4-fold, but has provided support for the concept of abnormal xenobiotic metabolism contributing to the etiology of PD. However, subsequent reports have failed to confirm these findings (Planté-Bordeneuve et al., 1994; Kurth et al., 1995; Bordet et al., 1996; Diederich et al., 1996). Additionally, there was no increase in CYP2D6B in patients with the PD–dementia complex (Chen et al., 1996). A more recent study compared 100 pathologically proven PD cases with matched controls and found the slow acetylator phenotype for *N*-acetyltransferase 2 was more common in the PD group (69% compared to a control value of 37%; $p = 0.000002$) (Bandmann et al., 1997). Although the CYP2D6B allele may be associated with a slightly increased risk of developing PD, the literature is confusing and does not provide a clear answer.

Kurth et al. (1993) found an increase in a specific monoamine oxidase B (MAO-B) allele in patients with PD, but a further study failed to replicate this (Kurth et al., 1995).

Many other candidates have been studied. Linkage studies of glutathione peroxidase, tyrosine hydroxylase, brain-derived neurotrophic factor (BDNF), superoxide dismutase (SOD), catalase, and amyloid precursor protein have all been negative.

ENVIRONMENT

A number of studies have looked for an environmental agent that is linked to the development of PD. Epidemiological studies have searched for "clusters" of PD in an attempt to identify these risk factors (Tanner, 1986).

Geographic/Occupational

Rural Environment

A rural residency appears to increase the risk of the development of PD and in particular young-onset PD (Rajput et al., 1986; Rajput and Uitti, 1988; Davanipour and Will, 1990; Semchuk et al., 1991), However, this finding has not been confirmed in all studies (Semchuk et al., 1991).

Rural living is associated with farming and pesticide use, and an association with the agricultural industry has been found with increased incidence in PD patients (Barbeau et al., 1987; Gorrell et al., 1996; Fall et al., 1999). In addition, a further lifestyle study showed increased herbicide exposure in patients with PD (Semchuk et al., 1992). Organochloride pesticides were identified as risk factors in a German case control study (Seidler et al., 1996) with the offending agent being identified as the organochloride dieldrin, which was found in 6 of 20 PD brains and none of 14 controls (Fleming et al., 1994). Another study identified dithiocarbamate as a risk factor for PD (Semchuk et al., 1991), a compound that has also been shown to enhance MPTP toxicity (Corsini et al., 1985). Some studies have found the significant association of PD with farming as an occupation cannot be accounted for by pesticide exposure alone (Gorrell et al., 1996). Another rural factor that has been linked to PD is the consumption of well water (Tanner, 1996), although this may simply be further evidence in support of herbicides or pesticides as etiological factors for PD.

There is also a north/south gradient, providing further evidence for an environmental factor (Tanner, 1996).

Wood Industries

Proximity to wood-related industries has also shown an association with PD (Barbeau et al., 1987; Sethi et al., 1989).

Specific Neurotoxins

1-Methyl-4-phenyl 1,2,3,6 tetrahydropyridine (MPTP)

The emergence of the neurotoxin MPTP, as an experimental tool to produce a model of PD, has provided some insight into the etiology and pathogenesis of idiopathic PD. MPTP was originally synthesized as a meperidine analog "designer drug" and sold as an injectable narcotic on the streets of northern California (Davis et al., 1979; Langston et al., 1983). It is likely that hundreds of people injected MPTP, but a few individuals who repeatedly used this drug developed an akinetic rigid syndrome with or without resting tremor within 7–14 d. These patients responded well to L-dopa, although subsequently went on to develop fluctuations and dyskinesia in response to treatment (Langston and Ballard, 1984). ^{18}F-dopa PET scans of some of these patients showed reduced uptake. Interestingly, a less severe reduction of ^{18}F-dopa uptake was also seen in individuals exposed to the drug who had not developed clinical features of parkinsonism (Calne et al., 1985; Martin et al., 1986), and subsequent repeat scans have shown further progressive loss of the nigrostriatal system, as determined by ^{18}F-dopa uptake over time.

Although MPTP-induced parkinsonism has clinical, pharmacological and ^{18}F-dopa uptake parallels with idiopathic PD, in the few cases that have so far come to postmortem, the pathological changes are similar but not identical to PD. Severe loss of nigral neurons in the substantia nigra is seen but without evidence of classic Lewy bodies (Langston et al., 1983).

Systemic injection of MPTP into nonhuman primates also produces parkinsonism with bradykinesia, rigidity, and freezing. Tremor may also develop but it is not the typical resting tremor of parkinsonism, usually being a postural or action tremor (Tetrud and Langston, 1992). Like humans, MPTP-treated monkeys develop severe dopamine depletion in the striatum with marked nigrostriatal neuronal loss. There is also cell loss in the locus coeruleus in MPTP-treated monkeys similar to that seen in idiopathic PD. In MPTP-treated monkeys, eosinophilic intraneuronal inclusions have been observed but they lack the structure typical of Lewy bodies (Forno et al., 1988).

Evidence now supports the view that MPTP causes cell death through inhibition of oxidative phosphorylation and the generation of free radicals. MPTP is a protoxin, being converted to its active form 1-methyl-4-phenylpyridinium (MPP$^+$) via the intermediate 2,3-dihydropyridinium (MPDP). Metabolism of MPTP to MPDP is by monoamine oxidase (MAO) A and B, preferentially by MAO-B (Singer et al., 1986). This conversion probably occurs in glia, as MAO-B is predominantly located in astrocytes and serotonergic neurons in the CNS, where it is located on the mitochondrial outer membrane (Levitt et al., 1982). Inhibition of this enzyme can prevent MPTP toxicity. MPDP is capable of passing across cell membranes and so the conversion of MPDP to MPP$^+$ probably occurs in the extracellular space by auto-oxidation (Di Monte et al., 1992).

MPP$^+$ is a substrate for the dopamine reuptake system (Chiba et al., 1984; Javitch and Synder, 1984). Blockade of MPP$^+$ uptake, by for instance nomifensine, can prevent MPTP toxicity (Melamed et al., 1985; Schultz et al., 1989). Thus MPP$^+$ is actively concentrated into dopaminergic neurons. Although taken up via the nerve terminals in the striatum, MPP$^+$ may be concentrated in the cell body through its affinity with neuromelanin (Lyden et al., 1983; D'Amato et al., 1987). Thus neuromelanin, which is an autooxidation product of L-dopa, may act as a toxic sink in the cell body.

Once inside the cell MPP$^+$ is concentrated to millimolar proportions by an energy-dependent mitochondrial ion concentrating system. Once within mitochondria, MPP$^+$ inhibits NADH CoQ$_{10}$ reductase (complex I), the first enzyme of the mitochondrial respiratory chain. There is also evidence that MPP$^+$ generates free radicals and that the nitric oxide synthase (NOS) inhibitor, 7-nitroindozole (7-NI), can prevent MPP$^+$ toxicity in monkeys (Hantraye et al., 1996). The current assumption is that MPP$^+$ induces cell death through a combination of inhibition of adenosine triphosphate (ATP) synthesis and free-radical generation. The interrelationship of mitochondrial inhibition and free-radical generation will be discussed in detail with reference to the pathogenesis of cell death in PD.

Although much is now known about the mechanism of action of MPTP, there are still several intriguing features that remain to be explained. For instance, it is estimated that possibly several hundred individuals were exposed to this chemical, yet only seven or eight developed parkinsonism. The simplest explanation for this phenomenon would be that these individuals were those who were exposed to the highest concentrations of the drug. However, an intriguing alternative is that they were genetically susceptible to the effects of this toxin. This could, for instance, be determined through an abnormality of mitochondrial function or free-radical metabolism that in itself was incapable of causing nigral cell death but, when combined with MPP^+, precipitated a specific cascade that resulted in severe cell loss and parkinsonism. It is interesting to note, therefore, that one of the sons of a man who developed MPTP-induced parkinsonism has subsequently developed clinical features of PD, suggesting that both may have some predisposing genetic factor (J. W. Langston, personal communication).

Carbon Monoxide

Like nitric oxide (NO), carbon monoxide is a common environmental pollutant but may also play an important role in cell signaling. Acute carbon monoxide poisoning results in complexing of carbon monoxide with the ferrous iron, protophorphyrin IX, and this prevents the carriage of oxygen. In addition, carbon monoxide is a potent inhibitor of cytochrome oxidase (complex IV) of the mitochondrial respiratory chain. Survivors of carbon monoxide poisoning have developed parkinsonism within a few days or weeks of exposure (Grinker, 1926; Gordon, 1965; Klawans et al., 1982). The affected patients show necrosis of the globus pallidus on computed tomography (CT) and magnetic resonance imaging (MRI).

Manganese

The first description of manganese inducing an akinetic rigid syndrome was in 1837 when Couper (1837) described a patient who developed stiffness and rigidity while working in a French manganese mill. Since that time, there have been numerous reports of manganism developing among individuals exposed to manganese dioxide ore, usually by inhalation of manganese dust. Exposure is typically chronic over 6 mo to 16 yr and the onset of manganism is slow, beginning with apathy, muscle weakness and cramps, and general irritability. Progression occurs and is characterized by dysarthria and psychosis followed by severe rigidity, anarthria, and some dystonia (Huang et al., 1993). Manganese predominantly appears to affect the striatum and globus pallidus with sparing of the substantia nigra (Mena, 1979).

Cadmium

Acute exposure to cadmium resulting in the development of parkinsonism 3 mo later has been reported (Okuda et al., 1997).

Diet

Because of the complex nature of individual diets, identifying dietary factors that may influence the development of PD is difficult. However, there are specific cases of dietary-induced parkinsonism, such as the Guam amyotrophic lateral sclerosis–parkinsonism–dementia complex, which some have suggested may be due to ingestion of a toxin from the cycad plant (Spencer, 1987). This has not subsequently been confirmed.

Niacin-containing foods (coffee, wine, meat, smoked ham, eggs, white bread, and tomatoes) may reduce the risk of PD (Fall et al., 1999), but others have not confirmed these findings (Anderson et al., 1999). The association with vitamin consumption is unclear, although vitamin A may be associated with an increased risk of PD (Cerhan et al., 1994; Anderson et al., 1999). There is no consensus regarding vitamins C or E (Golbe et al., 1990; Morens et al., 1996; Hellenbrand et al., 1996a, b; Logroscino et al., 1996; de Rijk et al., 1997; Anderson et al., 1999). However, one large prospective study found vitamin C to decrease the risk of developing PD (Cerhan et al., 1994). It has been suggested that PD is associated with diets high in animal fat (Anderson et al., 1999). Epidemiological data obtained in a prospective study of a cohort in New York City indicate that low calorie intake is associated with reduced risk for PD, and that a high calorie intake increases risk (Logroscino et al., 1996). Animal studies have shown that dietary restriction increases resistance of substantia nigra dopaminergic neurons to MPTP-induced damage and improves behavioral outcome (Duan and Mattson, 1999).

INFECTION

Encephalitis Lethargica

Between 1917 and 1919, there was an epidemic of an influenza-like illness starting in Austria and France but spreading throughout Europe and North America. The illness was characterized by fever, headache, lethargy, and paralysis, particularly of the extraocular muscles. Following this, stupor, coma, sleep disturbance, and seizures could occur. Ocular gyric crises were seen in a high proportion of patients. Mortality was 30–40% and parkinsonism developed in the majority of survivors over the next 10 yr (Ziegler, 1928; Von Economo, 1931; Anonymous, 1981). The specific agent causing encephalitis lethargica was never identified.

Although there has been no outbreak of encephalitis lethargica since the 1920s, infection as a cause for PD has still attracted some attention. There are numerous anecdotal reports of infections, particularly encephalitis, being associated with parkinsonism. These include a wide variety of viruses, bacteria—including *Borrelia burgdorferii* (Lyme disease)—and even fungi, such as a cryptococcus or aspergillus. However, there is no evidence to suggest that any of these are relevant to the vast majority of patients with idiopathic PD. For instance, patients who develop PD before the age of 40 have no greater history of CNS infection than do patients who develop the disease over the age of 60 (Dulaney et al., 1990). Intrauterine exposure to, for instance, the influenza virus pandemic from 1890 to 1930 has not been supported by any association with year of birth (Ebmeier et al., 1989). Searches for viral particles of antigens within the brains of patients with PD has not proved rewarding (Schwartz and Elizan, 1979; Mann et al., 1981).

INFLAMMATION AND GLIAL CELLS

The presence of activated microglia, present at the time of death in PD, has supported the possibility that there may be some immune component to the pathogenesis of PD (McGeer et al., 1988). Different subpopulations of glial cells may either be neuroprotected by mechanisms such as dopamine and free-radical scavenging or they may contribute to the neurotoxic process (Hirsch et al., 1998). Activated microglia are involved in inflammation of the nervous system by releasing proteinases and cytokines among other things (Benveniste, 1992). In PD cases, levels of factors such as interleukin 1-β (IL-1β), interferon-γ (IFN-γ), and tumor necrosis factor-α (TNF-α) are massively elevated compared to controls (Hirsch et al., 1998). However, these reactions are not specific to PD and are seen in other neurodegenerative disorders such as Alzheimer's disease (AD) and multiple sclerosis (MS) (Hopkins and Rothwell, 1995). It is likely that the inflammation observed forms part of the final common pathway of cell death seen in the neurodegenerative disorders.

PATHOGENESIS

Several biochemical abnormalities have now been identified in PD substantia nigra. These include abnormal iron accumulation, alteration in the concentration of iron-binding proteins, evidence for increased oxidative stress and oxidative damage, and mitochondrial complex I deficiency. There is also some preliminary evidence of increased NO formation and the generation of nitrotyrosine residues within PD substantia nigra. Each of these factors may form part of the pathogenesis of nigral cell death, as well as being potential etiological factors in themselves. Research has attempted to determine whether these biochemical abnormalities are primary or secondary and whether they may be related to each other in a single sequence of events.

All of these biochemical abnormalities have been, by necessity, identified in post-mortem brain. An advantage of this is that the pathological diagnosis of PD can be confirmed. However, it is important to distinguish between *bone fide* biochemical abnormalities and those that might be a consequence of postmortem changes. In addition, the PD brains that become available for such analyses are inevitably at the end stage of the disease process when the majority of the nigral neurons have already disappeared and gliosis is often widespread.

Iron

High iron concentrations are found in control substantia nigra, globus pallidus, and striatum. In PD, there is a 35% increase in substantia nigra iron levels (Dexter et al., 1987; Sofic et al., 1991). Other degenerative diseases involving cell loss in the basal ganglia also showed increased iron in these areas, e.g., progressive supranuclear palsy (PSP) and multiple system atrophy (MSA). These studies suggested that increased iron concentrations were a reflection of neuronal cell loss rather than any specific pathogenetic factor.

Perhaps not surprisingly, high concentrations of iron were found in macrophages, astrocytes, and reactive microglia in the PD substantia nigra (Jellinger et al., 1990). However, one study using X-ray microanalysis found increased levels of iron in neuromelanin. In this respect, neuromelanin could again act as a toxic sink (Jellinger et al., 1992). In contrast, another study found no difference in iron concentrations between melanized and nonmelanized cells in controls but a significant increase in the cytoplasm of dopaminergic neurons (Hirsch et al., 1991). However, there was no apparent correlation between the high concentrations of iron and morphological alterations in the neurons, which might suggest degeneration.

Iron is capable of catalyzing oxidative reactions that may generate hydrogen peroxide and the hydroxyl ion.

$$O_2 + Fe^{2+} \rightarrow O_2^{\cdot-} + Fe^{3+}$$

$$H_2O_2 + Fe^{2+} \rightarrow OH^{\cdot} + OH^- + Fe^{3+}$$

Thus, if iron is available in a free and reactive form, it has the potential for exacerbating oxidative stress and damage. Iron is normally bound to ferritin, which exists in two forms, H and L. Most brain ferritin is in the H form. Three studies have now been undertaken on ferritin concentrations in PD brain. One used a polyclonal antibody predominantly against L-ferritin and found a significant decrease in the concentration of this protein in PD substantia nigra and other areas (Dexter et al., 1990). This decrease was not seen in other parkinsonian syndromes such as PSP or MSA. Another study (Riederer et al., 1989), again using an antibody against mainly L-ferritin, found an increase in the number of ferritin-positive microglia in substantia nigra. This latter work used immunohistochemistry and therefore is not directly quantifiable. In addition, this study incorporated parkinsonian syndromes as well as idiopathic PD into the disease group (M. Youdim, personal communication).

A third and more comprehensive study involved monoclonal antibodies against both L and H ferritin together with a double-capture technique incorporated into an enzyme-linked immunosorbent assay (ELISA) study together with Western blotting studies of PD substantia nigra protein (Mann et al., 1994). The results did not identify any significant difference in ferritin levels between control and PD substantia nigra. Thus, there are no hard data that ferritin levels are abnormal in PD. Indeed ferritin has such a high iron-binding capacity that the increase of iron noticed in PD brain may not require any increased buffering capacity from ferritin.

Oxidative Stress and Damage

There are several lines of evidence that suggest increased oxidative stress and oxidative damage to biomolecules in PD substantia nigra.

1. Glutathione (GSH) in its reduced form is an important compound in antioxidant defense and in the repair of oxidized proteins. It is oxidized to its disulphide, GSSG. High GSH/GSSG ratios are maintained by glutathione (GSSG) reductase, which converts GSSG to GSH. One study reported a decrease in the activity of glutathione peroxidase in PD nigra, putamen, and globus pallidus (Kish et al., 1985), but others (Riederer et al., 1989) have not reproduced this.

   There is evidence that GSH levels are decreased in PD substantia nigra (Perry et al., 1982; Sian et al., 1994). Total GSH levels appear to be slightly lower in PD substantia nigra. This combination suggests enhanced free-radical generation in the PD nigra.

2. SOD exists in cytosolic (copper/zinc, Cu/Zn) SOD and mitochondrial manganese (Mn) SOD forms and is important in dismutating superoxide ions. Thus, levels of this enzyme are indicative of superoxide generation. Both copper/zinc and manganese SOD appear to be increased in PD substantia nigra (Marttila et al., 1988b; Saggu et al., 1989). High levels of copper/zinc SOD are expressed at the mRNA level in control and PD nigral pigmented neurons (Hirsch et al., 1989; Ceballos et al., 1990). Taken together, these observations suggest that PD nigral neurons, in particular, are exposed to increased superoxide generation.

3. Levels of polyunsaturated fatty acids (Dexter et al., 1986), malondialdehyde, and hydroperoxides (Jenner, 1991) are increased in PD substantia nigra. These are the products of free-radical damage to lipid membranes and imply oxidative damage in PD.

   Free-radical damage to DNA produces intracellular 8-hydroxydeoxyguanosine. Elevated concentrations of this product are seen in PD in the nuclear DNA and particularly in the mtDNA fractions from patients with PD (Sanchez-Ramos et al., 1994). Levels of these products are also particularly high in control brains in the substantia nigra and striatum confirming that, even in controls, this area of the brain is a site of high oxidative stress.

Nitric Oxide

The free-radical gas NO· is present in many tissues, including the CNS. NO· is generated by the conversion of L-arginine to L-citrulline by NOS. At least three NOS isoforms are recognized, and are all expressed within the brain. NO· acts as an atypical molecular messenger. At higher concentrations, it may have a toxic role, and has been implicated in the neurodegeneration that occurs in PD. As a free radical, NO· could potentially contribute to dopaminergic neuronal death by mechanisms such as increased lipid peroxidation, release of iron (II), and damage to DNA. It can also inhibit a number of enzymes such as cytochrome c oxidase and SOD. As discussed earlier, it also affects mitochondrial function by inhibiting complexes II, III, and IV. Animal studies have implicated NO· in the nigrostriatal neuronal loss. In addition to its possible neuroprotective effect with regard to MPP$^+$ toxicity, 7-nitroindazole (7-NI) also protects in the methamphetamine animal model of PD (Itzhak and Ali, 1996).

Clinical Evidence of NO· Involvement in PD

NOS activity is at its highest in the nigrostriatal system in nonhuman primates and humans. Attempts to demonstrate altered levels of NO· in the brains of PD patients have been inconclusive with both decreased and increased levels of cerebrospinal fluid nitrate, a marker of NOS activity, being seen (Kuiper et al., 1994, Molina et al., 1994, 1996; Qureshi et al., 1995).

Mitochondrial Deficiency

Mitochondria are responsible for producing ATP by oxidative phosphorylation (OXPHOS). This system is also responsible for producing 95% of the cell's superoxide ions during aerobic metabolism. Thus, mitochondria occupy a pivotal role in both energy supply and the potential for generating cell damage.

The mitochondrial respiratory chain and the OXPHOS system comprises five multimeric proteins (complexes I–V) located on the inner mitochondrial membrane. Reducing equivalents feed into the respiratory chain through complexes I or II, or from β-oxidation via the electron-transfer flavoprotein. An electrochemical gradient is generated across the inner membrane and this drives ATP synthesis through the reduction of molecular oxygen. Inhibition of the respiratory chain, particularly at complex I or complex III, results in the increased generation of free radicals (Takeshige and Minakimi, 1979; Hasegawa et al., 1990).

MPP$^+$ inhibits complex I in a concentration-dependent but reversible manner. Prolonged exposure of submitochondrial particles to MPP$^+$, however, results in more severe and irreversible inhibition of complex I when electron flow through the chain is prevented by inhibition of cytochrome oxidase (complex IV). This severe and irreversible inhibition can be prevented with free-radical scavengers (Cleeter et al., 1994). Free radicals can damage the respiratory chain. In vitro studies suggest that both complexes I and III are particularly affected whereas in vivo studies suggest that complex IV and, less so, complex I are inhibited (Hillered and Ernster, 1983; Zhang et al., 1990; Benzi et al., 1991; Thomas et al., 1993).

Mitochondria contain between 2 and 10 molecules of DNA. MtDNA is a circular double-stranded molecule, 16.5 kb long, and encodes its own 22 transfer RNAs, 2 ribosomal RNAs, and 13 proteins, all of which are part of the respiratory chain. This includes seven subunits of complex I, cytochrome *b* of complex III, three subunits of complex IV, and two subunits of complex V (ATPase). MtDNA is dependent on the nuclear genome for its replication, transcription, translation, and repair and is considered particularly susceptible to the damaging effects of free radicals because of its lack of a histone coat, as well as its rather limited repertoire of repair enzymes.

Numerous mutations of mtDNA have recently been associated with human diseases. These include a variety of ocular and limb myopathies, encephalomyopathies including mitochondrial encephalopathy, lactic acidosis, stroke-like episodes, and myoclonic epilepsy with ragged red fibers. MtDNA mutations responsible for these disorders include point mutations of tRNAs or protein-coding genes, and genomic rearrangements including deletions and duplications. MtDNA mutations may be somatic—such as occur with senescence—or inherited. MtDNA is inherited maternally and many primary mitochondrial disorders exhibit such inheritance patterns. However, approx 40% of patients with proven mtDNA mutations have no family history. Patients with mitochondrial myopathies have an increased incidence of movement disorders over and above that expected in controls (Truong et al., 1990). Also, certain families with point mutations in complex I coding genes have Leber's hereditary optic neuropathy (LHON), associated particularly with dystonia (Jun et al., 1994; De Vries et al., 1996). Furthermore, mitochondrial complex I deficiency has been identified in platelet mitochondria from patients with dystonia (Benecke et al., 1992; Schapira et al., 1997).

A mitochondrial defect in PD was first identified in 1989 in substantia nigra from patients with PD (Schapira et al., 1989a, b). This study has been expanded over the years and results to date show that there is a specific approx 35% complex I deficiency in PD nigra. This defect in complex I activity does not affect any other part of the respiratory chain. In addition, no defect in mitochondrial activity has been identifiable in any other part of PD brain, including caudate putamen, globus pallidus, tegmentum, cortex, cerebellum, or substantia innominata (Schapira et al., 1990a).

This observation, which provided a direct biochemical link between idiopathic PD and the MPTP model of this disorder, obviously raised many questions, specifically regarding its primary or secondary role in etiology/pathogenesis. As virtually all the brains taken from PD patients had been exposed to L-dopa, an important question related to whether L-dopa caused the complex I deficiency. However, there is no deficiency of complex I activity in PD striatum, which one might expect from the rat model (Cooper et al., 1995). L-dopa does not appear to cause any deficiency of complex I activity in platelet mitochondria, which again might be expected given the relatively high circulating levels of L-dopa in the periphery. It has been shown that patients with MSA who have take L-dopa in quantities and for duration comparable to patients with PD have no defect of mitochondrial activity in their substantia nigra (Gu et al., 1997). Furthermore, cell loss in the nigra is at least as severe in MSA as it is in PD and so one would expect that, if the mitochondrial defect in PD were simply a reflection of this degeneration, the same abnormality should be present in MSA. Its absence in MSA, therefore, suggests that its presence in PD is the result of a more specific cause than simple cell loss.

Following the report of complex I deficiency in PD substantia nigra, respiratory-chain abnormalities were described in skeletal-muscle mitochondria from PD patients. This particular area has proved very contentious, with several groups either describing similar defects or no abnormality whatsoever (see Schapira, 1994, for review). Two magnetic resonance spectroscopy studies on skeletal-muscle mitochondrial function in PD have shown conflicting results (Taylor et al., 1994; Penn et al., 1995). It is possible, however, to encompass these different observations into a single hypothesis (*see* below).

Finally, mitochondrial complex I deficiency was also identified in platelet mitochondria of PD patients (Parker et al., 1989). In contrast to skeletal muscle, there is a consensus among several laboratories that complex I deficiency does exist in PD platelet mitochondria. The majority of studies, however, suggest that this deficiency, as least based on a group to group analysis, is modest (circa 20–25%) (see Schapira, 1994, for review). The complex I deficiency in PD lacks the sensitivity to allow its use as a biomarker of PD.

The Mitochondria/Free-Radical Defect in PD

Studies investigating biochemical abnormalities of PD, whether using postmortem material or living patients, have inherent problems. One of the most important problems may be the assumption of common etiology among all PD patients. Thus, collecting PD material and comparing results of specific assays to a matched control population may obscure defects that may be present in only a subgroup. This may be particularly important with reference to mitochondrial dysfunction in PD. The complex I activity data, in the PD groups using both brain and platelet material, show considerable scatter and overlap with controls. The statistical difference is derived from the mean of the group-to-group analysis. A more pertinent analysis may be to use an additional factor to help segregate subgroups of PD patients and allow a clearer distinction with the control group. This may be possible, for instance in the platelet mitochondria studies, by using only those PD patients with complex I activities less than 75% of the control mean.

The observation of mitochondrial and free-radical abnormalities in PD is consistent with the hypothesis that these two defects may form part of a self-amplifying cycle of events. As has already been discussed, a defect of complex I activity may generate increased superoxide ions that may enhance oxidative stress and damage in the presence of elevated concentrations of iron. Increased free-radical generation itself may also amplify the complex I defect, and so on. Although this process may begin at a very low level precipitated by either an inborn error of metabolism or an environmental agent, their interrelationship will ensure the establishment of the cycle and its perpetuation. Whether these events occur at the proximal or distal ends of the cascade that terminates in dopaminergic cell death is at present unknown, although it would seem more logical that they are involved in the end-stage of the pathway.

Mitochondrial Genetics

Following the identification of the complex I defect in PD, several studies investigated the structure of mtDNA in tissues from PD patients. An early study suggested increased levels of the common deletion in PD brain (Ikebe et al., 1990). However, this study did not use age-matched controls and the increase in the common deletion was in fact no greater than would be expected from the aging process. Other studies using properly matched controls found no increase in this mitochondrial mutation in PD substantia nigra (Schapira et al., 1990b; Lestienne et al., 1991). Several studies have sequenced mtDNA in PD but these have all involved small numbers of unselected patients (Ozawa et al., 1991; Ikebe et al., 1995). Although some reports have suggested an increased frequency of certain mtDNA polymorphisms in PD, this has not been replicated in other studies (Shoffner et al., 1993; Lücking et al., 1995).

It is possible to use a cell complementation system to assess whether a respiratory-chain deficiency is determined by mitochondrial or nuclear DNA. This system makes use of ρ^0 cells that have no mtDNA but are capable of surviving in supplemented medium. These cells have a nucleus but no functioning respiratory chain. MtDNA taken from donor cells may then be fused into these ρ^0 cells. The production of a respiratory-chain defect in the recipient cells will indicate that this has derived from the donor mtDNA. This system has been used to show that the respiratory-chain defects of LHON 3460 (Cock et al., 1998) and mtDNA depletion syndrome (Bodnar et al., 1993) have a nuclear component. Others have shown that the biochemical deficiency associated with mtDNA deletions and the 3243 bp tRNA $^{Leu(UUR)}$ mutation associated with the MELAS phenotype follow the transfer of mtDNA from donor to recipient ρ^0 cells (Dunbar et al., 1996).

Two recent studies have used genetic transplantation to investigate the potential for PD mtDNA to determine the complex I defect. In one study, unselected PD patients' cells were fused and grown in mixed culture (Swerdlow et al., 1996). In another, PD patients were selected on the basis of demonstrating a peripheral complex I deficiency. These patients' cells were then fused with ρ^0 cells and grown both in mixed and clonal culture (Gu et al., 1998). In both, mtDNA transferred from the PD patients induced a complex I defect in the recipient cybrid cells. These results indicate that the mtDNA in these patients caused the complex I deficiency through either inherited or somatic mutations. Further experiments suggested that the recipient cells also developed abnormal calcium handling and a lower mitochondrial membrane potential.

APOPTOSIS

Apoptosis is a form of programmed cell death that involves a stereotyped sequence of biochemical and morphological changes. Evidence that dopaminergic neurons die by an apoptotic process in PD is accumulating. Analyses of postmortem brain tissue from PD patients have revealed evidence for neuronal apoptosis including nuclear condensation, chromatin fragmentation and formation of apoptotic bodies (Mochizuki et al., 1996; Tompkins et al., 1997; Tatton et al., 1998). In addition, aberrant expression of cell cycle-related genes in substantia nigra dopaminergic neurons has been documented in immunohistochemical analyses of PD brain tissue sections (Nakamura et al., 1997). Additional findings provide quite compelling evidence that dopaminergic neurons can die by apoptosis in experimental animal models of PD. Analyses of brain tissue from MPTP-treated mice revealed evidence for apoptosis-related DNA damage in substantia nigra dopaminergic neurons (Tatton and Kish, 1997). Moreover, it was reported that inhibitors of caspases, proteases that play a central role in apoptosis, can greatly improve survival of dopaminergic neurons following their transplantation into the substantia nigra of lesioned animals (Schierle et al., 1999). Levels of prostate apoptosis response-4 (Par-4), a protein that plays a central role in neuronal apoptosis (Duan et al., 1999a), increase dramatically in midbrain dopaminergic neurons of monkeys and mice after MPTP administration (Duan et al., 1999b). The increase in Par-4 levels occurs in both neuronal cell bodies in the substantia nigra and their axon terminals in the striatum, and precedes loss of tyrosine hydroxylase immunoreactivity and cell death. The latter studies further showed that blockade of Par-4 production by antisense treatment prevents apoptosis of cultured human dopaminergic cells, demonstrating that a critical role for Par-4 production by antisense treatment prevents apoptosis of cultured human dopaminergic cells, indicating a critical role for Par-4 in the cell death process. Thus, therapeutic approaches aimed at preventing apoptosis of dopaminergic neurons are being pursued.

CONCLUSIONS

There is increasing evidence that nigrostriatal cell death involves multiple processes that form a final pathway that may be common to many neurodegenerative diseases. Oxidative stress and free-radical generation involving NO occurs. Free-radical scavenger systems such as glutathione are suboptimal in the basal ganglia of PD patients, which enables an ever-increasing cycle of oxidative stress to occur. Mitochondrial function is further compromised, leading to decreased ATP production and subsequent cell death. Additional inflammatory changes occur, predominantly in the glia, which further accelerate these processes. Manipulation of these processes may lead to effective neuroprotective agents that can slow the rate of neuronal loss.

The key question is what triggers this final common pathway in PD. Available data imply that multiple etiologies are more likely than a single common factor. This hypothesis for etiology in PD may also include a combination of genetic and environmental factors where the contribution of each may vary between affected patients. Genetic susceptibility may be determined in part through impaired metabolism of free radicals or complex I activity, which in turn may be the product of nuclear or mitochondrial genomic defects. Environmental interactions may include exogenous toxin(s) with uptake and conversion characteristics similar to MPTP, such that they are targeted to the substantia nigra, or endogenously generated neurotoxin(s) such as the tetrohydroisoquinolines or β-carbolines. Evidence is emerging that genetic factors may be more important in idiopathic PD than previously thought.

The ability to identify subsets in patients with PD will accelerate research into etiology and pathogenesis and circumvent the problems of specific factors being obscured by group-to-group comparison.

REFERENCES

Abbas, N., Lücking, C. B., Ricard, S., Dürr, A., Bonifati, V., De Michele, G., et al. (1999) A wide variety of mutations in the parkin gene are responsible for autosomal recessive parkinsonism in Europe. *Hum. Mol. Genet.* **8,** 567–574.

Allen, W. (1937) Inheritance of the shaking palsy. *Arch. Int. Med.* **60,** 424–426.

Alonso, M. E., Otero, E., D'Regules, R., and Figueroa, H. H. (1986) Parkinson's disease: a genetic study. *Can. J. Neurol. Sci.* **13,** 248–251.

Anderson, C., Checkoway, H., Franklin, G. M., Beresford, S., Smith-Weller, T., and Swanson, P. D. (1999) Dietary factors in Parkinson's disease: the role of food groups and specific foods. *Mov. Disord.* **14,** 21–27.

Anonymous (1981) Encephalitis lethargica. *Lancet* **2,** 1396–1397.

Armstrong, M., Daly, A. K., Cholerton, S., Bateman, D. N., and Idle, J. R. (1992) Mutant debrisoquine hydroxylation genes in Parkinson's disease. *Lancet* **339,** 1017–1018.

Bandmann, O., Vaughan, J., Holmans, P., Marsden, C. D., and Wood, N. W. (1997) Association of slow acetylator genotype for N-acetyltransferase 2 with familial Parkinson's disease. *Lancet* **350**, 1136–1139.

Barbeau, A., Roy, M., Bernier, G., Campanella, G., and Paris, S. (1987) Ecogenetics of Parkinson's disease: prevalence and environmental aspects of rural areas. *Can. J. Neurol. Sci.* **14**, 36–41.

Benecke, R., Strümper, P., and Weiss H. (1992) Electron transfer complex I defect in idiopathic dystonia. *Ann. Neurol.* **32**, 683–686.

Benveniste, E. N. (1992) Inflammatory cytokines within the central nervous system: sources, function, and mechanism of action. *Am. J. Physiol.* **263**, 1–16.

Benzi, G., Curti, D., Pastoris, O., Marzatico,, F., Villa, R. F., and Dagani, F. (1991) Sequential damage in mitochondrial complexes by peroxidative stress. *Neurochem. Res.* **16**, 1295–1302.

Bharucha, N. E., Stokes, L., Schoenberg, B. S., Ward, C., Ince, S., Nutt, J. G., et al. (1986) A case-control study of twin pairs discordant for Parkinson's disease: a search for environmental risk factors. *Neurology* **36**, 284–288.

Bodnar, A. G., Cooper, J. M., Holt, I. J., Leonard, J. V., and Schapira, A. H. V. (1993) Nuclear complementation restores mtDNA levels in cultured cells from a patient with mtDNA depletion. *Am. J. Hum. Genet.* **53**, 663–669.

Bonifati, V., Fabrizio, E., Vanacore, N., De Mari, M., and Meco, G. (1995) Familial Parkinson's disease: a clinical genetic analysis. *Can. J. Neurol. Sci.* **22**, 272–279.

Borden, K. L. B. (1998) Structure/function in neuroprotection and apoptosis. *Ann. Neurol.* **44(Suppl. 1),** S65–S71.

Bordet, R., Broly, F., Destée, A., and Libersa, C. (1996) Debrisoquine hydroxylation genotype in familial forms of idiopathic Parkinson's disease. *Adv. Neurol.* **69**, 97–100.

Brooks, D. J., Ibanez, V., Sawle, G. V., Quinn, N., Lees, A. J., Mathias, C. J., et al. (1990) Differing patterns of structural ^{18}F-dopa uptake in Parkinson's disease, multiple system atrophy and progressive supranuclear palsy. *Ann. Neurol.* **28**, 547–555.

Burn, D. J., Mark, M. H., Playford, E. D., Maraganore, D. M., Zimmerman, T. R., Duvoison, R. C., et al. (1992) Parkinson's disease in twins studied with ^{18}F-dopa positron emission tomography. *Neurology* **42**, 1894–1900.

Butterfield, P. G., Valanis, B. G., Spencer, P. S., Lindemann, C. A., and Nutt, J. G. (1993) Environmental antecedents of young-onset Parkinson's disease. *Neurology* **43**, 1150–1158.

Calne, D. B., Langston, J. W., Martin, W. R., Stoessl, A. J., Ruth, T. J., Adam, M. J., et al. (1985) Positron emission tomography after MRI: observations relating to the cause of Parkinson's disease. *Nature* **317**, 246–248.

Ceballos, I., Lafon, M., Javoy-Agid, F., Hirsch, E., Nicole, A., Sinet, P. M., and Agid, Y. (1990) Superior dismutase and Parkinson's disease. *Lancet* **335**, 1035–1036.

Cerhan, J. R., Wallace, R. B., and Folsom, A. R. (1994) Antioxidant intake and risk of Parkinson's disease (PD) in older women. *Am. J. Epidemiol.* **139**, S65.

Charcot, J. (1867) Lecons sur les maladies du systeme nerveux. VA Delahaye, Paris.

Chan, P., Tanner, C. M., Jiang, X., and Langston, J. W. (1998) Failure to find the alpha-synuclein gene missense mutation (G209A) in 100 patients with younger onset Parkinson's disease. *Neurology* **50**, 513–514.

Chen, X., Xia, Y., Gresham, L. S., Molgarrd, C. A., Thomas, R. G., Galasko, D., et al. (1996) ApoE and CYP2D6 polymorphism with and without parkinsonism-dementia complex in the people of Chamorro, Guam. *Neurology* **47**, 779–784.

Chiba, K., Trevor, A. J., and Castagnoli, N. Jr. (1984) Active uptake of MPP$^+$ a metabolic of MPTP, by brain synaptosomes. *Biochem. Biophys. Res. Commun.* **128**, 1229–1232.

Cleeter, M. J. W., Cooper, J. M., Darley-Usmar, V. M., Moncada, S., and Schapira, A. H. V. (1994) Reversible inhibition of cytochrome c oxidase, the terminal enzyme of the mitochondrial respiratory chain, by nitric oxide. Implications for neurodegenerative diseases. *FEBS* **345**, 50–54.

Cock, H. C., Tabrizi, S. J., Cooper, J. M., and Schapira, A. H. V. (1998) The influence of nuclear background on the biochemical expression of 3460 Leber's hereditary optic neuropathy. *Ann. Neurol.* **44,** 187–193.

Cooper, J. M., Daniel, S. E., Marsden, C. D., and Schapira AHV (1995) L-dihydroxyphenylalanine and complex I deficiency in Parkinson's disease brain. *Mov. Disord.* **10,** 295–297.

Corsini, G. U., Pintus, S., Chiueh, C. C., Weiss, J. F., and Kopin, I. J. (1985) 1-Methyl-4-phenyl-1,2,3,6-tetrahydropyridine (MPTP) neurotoxicity in mice is enhanced by pretreatment with diethydithiocarbamate. *Eur. J. Pharmacol.* **119,** 127–128.

Couper, J. (1837) On the effects of black oxide of manganese when inhaled into the lungs. *Br. Ann. Med. Pharmacol. Vit. Stat. Gen. Sci.* **1,** 41–42.

D'Amato, R. J., Benham, D. F., and Snyder, S. H. (1987) Characterization of the binding of N-methyl-4-phenylpyridine, the toxic metabolite of the parkinsonian neurotoxin N-methyl-4-phenyl 1,2,3,6 tetrahydropyridine, to neuromelanin. *J. Neurochem.* **48,** 653–658.

Davanipour, Z. and Will, A. D. (1990) Residual histories in Parkinson's disease patients. *Ann. Neurol.* **28,** 295–303.

Davis, G. C., Williams, A. C., Markey, S. P., Ebert, M. H., Caine, E. D., Reichert, C. M., and Kopin, I. J. (1979) Chronic parkinsonism secondary to intravenous injection of meperidine analogues. *Psychiatry Res.* **1,** 249–254.

Degl'Innocenti, F., Maurello, M. T., and Marini, P. (1989) A parkinsonian kindred. *Ital. J. Neurol. Sci.* **10,** 307–310.

De Michele, G., Filla, A., Volpe, G., De Marco, V., Gogliettino, A., Ambrosio, G., et al. (1996) Environmental and genetic risk factors in Parkinson's disease: a case control study in southern Italy. *Mov. Disord.* **11,** 17–23.

De Rijk, M. C., Breteler, M. M., de Breeijen, J. H, Launer, L. J., Grobbee, D. E., van der Mechene, F. G., and Hofman, A. (1997) Dietary antioxidants and Parkinson disease. *Arch. Neurol.* **54,** 762–765.

De Vries, D. D., Went, L. N., Bruyn, G. W., Scholte, H. R., Hofstra, R. M. W., Bolhuis, P. A., and van Ooost, B. A. (1996) Genetic and biochemical impairment of mitochondrial complex I activity in a family with Leber hereditary optic neuropathy and hereditary spastic dystonia. *Am. J. Hum. Genet.* **58,** 703–711.

Dexter, D. T., Carayon, A., Vidailhet, M., Ruberg, M., Agid, F., Agid, Y., et al. (1990) Decreased ferritin levels in brain in Parkinson's disease. *J. Neurochem.* **55,** 16–20.

Dexter, D. T., Carter, C., and Agid, F. (1986) Lipid peroxidation as cause of nigral death in Parkinson's disease. *Lancet* **2,** 639–640.

Dexter, D. T., Wells, F. R., Agid, F., Agid, Y., Lees, A. J., Jenner, P., and Marsden, C. D. (1987) Increased nigral iron content in post-mortem parkinsonian brain. *Lancet* **2,** 1219–1220.

Diederich, N., Hilger, C., Goetz, C., Keipes, M., Hentges, F. Vieregge, P., and Metz, H. (1996) Genetic variability of the CYP 2D6 gene is not a risk factor for sporadic Parkinson's disease. *Ann. Neurol.* **40,** 463–465.

Di Monte, D. A., Wu, E. Y., Irwin, I., Delanney, L. E., and Langston, J. W. (1992) Production and disposition of 1-methyl-4-phenylpyridinium in primary cultures of mouse astrocytes. *Glia* **5,** 48–55.

Duan, W., Rangnekar, V., and Mattson, M. P. (1999a) Par-4 production in synaptic compartments following apoptotic and excitotoxic insults: evidence for a pivotal role in mitochondrial dysfunction and neuronal degeneration. *J. Neurochem.* **72,** 2312–2322.

Duan, W., Zhang, Z., Gash, D. M., and Mattson, M. P. (1999b) Participation of Par-4 in degeneration of dopaminergic neurons in models of Parkinson's disease. *Ann. Neurol.* **46,** 587–597.

Duan, W. and Mattson, M. P. (1999) Dietary restriction and 2-deoxyglucose administration improve behavioral outcome and reduce degeneration of dopaminergic neurons in models of Parkinson's disease. *J. Neurosci. Res.* **57,** 185–206.

Dulaney, E., Stern, M., and Hurtig, H. (1990) The epidemiology of Parkinson's disease: a case-control study of young-onset versus old-onset patients. *Mov. Disord.* **5,** 12.

Dunbar, D. R., Moonie, P. A., Zeviani, M., and Holt, I. J. (1996) Complex I deficiency is associated with 3243G:C mitochondrial DNA in osteosarcoma cell cybrids. *Hum. Mol. Genet.* **5,** 123–129.

Duvoisin, R. C., Gearing, I. R., Schweitzer, M. R., and Yahr, M. D. (1969) A family study of parkinsonism, in *Progress in Neurogenetics* (Barbeau, A. and Brunette, J. R., eds.), Excerpta Medica Foundation, Amsterdam, pp. 492–496.

Ebmeier, K. P., Mutch, W. J., Calder, S. A. , Crawford, J. R., Stewart, L., and Besson, J. O. A. (1989) Does idiopathic parkinsonism in Aberdeen follow intrauterine influenza? *J. Neurol. Neurosurg. Psychiatry* **52,** 911–913.

Fall, P. A., Fredrikson, M., Axelson, O., and Granerus, A. (1999) Nutritional and occupational factors influencing the risk of Parkinson's disease: a case-control study of southeastern Sweden. *Mov. Disord.* **14,** 28–37.

Fleming, L., Mann, J. B., Bean, J., Briggle, T., and Sanchez-Ramos, J. R. (1994) Parkinson's disease and brain levels of organochloride pesticides. *Ann. Neurol.* **36,** 100–103.

Forno, L. S., Langston, J. W., DeLanney, L. E., and Irwin, I. (1988) An electron microscopic study of MPTP-induced inclusion bodies in an old monkey. *Brain Res.* **448,** 150–157.

Gasser, T. (1998) Genetics of Parkinson's disease. *Ann. Neurol.* **44(Suppl. 1),** S53–S57.

Gasser, T., Wszolek, Z. K., Trofatter, J., Ozelius, L., Iutti, R. J., Lee, C. S., et al. (1994) Genetic linkage studies in autosomal dominant parkinsonism: evaluation of seven candidate genes. *Ann. Neurol.* **36,** 387–396.

George, J. M., Jin, H., Woods, W. S., and Clayton, D. F. (1995) Characterization of a novel protein regulated during the critical period of song learning in the zebra finch. *Neuron* **15,** 361–372.

Golbe, L. I. (1999) Alpha-synuclein and Parkinson's disease. *Mov. Disord.* **14,** 6–9.

Golbe, L. I., Di Iorio, G., Bonavita, V., Miller, D. C., and Duvoisin, R. C, (1990) A large kindred with autosomal dominant Parkinson's disease. *Ann. Neurol.* **27,** 276–282.

Golbe, L. I., Di Iorio, G., Sanges, G., Lazzarini, A. M., La Sala, S., Bonavita, V., and Duvoisin, R. C. (1996) Clinical genetic analysis of Parkinson's disease in the Contursi kindred. *Ann. Neurol.* **40,** 767–775.

Gordon, E. B. (1965) Carbon-monoxide encephalopathy. *Br. Med. J.* **1,** 1232–1233.

Gorrell, J. M., DiMonte, D., and Graham, D. (1996) The role if the environment in Parkinson's disease. *Environ. Health Perspect.* **104,** 652–654.

Gowers, W. R. (1888) *Diseases of the Nervous System.* Chapman, London, p. 996.

Grinker, R. R. (1926) Parkinsonism following carbon monoxide poisoning. *J. Nerv. Mental Disord.* **64,** 18–28.

Gu, M., Cooper, J. M., Taanman, J. W., and Schapira, A. H. V. (1998) Mitochondrial DNA transmission of the mitochondrial defect in Parkinson's disease. *Ann. Neurol.* **44,** 177–186.

Gu, M., Gash, M. T., Cooper, J. M., Wenning, G. K., Daniel, S. E., Quinn, N. P., et al. (1997) Mitochondrial respiratory chain function in multiple system atrophy. *Mov. Disord.* **12,** 418–422.

Hantraye, P., Brouillet, E., Ferrante, R., Palfi, S., Dolan, R., Matthews, R. T., and Beal, M. F. (1996) Inhibition of neuronal nitric oxide synthase prevents MPTP-induced parkinsonism in baboons. *Nat. Med.* **2,** 1017–1021.

Hasegawa, E., Takeshige, K., Oishi, T., Murai, Y., and Minikami, S. (1990) 1-Methyl-4-phenylpyridinium (MPP+) induces NADH dependent superoxide formation, and enhances NADH-dependent lipid peroxidation in bovine heart submitochondrial particles. *Biochem. Biophys. Res. Commun.* **170,** 1049–1055.

Hattori, N., Kitada, T., Matsumine, H., Asakawa, S., Yamamura, Y., Yoshino, H., et al. (1998) Molecular genetic analysis of a novel parkin gene in Japanese families with autosomal recessive juvenile parkinsonism: evidence for variable homozygous deletions in the parkin gene in affected individuals. *Ann. Neurol.* **44,** 935–941.

Hellenbrand, W., Boeing, H., Robra, B. P., Seidler, A., Vieregge, P., Nischan, P., et al. (1996a) Diet and Parkinson's disease. II. A possible role for the past intake of specific nutrients. Results from a self-administered food frequency questionnaire in a case-control study. *Neurology* **47**, 644–650.

Hellenbrand, W., Seidler, A., Boeing, H., Robra, B. P., Vieregge, P., Nischan, P., et al. (1996b) Diet and Parkinson's disease. I. A possible role for the past intake of specific foods and food group. Results from a self-administered food frequency questionnaire in a case-control study. *Neurology* **47**, 636–643.

Hillered, L. and Ernster, L. (1983) Respiratory activity of isolated rat brain mitochondria following in vitro exposure to oxygen radicals. *J. Cereb. Blood Flow Metab.* **3**, 207–214.

Hirsch, E. C., Brandel, J.-P., Galle, P., Javoy-Agid, F., and Agid, Y. (1991) Iron and aluminium increase in the substantia nigra of patients with Parkinson's disease: an x-ray microanalysis. *J. Neurochem.* **56**, 446–451.

Hirsch, E. C., Graybiel, A. M., and Agid, Y. (1989) Selective vulnerability of pigmented dopaminergic neurons in Parkinson's disease. *Acta Neurol. Scand. Suppl* **126**, 19–122.

Hirsch, E. C., Hunot, S., Damier, P., and Faucheux, B. (1998) Glial cells and inflammation in Parkinson's disease: A role in neurodegeneration? *Ann. Neurol.* **44(Suppl 1),** S115–S120.

Hopkins, S. J. and Rothwell, N. J. (1995) Cytokines and the nervous system. I: expression and recognition. *Trends Neurosci.* **18**, 83–88.

Hornykiewicz, O. (1975) Parkinson's disease and its chemotherapy. *Biochem. Pharmacol.* **24**, 1061–1065.

Huang, C. C., Lu, C. S., Chu, N. S., Hochberg, F., Lilienfeld, D., Olanow, W., and Calne, D. B. (1993) Progression after chronic manganese exposure. *Neurology* **43**, 1479–1483.

Ikebe, S. I., Tanaka, M., and Azawa, T. (1995) Point mutations of mitochondrial genome in Parkinson's disease. *Mol. Brain Res.* **28**, 281–295.

Ikebe, S. I., Tanaka, M., Ohno, K., Sato, W., Hattori, K., Kondo, T., et al. (1990) Increase of deleted mitochondrial DNA in the striatum in Parkinson's disease and senescence. *Biochem. Biophys. Res. Commun.* **170**, 1044–1048.

Itzhak, Y. and Ali, S. F. (1996) The neuronal nitric oxide synthase inhibitor, 7-nitroindazole, protects against methamphetamine-induced neurotoxicity in vivo. *J. Neurochem.* **67**, 1770–1773.

Javitch, J. A. and Snyder, S. H. (1984) Uptake of MPTP ($^+$) by dopamine neurons explains selectivity of parkinsonism-inducing neurotoxin, MPTP. *Eur. J. Pharmacol.* **104**, 455–456.

Jellinger, K., Kienzl, E., Rumpelmair, G., Riederer, P., Stachelberger, H., Ben-Shacher, D., and Youdim, M. B. H. (1992) Iron-melanin complex in substantia nigra of parkinsonian brains: an x-ray microanalysis. *J. Neurochem.* **59**, 1168–1171.

Jellinger, K., Paulus, W., Grundke-Iqbal, I., Riederer, P., and Youdim, M. B. (1990) Iron-melanin complex in substantia nigra of parkinsonian brains: an x-ray microanalysis. *J. Neural Trans. Parkinson's Dis. Diment. Sect.* **2**, 327–340.

Jenner, P. (1991) Oxidative stress as a cause of Parkinson's disease. *Acta Neurol. Scand.* **84**, 6–15.

Johnson, W. G., Hodge, S. E., and Duvoisin, R. (1990) Twin studies and the genetics of Parkinson's disease—a reappraisal. *Mov. Disord* **5**, 87–94.

Jun, A. S., Brown, M. D., and Wallace, D. C. (1994) A mitochondrial DNA mutation at nucleotide pair 14459 of the NADH dehydrogenase subunit 6 gene associated with maternally inherited Leber hereditary optic neuropathy and dystonia. *Proc. Natl. Acad. Sci. USA* **91**, 6202–6210.

Kish, S. J., Morito, C., and Hornykiewicz, O. (1985) Glutathione peroxidase activity in Parkinson's disease. *Neurosci. Lett.* **58**, 343–346.

Kitada, T., Asakawa, S., Hattori, N., Matsumine, H., Yamamura, Y., Minoshima, S., et al. (1998) Mutations in the parkin gene cause autosomal recessive juvenile parkinsonism. *Nature* **392**, 605–608.

Klawans, H. L., Stein, R. W., Tanner, C. M., and Goetz, C. G. (1982) A pure parkinsonian syndrome following acute carbon monoxide intoxication. *Arch. Neurol.* **39,** 302–304.

Kruger, R., Kuhn, W., Muller, T., Woitalla, D., Graber, M., Kosel, S., et al. (1998) Ala30Pro mutation in the gene encoding alpha-synuclein in Parkinson's disease. *Nature Genet.* **18,** 106–108.

Kuiper, M. A., Visser, J. J., Bergmans, P. L. M., Scheltens, P., and Wolters, E. C. (1994) Decreased cerebrospinal-fluid nitrate levels in Parkinson's-disease, Alzheimer's-disease and multiple system atrophy patients. *J. Neurol. Sci.* **121,** 46–49.

Kurth, J. H., Hubble, J. P., and Eggers, E. A. (1995) Lack of association of CYP 2D6 and MAO-B alleles with Parkinson's disease in a Kansas cohort. *Neurology* **45(Suppl. 4),** A429.

Kurth, J. H., Kurth, M. C., Poduslo, S. E., and Schwankhaus, J. D. (1993) Association of a monoamine oxidase B allele with Parkinson's disease. *Ann. Neurol.* **33,** 368–372.

Langston, J. W. and Ballard, P. A. (1984) Parkinsonism induced by 1-methyl-4-phenyl 1,2,3,6 tetrahydropyridine: implications for treatment and the pathophysiology of Parkinson's disease. *Can. J. Neurol. Sci.* **11,** 160–165.

Langston, J. W., Ballard, P., Tetrud, J. W., and Irwin, I. (1983) Chronic parkinsonism in humans due to a product of meperidine analog synthesis. *Science* **219,** 979–980.

Lazzarini, A. M., Myers, R. H., Zimmerman, T. R. Jr., Mark, M. H., Golbe, L. I., Sage, J. I., et al. (1994) A clinical genetic study of Parkinson's disease: evidence for dominant transmission. *Neurology* **44,** 499–506.

Leroy, E., Boyer, R., Auburger, G., Leube, B., Ulm, G., Mezey, E., et al. (1998) The ubiquitin pathway in Parkinson's disease. *Nature* **395,** 451–452.

Lestienne, P., Riederer, P., and Jellinger, K. (1991) Mitochondrial DNA in postmortem brain from patients with Parkinson's disease. *J. Neurochem.* **56,** 1819.

Levitt, P., Pintar, J. E., and Breakefield, X. O. (1982) Immunocytochemical demonstration of monoamine oxidase B in brain astrocytes and serotonergic neurons. *Proc. Natl. Acad. Sci. USA* **79,** 6385–6389.

Logroscino, G., Marder, K., Cote, L., Tang, M. X., Shea, S., and Mayeux, R. (1996) Dietary lipids and antioxidants in Parkinson's disease: a population-based, case-control study. *Ann. Neurol.* **39,** 89–94.

Lücking, C. B., Kösel, S., Mehraein, P., and Graeber, M. B. (1995) Absence of the mitochondrial A7237T mutation in Parkinson's disease. *Biochem. Biophys. Res. Commun.* **211,** 700–704.

Lyden, A., Bondesson, U., Larsson, B. S., and Lindquist, N. G. (1983) Melanin affinity of 1-methyl-4-phenyl 1,2,3,6 tetrahydropyridine and inducer of chronic parkinsonism in humans. *Acta Pharmacol.* **53,** 429–432.

Mann, V. M., Cooper, J. M., Daniel, S. E., Jenner, P., Marsden, C. D., and Schapira, A. H. V. (1994) Complex I, iron and ferritin in Parkinson's disease substantia nigra. *Ann. Neurol.* **36,** 876–881.

Mann, D. M. A., Yates, P. O., Davies, J. S., and Hawkes, J. (1981) Viruses, parkinsonism and Alzheimer's disease. *J. Neurol. Neurosurg. Psychiatry* **44,** 651.

Marder, K., Tang, M. X., Mejia, H., Alfaro, B., Cote, L., Louis, E., et al. (1996) Risk of Parkinson's disease among first-degree relatives: a community-based study. *Neurology* **47,** 155–160.

Maroteaux, L. and Scheller, R. H. (1991) The rat brain synucleins; family of proteins transiently associated with neuronal membrane. *Brain. Res. Mol. Brain. Res.* **11,** 335–343.

Maroteaux, L., Campanelli, J. T., and Scheller, R. H. (1988) Synuclein: a neuron-specific protein localized to the nucleus and presynaptic nerve terminal. *J. Neurosci.* **8,** 2804–2815.

Marsden, C. D. (1987) Parkinson's disease in twins. *J. Neurol. Neurosurg. Psychiatry* **50,** 105–106.

Martin, W. E., Young, W. E., and Anderson, V. E. (1973) Parkinson's disease. A genetic study. *Brain* **96,** 495–506.

Martin, W. R. W, Stoessel, A. J., and Adam, M. J. (1986) Imaging of dopamine systems in human subjects exposed to MPTP, in *MPTP: A Neurotoxin Producing Parkinsonian Syndrome* (Markey, S. P., Castagnoli, N. Jr., Trevor, A. J., and Kopin, I. J., eds.), Academic, New York, pp. 315–325.

Marttila, R. J. and Rinne, U. K. (1976) Arteriosclerosis, hereditary, and some previous infections in the etiology of Parkinson's disease. A case-control study. *Clin. Neurol. Neurosurg.* **79,** 46–56.

Marttila, R. J., Kaprio, J., Kopshenvuo, M., and Rinne, U. K. (1988a) Parkinson's disease in a nationwide twin cohort. *Neurology* **38,** 1217–1219.

Marttila, R. J., Lorentz, H., and Rinne, U. K. (1988b) Oxygen toxicity protecting enzymes in Parkinson's disease. Increase of superoxide dismutase-like activity in the substantia nigra and basal nucleus. *J. Neurol. Sci.* **86,** 321–331.

Matsumine, H., Saito, M., Shimodo-Matsubayashi, S., Tanaka, H., Ishihawa, A., Nakagawa-Hattori, Y., et al. (1997) Localization of a gene for an autosomal recessive form of juvenile parkinsonism to chromosome 6q25. 2–27. *Am. J. Hum. Genet.* **60,** 588–596.

McGeer, P. L., Itagaki, S., Boyes, B. E., and McGeer, E. G. (1988) Reactive microglia are positive for HLA-DR in the substantia nigra of Parkinson's and Alzheimer's disease brains. *Neurology* **38,** 1285–1291.

Melamed, E., Rosenthal, J., Globus, M., Cohen, O., and Uzzan, A. (1985) Suppression of MPTP-induced dopaminergic neurotoxicity in mice by nomifensine and L-dopa. *Brain Res.* **342,** 401–404.

Mena, I. (1979) Manganese poisoning, in *Handbook of Clinical Neurology Vol. 36. Intoxications of the Nervous System, Part I* (Vinken, P. J. and Bruyn, G. W., eds.), Amsterdam, North Holland, pp. 217–237.

Mjönes, H. (1949) Paralysis agitans: a clinical and genetic study. *Acta Psychiat. Neurol.* **54,** 1–195.

Mochizuki, H., Goto, K., Mori, H., and Mizuno, Y. (1996) Histochemical detection of apoptosis in Parkinson's disease. *J. Neurol. Sci.* **137,** 120–123.

Molina, J. A., Jiminez-Jiminez, F. J., Navarro, J. A., Ruiz, E., Arenas, J., Carrera-Valdivia, F., et al. (1994) Plasma-levels of nitrate in patients with Parkinson's disease. *J. Neurol. Sci.* **127,** 87–89.

Molina, J. A., Jiminez-Jiminez, F. J., Navarro, J. A., Vargas, C., Gomez, P., Benito-Leon, J., et al. (1996) Cerebrospinal-fluid nitrate levels in patients with Parkinson's disease. *Acta. Neurol. Scand.* **93,** 123–126.

Morens, D. M., Grandinetti, A., and Waslien, C. I. (1996) Case study of idiopathic Parkinson's disease and dietary vitamin intake. *Neurology* **47,** 644–650.

Nakamura, S., Kawamoto, Y., Nakano, S., Akiguchi, I., and Kimura, J. (1997) p35nck5a and cyclin-dependent kinase 5 co-localize in Lewy bodies of brains with Parkinson's disease. *Acta Neuropathol. (Berl.)* **94,** 153–157.

Nukada, H., Kowa, H., Saitoh, T., Tazaki, Y., and Miura, S. (1978) A big family of paralysis agitans. *Rinsho Shinkeigaju* **18,** 627–634.

Okuda, B., Iwamoto, Y., Tachibana, H., and Sugita, M. (1997) Parkinsonism after acute cadmium poisoning. *Clin. Neurol. Neurosurg.* **99,** 263–265.

Ozawa, T., Tanaka, M., Ino, H., Ohno, K., Sano, T., Wada, Y., et al. (1991) Distinct clustering of point mutations in mitochondrial DNA among patients with mitochondrial encephalomyopathies and with Parkinson's disease. *Biochem. Biophys. Res. Commun.* **176,** 938–946.

Parker, W. D., Boyson, S. J., and Parks, J. K. (1989) Abnormalities of the electron transport chain in idiopathic Parkinson's disease. *Ann. Neurol.* **26,** 719–723.

Payami, H., Larsen, K., Bernard, S., and Nutt, J. (1994) Increased risk of Parkinson's disease in parents and siblings of patients. *Ann. Neurol.* **36,** 659–661.

Penn, A. M. W., Roberts, T., Hodder, J., Allen, P. S., Zhu, G., and Martin, W. R. W. (1995) Generalized mitochondrial dysfunction in Parkinson's disease detected by magnetic resonance spectroscopy of muscle. *Neurology* **45,** 2097–2099.

Perry, T. L., Godin, D. V., and Hansen, S. (1982) Parkinson's disease: a disorder due to nigral glutathione deficiency. *Neurosci. Lett.* **33,** 305–310.

Piccini, P., Burn, D. J., Ceravolo, R., Maraganore, D., and Brooks, D. J. (1999) The role of inheritance in sporadic Parkinson's disease: evidence from a longitudinal study of dopaminergic function in twins. *Ann. Neurol.* **4,** 577–582.

Planté-Bordeneuve, V., Davis, M. B., Maraganore, D. M., Marsde, C. D., and Harding, A. E. (1994) Debrisoquine hydroxylase gene polymorphism in familial Parkinson's disease. *J. Neurol. Neurosurg. Psychiatry* **57,** 911–913.

Polymeropoulos, M. H., Higgins, J. J., Golbe, L. I., Johnson, W. G., Ide, S. E., Di Ioro, G., et al. (1996) Mapping of a gene for Parkinson's disease to chromosome 4q21-q23. *Science* **274,** 1197–1199.

Polymeropoulos, M. H., Lavedan, C., Leroy, E. , Ide, S. E., Dehejia, A., Dutra, A., et al. (1997) Mutation in the α-synuclein gene identified in families with Parkinson's disease. *Science* **276,** 2045–2047.

Qureshi, G. A., Baig, S., Bednar, I., Sodersten, P., Forsberg, G., and Siden, A. (1995) Increased cerebrospinal-fluid concentration of nitrate in Parkinson's disease. *Neuroreport* **6,** 1642–1644.

Rajput, A. H. and Uitti, R. J. (1988) Neurological disorders and services in Saskatchewan: a report based on provincial health care records. *Neuroepidemiology* **7,** 145–151.

Rajput, A. H., Uitti, R. J., Stern, W., and Laverty, W. (1986) Early onset Parkinson's disease in Saskatchewan: environmental considerations for etiology. *Can. J. Neurol. Sci.* **13,** 312–316.

Riederer, P. and Wuketich, S. (1976) Time course of nigrostriatal degeneration in Parkinson's disease: a detailed study of influential factors in human brain analogues. *J. Neural. Trans. Park. Dis. Dement. Sec.* **38,** 277–301.

Riederer, P., Sofic, E., Rausch, W. D., Schmidt, B., Reynolds, G. P., Jellinger, K., and Youdim, M. B. H. (1989) Transition metals, ferritin, glutathione and ascorbic acid in parkinsonian brains. *J. Neurochem.* **52,** 515–520.

Roman, G. C., Zhang, Z.-X., and Ellenberg, J. H. (1995) The neuroepidemiology of Parkinson's disease, in *Etiology of Parkinson's Disease* (Ellenberg, J. H., Koller, W. C., and Langston, J. W., eds.), Marcel Dekker, New York, pp. 203–343.

Saggu, H., Cooksey, J., Dexter, D., Wells, F. R., Lees, A., Jenner, P., and Marsden, C. D. (1989) A selective increase in particulate superoxide dismutase activity in Parkinson's substantia nigra. *J. Neurochem.* **53,** 692–697.

Sanchez-Ramos, J., Övervik, E., and Ames, B. N. (1994) A marker of oxyradical-mediated DNA damage (8-hydroxy-2'-deoxyguanosine) is increased in nigro-striatum of Parkinson's disease brain. *Neurodegeneration* **3,** 197–204.

Schapira, A. H. V. (1994) Evidence for mitochondrial dysfunction in Parkinson's disease: a critical appraisal. *Mov. Disord.* **9,** 125–138.

Schapira, A. H. V, Cooper, J. M., Dexter, D., Jenner, P., Clark, J. B., and Marsden, C. D. (1989a) Mitochondrial complex I deficiency in Parkinson's disease. *Lancet* **1,** 1269.

Schapira, A. H. V, Cooper, J. M., and Dexter, D. (1989b) Mitochondrial complex I deficiency in Parkinson's disease. *Ann. Neurol.* **26,** 122–123.

Schapira, A. H. V., Mann, V. M., Cooper, J. M., Dexter, D., Daniel, S. E., Jenner, P., et al. (1990a) Anatomic and disease specificity of NADH CoQ$_1$ reductase (complex I) deficiency in Parkinson's disease. *J. Neurochem.* **55,** 2142–2145.

Schapira, A. H. V., Holt, I. J., Sweeney, M., Harding, A. E., Jenner, P., and Marsden, C. D. (1990b) Mitochondrial DNA analysis in Parkinson's disease. *Mov. Disord.* **5,** 294–297.

Schapira, A. H. V., Warner, T., Gash, M. T., Cleeter, M. J. W., Marinho, C. F. M., and Cooper, J. M. (1997) Complex I function in familial and sporadic dystonia. *Ann. Neurol.* **41,** 556–559.

Schierle, G. S., Hansson, O., Leist, M., Nicotera, P., Widner, H., and Brundin, P. (1999) Caspase inhibition reduces apoptosis and increases survival of nigral transplants. *Nat. Med.* **5,** 97–100.

Schultz, W., Scarnati, E., Sundstrom, E., and Romo, R. (1989) Protection against 1-methyl-4-phenyl 1,2,3,6 tetrahydropyridine-induced parkinsonism by the catecholamine uptake inhibitor nomifensine: behavioural analysis in monkeys with partial striatal dopamine depletions. *Neuroscience* **31,** 219–230.

Schwartz, J. and Elizan, T. S. (1979) Search for viral articles and virus-specific products in idiopathic Parkinson disease brain material. *Ann. Neurol.* **6,** 261–263.

Seidler, A., Hellenbrand, W., Robra, B. P., Vieregge, P., Nischan, P., Joeg, J., et al. (1996) Possible environmental, occupational, and other etiologic factors for Parkinson's disease: a case-control study in Germany. *Neurology* **46,** 1275–1284.

Semchuk, K. M., Love, E. J., and Lee, R. F. (1991) Parkinson's disease and exposure to rural environmental factors: a population based case-control study. *Can. J. Neurol. Sci.* **18,** 279–286.

Semchuk, K. M., Love, E. J., and Lee, R. G. (1992) Parkinson's disease and exposure to agricultural work and pesticide chemicals. *Neurology* **42,** 1328–1335.

Semchuk, K. M., Love, E. J., and Lee, R. G. (1993) Parkinson's disease: a test of the multifactorial etiologic hypothesis. *Neurology* **43,** 1173–1180.

Sethi, K. D., Meador, K. J., Loring, D., and Meador, M. P. (1989) Neuroepidemiology of Parkinson's disease: analysis of mortality data for the U. S. A. and Georgia. *Intl. J. Neurosci.* **46,** 87–92.

Shoffner, J. M., Brown, M. D., Torrino, A., Lott, M. T., Cabell, M. F., Mirra, S. S., et al. (1993) Mitochondrial DNA variants observed in Alzheimer's disease and Parkinson's disease patients. *Genomics* **17,** 171–184.

Sian, J., Dexter, D. T., Lees, A. J., Daniel, S., Agid, Y., Javoy-Agid, F., et al. (1994) Alterations in glutathione levels in Parkinson's disease and other neurodegenerative disorders affecting basal ganglia. *Ann. Neurol.* **36,** 348–355.

Singer, T. P., Salach, J. I., Castognoli, N., Jr., and Trevor, A. J. (1986) Interactions of the neurotoxic amine 1-methyl-4-phenyl 1,2,3,6 tetrahydropyridine with monoamine oxidase. *Biochem. J.* **235,** 785–789.

Smith, C. A. D., Gough, A. C., Leigh, P. N., Summers, B. A., Harding, A. E., Maranganore, D. M., et al. (1992) Debrisoquine hydroxylase gene polymorphism and susceptibility to Parkinson's disease. *Lancet* **339,** 1375–1377.

Sofic, E., Paulus, W., Jellinger, K., Riederer, P., and Youdim, M. B. H. (1991) Selective increase of iron in substantia nigra zona compacta in parkinsonian brains. *J. Neurochem.* **56,** 978–982.

Spencer, P. (1987) Guam ALS/parkinsonism-dementia: a long-latency neurotoxic disorder caused by a "slow" toxin(s) in food? *Can. J. Neurol. Sci.* **14,** 347–357.

Spillantini, M. G., Schmidt, M. L., Lee, V. M., Trojanowski, J. Q., Jakes, R., and Goedert, M. (1997) Alpha-synuclein in Lewy bodies. *Nature* **388,** 839–840.

Swerdlow, R. H., Parks, J. K., Miller, S. W., Tuttle, J. B., Trimmer, P. A., Sheehan, J. P., et al. (1996) Origin and functional consequences of the complex I defect in Parkinson's disease. *Ann. Neurol.* **40,** 663–671.

Takahashi, H., Ohama, E., Suzuki, S., Horokawa, Y., Ishikawa, A., Morita, T., et al. (1994) Familial juvenile parkinsonism: clinical and pathological study in a family. *Neurology* **44,** 437–441.

Takeshige, K. and Minakami, S. (1979) NADH- and NADPH-dependent formation of superoxide anions by bovine heart submitochondrial particles. *Biochem. J.* **180,** 129–135.

Tanner, C. M. (1986) Influence of environmental factors on the onset of Parkinson's disease (PD). *Neurology* **36,** 215–224.

Tanner, C. M. (1996) Epidemiology of Parkinson's disease. *Neuroepidemiology* **14**, 317–335.

Tanner, C. M. and Ben-Shlomo, Y. (1999) Epidemiology of Parkinson's disease, in *Parkinson's Disease: Advances in Neurology, vol. 80.* Lippincott Williams and Wilkins, Philadelphia, pp. 153–159.

Tanner, C. M., Ottman, R., Goldman, S. M., Ellenberg, J., Chan, P., Mayeux, R., and Langston, J. W. (1999) Parkinson's disease in twins: an etiologic study. *JAMA* **281**, 341–346.

Tatton, N. A. and Kish, S. J. (1997) In situ detection of apoptotic nuclei in the substantia nigra compacta of 1-methyl-phenyl-1,2,3,6-tetrahydropyridine-treated mice using terminal deoxynucleotidyl transferase labeling and acridine orange staining. *Neuroscience* **77**, 1037–1048.

Tatton, N. A., Maclean-Fraser, A., Tatton, W. G., Perl, D. P., and Olanow, C. W. (1998) A fluorescent double-labeling method to detect and confirm apoptotic nuclei in Parkinson's disease. *Ann. Neurol.* **44**, S142–S148.

Taylor, D. J., Krige, D., Barnes, P. R. J., Kemp, G. J., Carroll, M. T., Mann, V. M., et al. (1994) A ^{31}P magnetic resonance spectroscopy study of mitochondrial function in skeletal muscle of patients with Parkinson's disease. *J. Neurol. Sci.* **125**, 77–81.

Tetrud, J. W. and Langston, J. W. (1992) Tremor in MPTP-induced parkinsonism. *Neurology* **42**, 407–410.

Thomas, P. K., Cooper, J. M., King, R. H. M., Workman, J. M., Schapira, A. H. V., Goss-Sampson, M. A., and Muller, D. P. R. (1993) Myopathy in vitamin E deficient rats: muscle fibre necrosis associated with disturbances of mitochondrial function. *J. Anat.* **183**, 451–461.

Tompkins, M. M., Basgall, E. J., Zamrini, E., and Hill, W. D. (1997) Apoptotic-like changes in Lewy-body-associated disorders and normal aging in substantia nigra neurons. *Am. J. Pathol.* **150**, 119–131.

Truong, D. D., Harding, A. E., Scaravilli, F., Smith, J. M., Morgan-Hughes, J. A., and Marsden, C. D. (1990) Movement disorders in mitochondrial myopathies: a report of nine cases with two autopsy studies. *Mov. Disord.* **5**, 109–117.

Ueda, K., Fukushima, H., Masliah, E., Xia, Y., Iwai, A., Yoshimoto, M., et al. (1993) Molecular cloning of cDNA encoding an unrecognized component of amyloid in Alzheimer disease. *Proc. Natl. Acad. Sci. USA* **90**, 11,282–11,286.

Vaughan, J., Durr, A., Tassin, J., Bereznai, B., Gasser, T., Bonifati, V., et al. (1998) The alpha-synuclein Ala53Thr mutation is not a common cause of familial Parkinson's disease: a study of 230 European cases. *Ann. Neurol.* **44**, 270–273.

Vieregge, P. and Heberlein, I. (1995) Increased risk of Parkinson's disease in relatives of patients. *Ann. Neurol.* **37**, 685.

Vieregge, P., Glaese, A., Ulm, G., and Kompf, D. (1992a) Familial Parkinson's disease. *Mov. Disord.* **7(Suppl. 1)**, 23–32.

Vieregge, P., Schiffke, K. A., Friedrich, H. J., Muller, B., and Ludin, H. P. (1992b) Parkinson's disease in twins. *Neurology* **42**, 1453–1461.

Von Economo, C. (1931) The sequelae of encephalitis lethargica, in *Encephalitis Lethargica: Its Sequelae and Treatment* (Newman, K., ed.), Oxford University Press, London, pp. 105–106.

Ward, C. D., Duvoisin, R. C., Ince, S. E., Nutt, J. D., Eldridge, R., and Calne, D. B. (1983) Parkinson's disease in 65 pairs of twins and in a set of quadruplets. *Neurology* **33**, 815–824.

Warner, T. T. and Schapira, A. H. V. (1998) The role of alpha synuclein gene mutation in patients with sporadic Parkinson's disease in the UK. *J. Neurol. Neurosurg. Psychiatry* **65**, 378–379.

Wooten, G. F., Currie, L. J., Bennett, J. P., Harrison, M. B., Trugman, J. M., and Parker, W. D. (1997) Maternal inheritance in Parkinson's disease. *Ann. Neurol.* **41**, 265–268.

Yamamura, Y., Sobue, I., Ando, K., Iida, M., and Yanagi, T. (1973) Paralysis agitans of early onset with marked diurnal fluctuation of symptoms. *Neurology* **23,** 239–244.

Zhang, Y., Marcillat, O., Giulivi, C., Ernster, I., and Davies, K. J. (1990) The oxidative inactivation of mitochondrial electron transport chain components and ATP. *J. Biol. Chem.* **265,** 16,330–16,336.

Ziegler, L. (1928) Follow-up studies in persons who have had epidemic encephalitis. *JAMA* **July 21,** 138–141.

4

Scope of Trinucleotide Repeat Disorders

Shoji Tsuji

INTRODUCTION

Expansions of CAG trinucleotide repeats coding for polyglutamine (polyQ) stretches have been found to be the causative mutation in at least eight hereditary neuro-degenerative disorders, which include spinal and bulbar muscular atrophy (SBMA) (La Spada et al., 1991); Huntington's disease (HD) (The Huntington's Disease Collaborative Research Group, 1993); spinocerebellar ataxia type 1 (SCA1) (Orr et al., 1993); dentatorubral-pallidoluysian atrophy (DRPLA) (Koide et al., 1994; Nagafuchi et al., 1994b); Machado-Joseph disease (Kawakami et al.,1995); spinocerebellar ataxia type 2 (SCA2) (Imbert et al., 1996; Pulst et al., 1996; Sanpei et al., 1996); spinocerebellar ataxia type 6 (SCA6) (Zhuchenko et al., 1997); and spinocerebellar ataxia type 7 (SCA7) (David et al., 1997).

A number of characteristic clinical and molecular genetic features are shared among these diseases:

1. The diseases are caused by expanded CAG repeats coding for polyglutamine stretches.
2. The mode of inheritance is basically autosomal dominant except for SBMA, which is transmitted as an X-linked recessive trait.
3. There is a threshold repeat length, beyond which individuals become affected. Generally, CAG repeats greater than 35–40 repeats are associated with diseases, except SCA6, for which the threshold length is 20 repeats.
4. There is an inverse correlation between the size of the expanded CAG repeats and the age at onset for all the diseases caused by expansion of CAG repeats, suggesting that the length of the polyQ stretches is intimately related to the pathogenic mechanisms.
5. There is a prominent genetic anticipation with paternal transmission resulting in more prominent anticipation than maternal transmission.
6. Despite the ubiquitous expression of the causative genes, selected regions of the central nervous system (CNS) are involved with the distribution unique to each disease. These suggest that common pathogenetic mechanisms underlie the neurodegeneration caused by expanded polyglutamine stretches.

This chapter will discuss the molecular mechanisms of neurodegeneration caused by expanded polyglutamine stretches with a focus on our research on DRPLA (Koide et al., 1994; Nagafuchi et al., 1994b).

From: *Pathogenesis of Neurodegenerative Disorders* Edited by: M. P. Mattson © Humana Press Inc., Totowa, NJ

POPULATION GENETICS OF CAG REPEAT DISEASES

Recent studies suggest that the prevalence rates of dominant SCAs, including DRPLA, are considerably different among different populations (Cancel et al., 1997; Geschwind et al., 1997a, b; Illarioshkin et al., 1996; Lorenzetti et al., 1997). On the other hand, strong linkage disequilibria have been demonstrated in expanded alleles (AE) of SCA1 (Wakisaka et al., 1995), SCA2 (Hernandez et al., 1995), MJD/SCA3 (Endo et al., 1996; Stevanin et al., 1995; Takiyama et al., 1995), and DRPLA (Yanagisawa et al., 1996) in particular populations. Haplotype analyses have also demonstrated that founder chromosomes are present in HD (Kremer et al., 1995; Rubinsztein and Leggo, 1997; Squitieri et al., 1994), MJD (Endo et al., 1996; Stevanin et al., 1995; Takiyama et al., 1995), DRPLA (Yanagisawa et al., 1996), and SCA7 (Johansson et al., 1998) chromosomes. These observations raise the possibility that *cis*-elements in the genomic structure are presumed to be associated with CAG repeat instability. Furthermore, in French MJD/SCA3 families, a close association was observed between AE and a particular haplotype that was also found in all normal alleles (AN) with larger than 33 repeats (Stevanin et al., 1997). In Japanese DRPLA patients, a particular haplotype has been found to be associated with AE, which was also exclusively associated with AN, with larger than 17 CAG repeats (Yanagisawa et al., 1996). These results raise the possibility that AE of the dominant SCAs are also generated from intermediate alleles (IA) associated with particular haplotypes, and that the prevalence rates of the dominant SCAs in individual populations correlate with the frequencies of IA of the corresponding genes.

Our initial analysis on the distribution of CAG repeats of DRPLA gene demonstrated that AN with larger than 17 repeats was overrepresented in Japanese populations in comparison to that observed in Caucasians (Burke et al., 1994). Furthermore, we have recently analyzed the relative prevalence rates of the dominant SCAs in large populations, based on data sets from Japanese ($n = 202$) and Caucasian ($n = 177$) pedigrees with dominant SCAs, and the distribution of the sizes of AN of the corresponding genes in both populations (Takano et al., 1998). The relative prevalence rates of SCA1 and SCA2 were higher in Caucasian pedigrees (15% and 14%, respectively) than in Japanese pedigrees (3% and 5%, respectively), and the differences were statistically significant (SCA1; $\chi^2 = 13.58$, df = 1, $p = 0.0002$, SCA2; $\chi^2 = 8.41$, df = 1, $p = 0.0037$). The relative prevalence rates of MJD/SCA3, SCA6, and DRPLA were higher in Japanese pedigrees (43%, 11%, and 20%, respectively) than in Caucasian pedigrees (30%, 5%, and 0%, respectively), and the differences were also statistically significant (MJD/SCA3; $\chi^2 = 5.05$, df = 1, $p = 0.024$, SCA6; $\chi^2 = 5.05$, df = 1, $p = 0.015$, DRPLA; $\chi^2 = 38.21$, df = 1, $p < 0.0001$) (Fig. 1).

The frequencies of large AN in SCA1 (larger than 30 repeats) and SCA2 (larger than 22 repeats) were significantly higher in Caucasians compared to Japanese (SCA1; $\chi^2 = 22.23$, df = 1, $p < 0.0001$, SCA2; $\chi^2 = 14.84$, df = 1, $p = 0.0001$). Cut-off values of 31 or 32 repeats for SCA1, and of 23 or 24 repeats for SCA2, also gave significantly higher frequencies of large AN of SCA1 and SCA2 genes in Caucasian than in Japanese populations. These results were in good accordance with the relatively higher prevalence rates of SCA1 and SCA2 in Caucasians than in Japanese. The frequencies of large AN in MJD/SCA3 (larger than 27 repeats), SCA6 (larger than 13 repeats), and

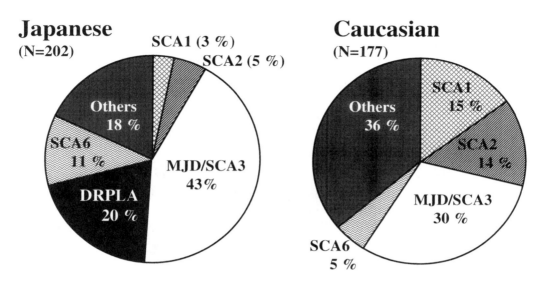

Fig. 1. Relative prevalence rate of dominant SCAs (Takano et al., 1998).

DRPLA (larger than 17 repeats) genes were significantly higher in Japanese than in Caucasians (MJD/SCA3; $\chi^2 = 24.16$, df = 1, $p < 0.0001$, SCA6; $\chi^2 = 38.64$, df = 1, $p < 0.0001$, DRPLA; $\chi^2 = 11.80$, df = 1, $p = 0.0006$). These results are also in accordance with the relatively higher prevalence rates of MJD/SCA3, SCA6, and DRPLA in Japanese than in Caucasians.

Thus, we found a close association between the relative prevalence rates of the dominant SCAs in Japanese and Caucasian pedigrees and the frequencies of large AN of the corresponding genes. The results suggest that the relative prevalence rates of these dominant SCAs are determined by the balance between continuous generation of new AE and loss of AE due to the impaired reproductive fitness of severely affected patients. Recent observations that particular haplotypes of large ANs of MJD/SCA3 and DRPLA genes are commonly shared with MJD/SCA3 (Stevanin et al., 1997) and DRPLA (Yanagisawa et al., 1996), respectively, strongly suggest that large AN with particular haplotypes are prone to further expansion to the disease range of CAG repeats.

MOLECULAR BASIS OF GENETIC ANTICIPATION AND MOLECULAR MECHANISMS OF INSTABILITY OF CAG REPEATS

As described above, DRPLA is characterized by a prominent genetic anticipation with a mean acceleration of age at onset of 25.6 ± 2.4 yr in paternal transmission and 14.0 ± 4.0 yr in maternal transmission (Ikeuchi et al., 1995a, b, c; Koide et al., 1994; Nagafuchi et al., 1994b). In accordance with the strong parental bias for genetic anticipation, a much larger intergenerational increase was observed for paternal transmission (5.8 ± 0.9 repeat units/generation, $n = 16$) compared to maternal transmission (1.3 ± 1.6 repeat units/generation, $n = 4$) (Ikeuchi et al., 1995a, b, c; Koide et al., 1994). This phenomenon has also been described for HD (Andrew et al., 1993; Duyao et al., 1993; Snell et al., 1993; The Huntington's Disease Collaborative Research Group et al., 1993), SCA1 (Chung et al., 1993; Orr et al., 1993), and SCA7 (David et al., 1997).

These results strongly indicate that similar mechanisms must underlie the intergenerational instability of the expanded CAG repeats during male gametogenesis. In fact, it has been demonstrated that DNA from the sperm of patients with HD show considerable variations in the size of expanded CAG repeats compared to DNA from somatic cells (Telenius et al., 1994).

With this background, we considered that transgenic mice harboring an entire mutant gene including flanking regions derived from a mutant allele would be required to investigate the molecular mechanisms of CAG repeat instability. The DRPLA gene spans only 20 kbp (Nagafuchi et al., 1994a), and the disease is among those characterized by large intergenerational changes including SCA7 (David et al., 1997, 1998; Ikeuchi et al., 1995b; Koide et al., 1994; Gouw et al., 1998), suggesting that the DRPLA gene is quite suitable for investigating the molecular mechanisms of CAG repeat instability. We generated transgenic mice harboring a single copy of a mutant DRPLA gene (Sato et al., 1999b). The transgenic mice, in fact, exhibited an age-dependent increase (+0.31/yr) in male transmission and an age-dependent contraction in female transmission (−1.21/yr). Such age-dependent increase in the intergenerational changes in the sizes of expanded CAG repeats in paternal transmission and age-dependent contraction in maternal transmission were also observed in 83 parent–offspring pairs of DRPLA patients (56 paternal and 27 maternal transmissions).

Based on a linear regression model and the continuous cell divisions required for spermatogenesis throughout adult life, the mean increase in the size of CAG repeats in male transmission in mice was calculated to be +0.31/yr and +0.0073/spermatogenesis cycle. These values were comparable to those observed in DRPLA patients, which were calculated to be +0.27 and +0.012, respectively. These results strongly indicate that the difference in the actual intergenerational changes between humans and mice is due to the reproductive life-span variations and that a common mechanism underlies the age-dependent increase in the sizes of CAG repeats both in humans and in mice.

In contrast to spermatogenesis, oogenesis occurs only during fetal life, and ceases at the diploten stage of the first meiotic prophase by 5 d after birth, suggesting that age-dependent contraction of CAG repeats occurs after the cessation of meiotic DNA replication. Similar observations have been made in transgenic mice for SCA1 and SBMA (La Spada et al., 1998; Kaytor et al., 1997). These results strongly suggest that contraction of the CAG repeats occurs during the prolonged resting stage, and mechanisms such as repair of damaged DNA or selective degeneration of the primary oocyte with larger CAG repeats might be involved in the contraction process.

The transgenic mice also exhibited somatic instabilities of CAG repeats, similar to those observed in DRPLA patients (Sato et al., 1999b). The size range of the CAG repeats was smallest in the cerebellum compared to that in the cerebrum and various somatic tissues. This observation has been well-documented in HD, SCA1, MJD, and DRPLA (Chong et al., 1995; Hashida et al., 1997; Takano et al., 1996; Telenius et al., 1994; Ueno et al., 1995). Because the cerebellum contains a dense population of granule cells, which are neuronal cells, it is assumed that neuronal cells exhibit the least instability because they do not undergo cell divisions and that cell divisions are required for the development of somatic instabilities of CAG repeats (Takano et al., 1996). Similar phenomena were observed in the granular layers of the cerebellar cortex and hippocampal formation in autopsied DRPLA brains (Hashida et al., 1997). Another interesting finding of this study is the age-dependent increase in the degree of somatic mosaicism. The size ranges of CAG repeats were much larger at 64 wk compared to those at 3 wk. These data strongly support the fact that the degree of somatic mosaicism increases with age. However, it remains to be elucidated how the age-dependent increase in the degree of the somatic mosaicism is involved in the pathogenesis of DRPLA.

MECHANISMS OF NEURODEGENERATION
CAUSED BY CAG REPEAT EXPANSION

There is increasing evidence suggesting "gain of toxic functions" of mutant proteins with expanded polyglutamine stretches, in particular, truncated mutant proteins containing expanded polyglutamine stretches. Such toxicities have been demonstrated not only in transient expression systems (Cooper et al., 1998; Igarashi et al., 1998; Ikeda et al., 1996; Martindale et al., 1998; Paulson et al., 1997; Skinner et al., 1997), but also in transgenic mice (Hodgson et al., 1999; Ikeda et al., 1996; Mangiarini et al., 1996).

To investigate whether full-length or truncated mutant DRPLA proteins with expanded polyglutamine stretches are toxic to cells, we created deletions with various lengths, either upstream or downstream of the expanded CAG repeats, and inserted this gene into a mammalian expression vector, pEF-BOS (Mizushima and Nagata, 1990), along with a DNA segment coding for an FLAG tag (Igarashi et al., 1998). COS7 cells were transfixed with these plasmid constructs, and the expression patterns of the proteins were investigated by immunohistochemical techniques using an M5 anti-FLAG monoclonal antibody (MAb). Formation of peri- and intranuclear aggregates was observed in COS7 cells expressing truncated mutant DRPLA proteins with an expanded polyglutamine stretch (Q82), but not in COS7 cells expressing truncated wild-type DRPLA protein (Q19). The cells with the aggregate bodies were shown to undergo apoptotic cell death as assayed by TUNEL. Electron microscopic observation demonstrated that the aggregate bodies were composed of fibrous structures 10–12 nm in diameter. Furthermore, these aggregates were not observed either in cells expressing full-length mutant DRPLA protein or in cells expressing full-length wild-type DRPLA protein, raising the possibility that the processing of mutant proteins carrying expanded polyglutamine stretches is important for the toxicity caused by expanded polyglutamine stretches (Goldberg et al., 1996).

Recent discovery of neuronal intranuclear inclusions (NIIs) containing mutant proteins has brought new controversial issues as to the mechanisms of neurodegeneration caused by expanded polyglutamine stretches. Although NIIs were first identified in HD and SCA1 transgenic mice (Davies et al., 1997; Hodgson et al., 1999; Skinner et al., 1997), subsequent intensive studies revealed NIIs in postmortem human brains, including cases of HD (Davies et al., 1997; Difiglia et al., 1997), SCA1 (Skinner et al., 1997), MJD/SCA3 (Paulson et al., 1997), DRPLA (Hayashi et al., 1998; Igarashi et al., 1998), SCA7 (Holmberg et al., 1998), and SBMA (Li et al., 1998). It has been controversial whether NIIs are toxic to neuronal cells (Sisodia, 1998; Zoghbi and Orr, 1999).

Various hypotheses have been proposed as to the mechanisms of aggregate formation of proteins containing expanded polyglutamine stretches. Perutz and colleagues proposed that polyglutamine stretches may function as polar zippers by joining complementary proteins through hydrogen bonding, and that extensions of the polyglutamine stretches may result in tight joining and aggregation of the affected proteins (Perutz et al., 1994, 1995, 1996). Another intriguing hypothesis has recently been proposed by Kahlem et al. (1996). They proposed that proteins with expanded polyglutamine stretches may serve as better substrates for transglutaminase than wild-type proteins, and that expanded polyglutamine stretches preferentially become crosslinked with polypeptides containing lysyl groups to form covalently bonded aggregates. We have demonstrated that the aggregate formation and apoptosis are partially suppressed by transglutaminase inhibitors such as cystamine and mono-dansylcadaverine (Igarashi et al., 1998). The results opened new prospects for developing therapeutic measures for the polyglutamine diseases.

To explore the mechanisms of neurodegeneration caused by expanded polyglutamine stretches, much effort has been made to identify proteins that interact with mutant proteins with expanded polyglutamine stretches. A number of proteins have so far been demonstrated to interact with the gene products (Bao et al., 1996; Boutell et al., 1998; Burke et al., 1996; Faber et al., 1998; Kalchman et al., 1996, 1997; Koshy et al., 1996; Li et al., 1995; Matilla et al., 1997; Nagai et al., 1999; Onodera et al., 1997; Sittler et al., 1998; Skinner et al., 1997; Wanker et al., 1997; Wood et al., 1998). Despite these efforts, however, it still remains unclear whether these interactions are involved in the pathogenic mechanism.

Although the physiological functions remain unclear, full-length wild-type DRPLA protein has recently been demonstrated to be localized predominantly in the nuclei of cultured cells (Miyashita et al., 1998; Sato et al., 1999a). Such nuclear localization is presumably mediated by putative nuclear localization signals (NLS) in these. In fact, we found that full-length DRPLA protein is expressed predominantly in the nucleus of neuronally differentiated PC12 cells using an adenovirus expression system. Furthermore, we demonstrated that intranuclear aggregate bodies are preferentially formed in neuronally differentiated PC12 cells and that these cells are more vulnerable than fibroblasts to the toxic effects of expanded polyglutamine stretches of the DRPLA protein (Sato et al., 1999a). These observations emphasize the importance of nuclear translocation of full-length or truncated DRPLA proteins with expanded polyglutamine stretches. The observations that transgenic mice expressing mutant ataxin-1 with a mutated NLS did not develop ataxia (Klement et al., 1998), and that addition of a nuclear export signal (NES) to mutant huntingtin suppressed the formation of NIIs and apoptosis (Saudou et al., 1998), further strengthens the role of nuclear translocation of mutant proteins with expanded polyglutamine stretches. Interaction of polyglutamine stretches and some nuclear proteins may be involved in the cytotoxicity caused by expanded polyglutamine stretches.

As reviewed above, recent advances have clearly shown the future directions, which include: (1) molecular mechanisms of nuclear translocation of mutant proteins with expanded polyglutamine stretches; (2) molecular mechanisms of aggregate formation, in particular, neuronal intranuclear inclusions; and (3) molecular mechanisms of apoptosis induced by expanded polyglutamine stretches (Fig. 2). Elucidation of these mechanisms and development of methods to interfere with these processes will no doubt make it possible to develop powerful new therapeutic strategies.

Mechanisms of aggregate formation and nuclear transport?

Mechanisms of apoptosis induced by aggregate bodies?

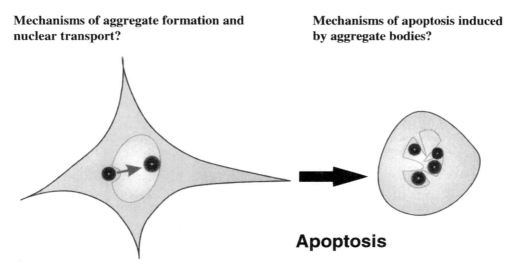

Apoptosis

Fig. 2. Proposed mechanisms of neurodegeneration caused by proteins with expanded polyglutamine stretches.

ACKNOWLEDGMENTS

This study was supported in part by: the Research for the Future Program from the Japan Society for the Promotion of Science (JSPS-RFTF96L00103); a Grant-in-Aid for Scientific Research on Priority Areas (Human Genome Program) from the Ministry of Education, Science, Sports and Culture, Japan; a grant from the Research Committee for Ataxic Diseases, the Ministry of Health and Welfare, Japan; a grant for Surveys and Research on Specific Diseases, the Ministry of Health and Welfare, Japan; and special coordination funds from the Japanese Science and Technology Agency.

REFERENCES

Andrew, S. E., Goldberg, Y. P., Kremer, B., Telenius, H., Theilmann, J., Adam, S., et al. (1993) The relationship between trinucleotide (CAG) repeat length and clinical features of Huntington's disease. *Nat. Genet.* **4,** 398–403.

Bao, J., Sharp, A. H., Wagster, M. V., Becher, M., Schilling, G., Ross, C. A., et al. (1996) Expansion of polyglutamine repeat in huntingtin leads to abnormal protein interactions involving calmodulin. *Proc. Natl. Acad. Sci. USA* **93,** 5037–5042.

Boutell, J. M., Wood, J. D., Harper, P. S., and Jones, A. L. (1998) Huntingtin interacts with cystathionine beta-synthase. *Hum. Mol. Genet.* **7,** 371–378.

Burke, J. R., Enghild, J. J., Martin, M. E., Jou, Y. S., Myers, R. M., Roses, A. D., et al. (1996) Huntingtin and DRPLA proteins selectively interact with the enzyme GAPDH. *Nat. Med.* **2,** 347–350.

Burke, J. R., Ikeuchi, T., Koide, R., Tsuji, S., Yamada, M., Pericak-Vance, M. A., and Vance, J. M. (1994) Dentatorubral-pallidoluysian atrophy and Haw River syndrome. *Lancet* **344,** 1711–1712.

Cancel, G., Durr, A., Didierjean, O., Imbert, G., Burk, K., Lezin, A., et al. (1997) Molecular and clinical correlations in spinocerebellar ataxia 2—a study of 32 families. *Hum. Mol. Genet.* **6,** 709–715.

Chong, S. S., McCall, A. E., Cota, J., Subramony, S. H., Orr, H. T., Hughes, M. R., and Zoghbi, H. Y. (1995) Gametic and somatic tissue-specific heterogeneity of the expanded SCA1 CAG repeat in spinocerebellar ataxia type 1. *Nat. Genet.* **10,** 344–350.

Chung, M. Y., Ranum, L. P., Duvick, L. A., Servadio, A., Zoghbi, H. Y., and Orr, H. T. (1993) Evidence for a mechanism predisposing to intergenerational CAG repeat instability in spinocerebellar ataxia type I. *Nat. Genet.* **5,** 254–258.

Cooper, J. K., Schilling, G., Peters, M. F., Herring, W. J., Sharp, A. H., Kaminsky, Z., et al. (1998) Truncated N-terminal fragments of huntingtin with expanded glutamine repeats form nuclear and cytoplasmic aggregates in cell culture. *Hum. Mol. Genet.* **7,** 783–790.

David, G., Abbas, N., Stevanin, G., Durr, A., Yvert, G., Cancel, G., et al. (1997) Cloning of the SCA7 gene reveals a highly unstable CAG repeat expansion. *Nat. Genet.* **17,** 65–70.

David, G., Durr, A., Stevanin, G., Cancel, G., Abbas, N., Benomar, A., et al. (1998) Molecular and clinical correlations in autosomal dominant cerebellar ataxia with progressive macular dystrophy (SCA7). *Hum. Mol. Genet.* **7,** 165–170.

Davies, S. W., Turmaine, M., Cozens, B. A., DiFiglia, M., Sharp, A. H., Ross, C. A., et al. (1997) Formation of neuronal intranuclear inclusions underlies the neurological dysfunction in mice transgenic for the HD mutation. *Cell* **90,** 537–548.

Difiglia, M., Sapp, E., Chase, K. O., Davies, S. W., Bates, G. P., Vonsattel, J. P., and Aronin, N. (1997) Aggregation of huntingtin in neuronal intranuclear inclusions and dystrophic neurites in brain. *Science* **277,** 1990–1993.

Duyao, M., Ambrose, C., Myers, R., Novelletto, A., Persichetti, F., Frontali, M., et al. (1993) Trinucleotide repeat length instability and age of onset in Huntington's disease. *Nat. Genet.* **4,** 387–392.

Endo, K., Sasaki, H., Wakisaka, A., Tanaka, H., Saito, M., Igarashi, S., et al. (1996) Strong linkage disequilibrium and haplotype analysis in Japanese pedigrees with Machado-Joseph disease. *Am. J. Med. Genet.* **67,** 437–444.

Faber, P. W., Barnes, G. T., Srinidhi, J., Chen, J., Gusella, J. F., and MacDonald, M. E. (1998) Huntingtin interacts with a family of WW domain proteins. *Hum. Mol. Genet.* **7,** 1463–1474.

Geschwind, D. H., Perlman, S., Figueroa, C. P., Treiman, L. J., and Pulst, S. M. (1997a) The prevalence and wide clinical spectrum of the spinocerebellar ataxia type 2 trinucleotide repeat in patients with autosomal dominant cerebellar ataxia. *Am. J. Hum. Genet.* **60,** 842–850.

Geschwind, D. H., Perlman, S., Figueroa, K. P., Karrim, J., Baloh, R. W., and Pulst, S. M. (1997b) Spinocerebellar ataxia type 6. Frequency of the mutation and genotype-phenotype correlations. *Neurology* **49,** 1247–1251.

Goldberg, Y. P., Nicholson, D. W., Rasper, D. M., Kalchman, M. A., Koide, H. B., Graham, R. K., et al. (1996) Cleavage of huntingtin by apopain, a proapoptotic cysteine protease, is modulated by the polyglutamine tract. *Nat. Genet.* **13,** 442–449.

Gouw, L. G., Castaneda, M. A., McKenna, C. K., Digre, K. B., Pulst, S. M., Perlman, S., et al. (1998) Analysis of the dynamic mutation in the SCA7 gene shows marked parental effects on CAG repeat transmission. *Hum. Mol. Genet.* **7,** 525–532.

Hashida, H., Goto, J., Kurisaki, H., Mizusawa, H., and Kanazawa, I. (1997) Brain regional differences in the expansion of a CAG repeat in the spinocerebellar ataxias: dentatorubral-pallidoluysian atrophy, Machado-Joseph disease, and spinocerebellar ataxia type 1. *Ann. Neurol.* **41,** 505–511.

Hayashi, Y., Kakita, A., Yamada, M., Egawa, S., Oyanagi, S., Naito, H., et al. (1998) Hereditary dentatorubral-pallidoluysian atrophy: ubiquitinated filamentous inclusions in the cerebellar dentate nucleus neurons. *Acta Neuropathol. (Berl.)* **95,** 479–482.

Hernandez, A., Magarino, C., Gispert, S., Santos, N., Lunkes, A., Orozco, G., et al. (1995) Genetic mapping of the spinocerebellar ataxia 2 (SCA2) locus on chromosome 12q23-q24. 1. *Genomics* **25,** 433–435.

Hodgson, J. G., Agopyan, N., Gutekunst, C. A., Leavitt, B. R., LePiane, F., Singaraja, R., et al. (1999) A YAC mouse model for Huntington's disease with full-length mutant huntingtin, cytoplasmic toxicity, and selective striatal neurodegeneration. *Neuron* **23**, 181–192.

Holmberg, M., Duyckaerts, C., Durr, A., Cancel, G., Gourfinkel-An, I., Damier, P., et al. (1998) Spinocerebellar ataxia type 7 (SCA7): a neurodegenerative disorder with neuronal intra-nuclear inclusions. *Hum. Mol. Genet.* **7**, 913–918.

Igarashi, S., Koide, R., Shimohata, T., Yamada, M., Hayashi, Y., Takano, H., et al. (1998) Suppression of aggregate formation and apoptosis by transglutaminase inhibitors in cells expressing truncated DRPLA protein with an expanded polyglutamine stretch. *Nat. Genet.* **18**, 111–117.

Ikeda, H., Yamaguchi, M., Sugai, S., Aze, Y., Narumiya, S., and Kakizuka, A. (1996) Expanded polyglutamine in the Machado-Joseph disease protein induces cell death in vitro and in vivo. *Nat. Genet.* **13**, 196–202.

Ikeuchi, T., Koide, R., Onodera, O., Tanaka, H., Oyake, M., Takano, H., and Tsuji, S. (1995a) Dentatorubral-pallidoluysian atrophy (DRPLA). Molecular basis for wide clinical features of DRPLA. *Clin. Neurosci.* **3**, 23–27.

Ikeuchi, T., Koide, R., Tanaka, H., Onodera, O., Igarashi, S., Takahashi, H., et al. (1995b) Dentatorubral-pallidoluysian atrophy (DRPLA): clinical features are closely related to unstable expansions of trinucleotide (CAG) repeat. *Ann. Neurol.* **37**, 769–775.

Ikeuchi, T., Onodera, O., Oyake, M., Koide, R., Tanaka, H., and Tsuji, S. (1995c) Dentatorubral-pallidoluysian atrophy (DRPLA): close correlation of CAG repeat expansions with the wide spectrum of clinical presentations and prominent anticipation. *Semin. Cell Biol.* **6**, 37–44.

Illarioshkin, S. N., Slominsky, P. A., Ovchinnikov, I. V., Markova, E. D., Miklina, N. I., Klyushnikov, S. A., et al. (1996) Spinocerebellar ataxia type 1 in Russia. *J. Neurol.* **243**, 506–510.

Imbert, G., Saudou, F., Yvert, G., Devys, D., Trottier, Y., Garnier, J. M., et al. (1996) Cloning of the gene for spinocerebellar ataxia 2 reveals a locus with high sensitivity to expanded CAG/glutamine repeats. *Nat. Genet.* **14**, 285–291.

Johansson, J., Forsgren, L., Sandgren, O., Brice, A., Holmgren, G., and Holmberg, M. (1998) Expanded CAG repeats in Swedish spinocerebellar ataxia type 7 (SCA7) patients: effect of CAG repeat length on the clinical manifestation. *Hum. Mol. Genet.* **7**, 171–176.

Kahlem, P., Terre, C., Green, H., and Djian, P. (1996) Peptides containing glutamine repeats as substrates for transglutaminase-catalyzed cross-linking: relevance to diseases of the nervous system. *Proc. Natl. Acad. Sci. USA* **93**, 14,580–14,585.

Kalchman, M. A., Graham, R. K., Xia, G., Koide, H. B., Hodgson, J. G., Graham, K. C., et al. (1996) Huntingtin is ubiquitinated and interacts with a specific ubiquitin-conjugating enzyme. *J. Biol. Chem.* **271**, 19,385–19,394.

Kalchman, M. A., Koide, H. B., McCutcheon, K., Graham, R. K., Nichol, K., Nishiyama, K., et al. (1997) HIP1, a human homologue of s-cerevisiae sla2p, interacts with membrane-associated huntingtin in the brain. *Nat. Genet.* **16**, 44–53.

Kawakami, H., Maruyama, H., Nakamura, S., Kawaguchi, Y., Kakizuka, A., Doyu, M., and Sobue, G. (1995) Unique features of the CAG repeats in Machado-Joseph disease. *Nat. Genet.* **9**, 344–345.

Kaytor, M. D., Burright, E. N., Duvick, L. A., Zoghbi, H. Y., and Orr, H. T. (1997) Increased trinucleotide repeat instability with advanced maternal age. *Hum. Mol. Genet.* **6**, 2135–2139.

Klement, I. A., Skinner, P. J., Kaytor, M. D., Yi, H., Hersch, S. M., Clark, H. B., et al. (1998) Ataxin-1 nuclear localization and aggregation: role in polyglutamine-induced disease in SCA1 transgenic mice. *Cell* **95**, 41–53.

Koide, R., Ikeuchi, T., Onodera, O., Tanaka, H., Igarashi, S., Endo, K., et al. (1994) Unstable expansion of CAG repeat in hereditary dentatorubral-pallidoluysian atrophy (DRPLA). *Nat. Genet.* **6**, 9–13.

Koshy, B., Matilla, T., Burright, E. N., Merry, D. E., Fischbeck, K. H., Orr, H. T., and Zoghbi, H. Y. (1996) Spinocerebellar ataxia type-1 and spinobulbar muscular atrophy gene products interact with glyceraldehyde-3-phosphate dehydrogenase. *Hum. Mol. Genet.* **5,** 1311–1318.

Kremer, B., Almqvist, E., Theilmann, J., Spence, N., Telenius, H., Goldberg, Y. P., and Hayden, M. R. (1995) Sex-dependent mechanisms for expansions and contractions of the cag repeat on affected huntington disease chromosomes. *Am. J. Hum. Genet.* **57,** 343–350.

La Spada, A. R., Peterson, K. R., Meadows, S. A., McClain, M. E., Jeng, G., Chmelar, R. S., et al. (1998) Androgen receptor YAC transgenic mice carrying CAG 45 alleles show trinucleotide repeat instability. *Hum. Mol. Genet.* **7,** 959–967.

La Spada, A. R., Wilson, E. M., Lubahn, D. B., Harding, A. E., and Fischbeck, K. H. (1991) Androgen receptor gene mutations in X-linked spinal and bulbar muscular atrophy. *Nature* **352,** 77–79.

Li, M., Miwa, S., Kobayashi, Y., Merry, D. E., Yamamoto, M., Tanaka, F., et al. (1998) Nuclear Inclusions of the androgen receptor protein in spinal and bulbar muscular atrophy. *Ann. Neurol.* **44,** 249–254.

Li, X. J., Li, S. H., Sharp, A. H., Nucifora Jr, F. C., Schilling, G., Lanahan, A., et al. (1995) A huntingtin-associated protein enriched in brain with implications for pathology. *Nature* **378,** 398–402.

Lorenzetti, D., Bohlega, S., and Zoghbi, H. Y. (1997) The expansion of the CAG repeat in ataxin-2 is a frequent cause of autosomal dominant spinocerebellar ataxia. *Neurology* **49,** 1009–1013.

Mangiarini, L., Sathasivam, K., Seller, M., Cozens, B., Harper, A., Hetherington, C., Lawton, M., et al. (1996) Exon 1 of the HD gene with an expanded CAG repeat is sufficient to cause a progressive neurological phenotype in transgenic mice. *Cell* **87,** 493–506.

Martindale, D., Hackam, A., Wieczorek, A., Ellerby, L., Wellington, C., McCutcheon, K., et al. (1998) Length of huntingtin and its polyglutamine tract influences localization and frequency of intracellular aggregates. *Nat. Genet.* **18,** 150–154.

Matilla, A., Koshy, B. T., Cummings, C. J., Isobe, T., Orr, H. T., and Zoghbi, H. Y. (1997) The cerebellar leucine-rich acidic nuclear protein interacts with ataxin-1. *Nature* **389,** 974–978.

Miyashita, T., Nagao, K., Ohmi, K., Yanagisawa, H., Okamura-Oho, Y., and Yamada, M. (1998) Intracellular aggregate formation of dentatorubral-pallidoluysian atrophy (DRPLA) protein with the extended polyglutamine. *Biochem. Biophys. Res. Commun.* **249,** 96–102.

Mizushima, S. and Nagata, S. (1990) pEF-BOS, a powerful mammalian expression vector. *Nucleic Acids Res.* **18,** 5322.

Nagafuchi, S., Yanagisawa, H., Ohsaki, E., Shirayama, T., Tadokoro, K., Inoue, T., and Yamada, M. (1994a) Structure and expression of the gene responsible for the triplet repeat disorder, dentatorubral and pallidoluysian atrophy (DRPLA). *Nat. Genet.* **8,** 177–182.

Nagafuchi, S., Yanagisawa, H., Sato, K., Sato, K., Shirayama, T., Ohsaki, E., et al. (1994b) Expansion of an unstable CAG trinucleotide on chromosome 12p in dentatorubral and pallidoluysian atrophy. *Nat. Genet.* **6,** 14–18.

Nagai, Y., Onodera, O., Chun, J., Strittmatter, W. J., and Burke, J. R. (1999) Expanded polyglutamine domain proteins bind neurofilament and alter the neurofilament network. *Exp. Neurol.* **155,** 195–203.

Onodera, O., Burke, J. R., Miller, S. E., Hester, S., Tsuji, S., Roses, A. D., and Strittmatter, W. J. (1997) Oligomerization of expanded-polyglutamine domain fluorescent fusion proteins in cultured mammalian cells. *Biochem. Biophys. Res. Commun.* **238,** 599–605.

Orr, H. T., Chung, M. Y., Banfi, S., Kwiatkowski Jr, T. J., Servadio, A., Beaudet, A. L., et al. (1993) Expansion of an unstable trinucleotide CAG repeat in spinocerebellar ataxia type 1. *Nat. Genet.* **4,** 221–226.

Paulson, H. L., Perez, M. K., Trottier, Y., Trojanowski, J. Q., Subramony, S. H., Das, S. S., et al. (1997) Intranuclear inclusions of expanded polyglutamine protein in spinocerebellar ataxia type 3. *Neuron* **19,** 333–344.

Perutz, M. F. (1995) Glutamine repeats as polar zippers: their role in inherited neuro-degenerative disease. *Mol. Med.* **1,** 718–721.

Perutz, M. F. (1996) Taking the pressure off. *Nature* **380,** 205–206.

Perutz, M. F., Johnson, T., Suzuki, M., and Finch, J. T. (1994) Glutamine repeats as polar zippers: their possible role in inherited neurodegenerative diseases. *Proc. Natl. Acad. Sci. USA* **91,** 5355–5358.

Pulst, S. M., Nechiporuk, A., Nechiporuk, T., Gispert, S., Chen, X. N., Lopes-Cendes, I., et al. (1996) Moderate expansion of a normally biallelic trinucleotide repeat in spinocerebellar ataxia type 2. *Nat. Genet.* **14,** 269–276.

Rubinsztein, D. C. and Leggo, J. (1997) Non-mendelian transmission at the Machado-Joseph disease locus in normal females: preferential transmission of alleles with smaller CAG repeats. *J. Med. Genet.* **34,** 234–236.

Sanpei, K., Takano, H., Igarashi, S., Sato, T., Oyake, M., Sasaki, H., et al. (1996) Identification of the spinocerebellar ataxia type 2 gene using a direct identification of repeat expansion and cloning technique, DIRECT. *Nat. Genet.* **14,** 277–284.

Sato, A., Shimohata, T., Koide, R., Takano, H., Sato, T., Oyake, M., et al. (1999a) Adenovirus-mediated expression of mutant DRPLA proteins with expanded polyglutamine stretches in neuronally differentiated PC12 cells. *Hum. Mol. Genet.* **8,** 997–1006.

Sato, T., Oyake, M., Nakamura, K., Nakao, K., Fukusima, Y., Onodera, O., et al. (1999b) Transgenic mice harboring a full length human mutant DRPLA gene reveal CAG repeat instability. *Hum. Mol. Genet.* **8,** 99–106.

Saudou, F., Finkbeiner, S., Devys, D., and Greenberg, M. E. (1998) Huntingtin acts in the nucleus to induce apoptosis but death does not correlate with the formation of intranuclear inclusions. *Cell* **95,** 55–66.

Sisodia, S. S. (1998) Nuclear inclusions in glutamine repeat disorders: are they pernicious, coincidental, or beneficial? *Cell* **95,** 1–4.

Sittler, A., Walter, S., Wedemeyer, N., Hasenbank, R., Scherzinger, E., Eickhoff, H., et al. (1998) SH3GL3 associates with the Huntingtin exon 1 protein and promotes the formation of polygln-containing protein aggregates. *Mol. Cell* **2,** 427–436.

Skinner, P. J., Koshy, B. T., Cummings, C. J., Klement, I. A., Helin, K., Servadio, A., et al. (1997) Ataxin-1 with an expanded glutamine tract alters nuclear matrix-associated structures. *Nature* **389,** 971–974.

Snell, R. G., MacMillan, J. C., Cheadle, J. P., Fenton, I., Lazarou, L. P., Davies, P., et al. (1993) Relationship between trinucleotide repeat expansion and phenotypic variation in Huntington's disease. *Nat. Genet.* **4,** 393–397.

Squitieri, F., Andrew, S. E., Goldberg, Y. P., Kremer, B., Spence, N., Zeisler, J., et al. (1994) DNA haplotype analysis of Huntington disease reveals clues to the origins and mechanisms of CAG expansion and reasons for geographic variations of prevalence. *Hum. Mol. Genet.* **3,** 2103–2114.

Stevanin, G., Cancel, G., Didierjean, O., Durr, A., Abbas, N., Cassa, E., Feingold, J., et al. (1995) Linkage disequilibrium at the Machado-Joseph disease/spinal cerebellar ataxia 3 locus: evidence for a common founder effect in French and Portuguese-Brazilian families as well as a second ancestral Portuguese-Azorean mutation. *Am. J. Hum. Genet.* **57,** 1247–1250.

Stevanin, G., Lebre, A. S., Mathieux, C., Cancel, G., Abbas, N., Didierjean, O., et al. (1997) Linkage disequilibrium between the spinocerebellar ataxia 3/Machado-Joseph disease mutation and two intragenic polymorphisms, one of which, x359y, affects the stop codon. *Am. J. Hum. Genet.* **60,** 1548–1552.

Takano, H., Cancel, G., Ikeuchi, T., Lorenzetti, D., Mawad, R., Stevanin, G., et al. (1998) Close associations between the prevalence rates of dominantly inherited spinocerebellar ataxias with CAG repeat expansions and the frequencies of large normal CAG alleles in Japanese and Caucasian populations. *Am. J. Hum. Genet.* **63,** 1060–1066.

Takano, H., Onodera, O., Takahashi, H., Igarashi, S., Yamada, M., Oyake, M., et al. (1996) Somatic mosaicism of expanded CAG repeats in brains of patients with dentatorubral-pallidoluysian atrophy: cellular population-dependent dynamics of mitotic instability. *Am. J. Hum. Genet.* **58,** 1212–1222.

Takiyama, Y., Igarashi, S., Rogaeva, E. A., Endo, K., Rogaev, E. I., Tanaka, H., et al. (1995) Evidence for inter-generational instability in the CAG repeat in the MJD1 gene and for conserved haplotypes at flanking markers amongst Japanese and Caucasian subjects with Machado-Joseph disease. *Hum. Mol. Genet.* **4,** 1137–1146.

Telenius, H., Kremer, B., Goldberg, Y. P., Theilmann, J., Andrew, S. E., Zeisler, J., et al. (1994) Somatic and gonadal mosaicism of the Huntington disease gene CAG repeat in brain and sperm. *Nat. Genet.* **6,** 409–414.

The Huntington's Disease Collaborative Research Group. (1993) A novel gene containing a trinucleotide repeat that is expanded and unstable on Huntington's disease chromosomes. *Cell* **72,** 971–983.

Ueno, S., Kondoh, K., Kotani, Y., Komure, O., Kuno, S., Kawai, J., et al. (1995) Somatic mosaicism of CAG repeat in dentatorubral-pallidoluysian atrophy (DRPLA). *Hum. Mol. Genet.* **4,** 663–666.

Wakisaka, A., Sasaki, H., Takada, A., Fukazawa, T., Suzuki, Y., Hamada, T., et al. (1995) Spinocerebellar ataxia 1 (SCA1) in the Japanese in Hokkaido may derive from a single common ancestry. *J. Med. Genet.* **32,** 590–592.

Wanker, E. E., Rovira, C., Scherzinger, E., Hasenbank, R., Walter, S., Tait, D., et al. (1997) HIP-I: A huntingtin interacting protein isolated by the yeast two-hybrid system. *Hum. Mol. Genet.* **6,** 487–495.

Wood, J. D., Yuan, J., Margolis, R. L., Colomer, V., Duan, K., Kushi, J., et al. (1998) Atrophin-1, the DRPLA gene product, interacts with two families of ww domain-containing proteins. *Mol. Cell. Neurosci.* **11,** 149–160.

Yanagisawa, H., Fujii, K., Nagafuchi, S., Nakahori, Y., Nakagome, Y., Akane, A., et al. (1996) A unique origin and multistep process for the generation of expanded DRPLA triplet repeats. *Hum. Mol. Genet.* **5,** 373–379.

Zhuchenko, O., Bailey, J., Bonnen, P., Ashizawa, T., Stockton, D. W., Amos, C., et al. (1997) Autosomal dominant cerebellar ataxia (SCA6) associated with small polyglutamine expansions in the alpha 1a-voltage-dependent calcium channel. *Nat. Genet.* **15,** 62–69.

Zoghbi, H. Y. and Orr, H. T. (1999) Polyglutamine diseases: protein cleavage and aggregation. *Curr. Opin. Neurobiol.* **9,** 566–570.

5

Mechanisms of Neuronal Death in Huntington's Disease

Vassilis E. Koliatsos, Carlos Portera-Cailliau, Gabrielle Schilling, David B. Borchelt, Mark W. Becher, and Christopher A. Ross

INTRODUCTION

Huntington's disease (HD) is an inherited, progressive, and always fatal, neuro-degenerative disorder characterized by degeneration of neurons in the striatum and cerebral cortex, which results in involuntary motor movements. HD is one of several neurodegenerative diseases found to be caused by an expansion of CAG repeats encoding glutamine in specific proteins. The gene affected in HD is located on chromosome 4p and encodes a 3144 amino acid protein called huntingtin (Huntington's Disease Collaborative Research Group, 1993). Other diseases known to be caused by CAG repeat expansions of greater than 35 repeats include spinal cerebellar ataxia (CAG repeat expansion in genes SCA1, SCA2, SCA3, SCA6, and SCA7), spinobulbar muscular atrophy (CAG repeat expansion in AR), and dentato-rubro-pallido-luyisian atrophy (DRPLA gene). These conditions are a subset of a larger class of diseases (Fragile X syndrome, Friederich's ataxia, and myotonic dystrophy) associated with expansion of trinucleotide repeats either in coding or in noncoding regions of the DNA. Glutamine repeat (polyglutamine) diseases are the only trinucleotide expansion diseases in which triplet repeat expansion occurs within the coding DNA region, thus leading to protein changes. It is expected that the list of these diseases, all of which target excitable tissues in the human body (nerve and muscle cells), will grow in the coming years.

As is the case with other neurodegenerative disorders, polyglutamine repeat diseases are characterized by selective vulnerability of neurons; in the case of HD, it is primarily striatal neurons that degenerate. In HD and other polyglutamine repeat diseases, the responsible genes are expressed widely in nearly all organ systems, and yet pathology appears to be restricted to the central nervous system (CNS). Interestingly, affected CNS regions overlap considerably in different polyglutamine diseases, and consistently include areas such as the neostriatum, globus pallidus, the subthalamic and red nuclei, basis pontis, cerebellar Purkinje cells and dentate nucleus, and motor neurons in the brainstem and spinal cord (Warren and Nelson, 1993; LaSpada et al., 1994; Housman, 1995; Ross, 1995; Nasir et al. 1996; Bates, 1996; MacDonald and Gusella, 1997). Other regions and neuronal populations remain free of pathology; for

From: *Pathogenesis of Neurodegenerative Disorders* Edited by: M. P. Mattson © Humana Press Inc., Totowa, NJ

example, the temporal limbic system and the nucleus basalis of Meynert, which degenerate in Alzheimer's disease (AD), are spared in CAG repeat diseases.

The normal function of proteins affected in polyglutamine diseases is typically not compromised; instead, the pathogenic mechanism involves a gain-of-function in which the expanded protein acquires a novel adverse property (Ross et al. 1997). The commonalties among the various glutamine repeat diseases suggest that the glutamine repeat expansion itself may cause cell injury; the protein containing the glutamine repeat may act as a modifier of the degree and location of the injury, but when the repeat is long enough, then the common toxic effect predominates and leads to a severe and uniform neurodegeneration. On the other hand, the variance of neuropathologies among various glutamine repeat diseases suggests that the expanded glutamine repeat may alter the protein involved in a specific way irrespective of repeat length and, depending on the role of this protein in cells, the resulting conformational change may activate a unique cascade of injury. Of course, these two approaches in the understanding of the basic mechanisms of polyglutamine diseases have strengths and weaknesses and, overall, they represent complementary, rather then contrasting, neurobiological processes.

This chapter summarizes the neuropathology of HD and its principal animal models and outlines evidence for programmed cell death in the human disease and experimental animal models of HD. While our discussion of apoptosis in HD is self-explanatory, our inclusion, among the various models, of the excitotoxic paradigm in the discussion of apoptosis is based on the extensive characterization of this model by other investigators and us. Metabolic models (see below) generally match the excitotoxic models in terms of patterns of cell death. The extent of neuronal cell death in transgenic models varies and does not appear to be apoptotic when present (*see* below). The excitotoxic model allows us the additional opportunity of presenting our more generic notions on mechanisms of neuronal cell death, which transcend the largely artificial boundaries between apoptosis and necrosis.

NEUROPATHOLOGY OF HD

Adult-onset HD is characterized by loss of neostriatal projection (medium spiny) neurons accompanied by reactive astrocytosis and a less severe, but prominent degeneration of the globus pallidus (Vonsattel et al., 1985; Albin and Greenamyre, 1992; Oppenheimer and Esiri, 1992; Sharp and Ross, 1996) (Fig. 1). Variable degrees of neurodegeneration can occur in the claustrum, subthalamic nucleus, amygdala, neocortex, pons, olivary complex, and cerebellar cortex (i.e., Purkinje cells) (Rodda, 1981; Vonsattel et al., 1985; Hedreen et al., 1991; Jackson et al., 1995). Pathologic changes in nonstriatal areas consist of astrocytosis with or without mild neuronal loss.

Pediatric-onset (juvenile) HD, a relatively rare entity representing less than 1% of all HD cases, presents with rigidity and seizures rather than chorea, and is associated with more widespread brain pathology than adult-onset HD (Byers and Dodge, 1967; Jervis, 1963; Rodda, 1981; Nelson, 1995). In addition to caudate, putamen, and globus pallidus, most pediatric HD cases show cerebellar atrophy with loss of Purkinje and granule cells. There is also variable astrocytosis or loss of neurons in the dentate nucleus of the cerebellum, thalamus, hippocampus, and lateral vestibular nuclei (Byers et al., 1973). Many cases show marked loss of neurons in neocortex, including focal laminar

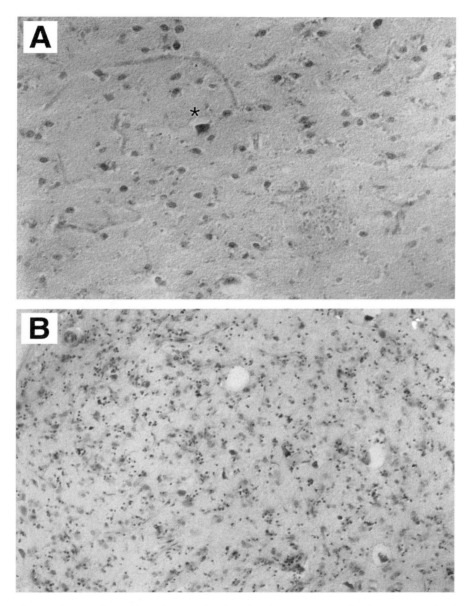

Fig. 1. The devastating effect of expanded huntingtin on small projection neurons in the neo-striatum is exemplified by the juxtaposition of a normal putamen from a control (**A**) and a patient with grade 3 HD (**B**). In HD, the neuronal cytoarchitecture appears to have been replaced by phagocytic glia. Large interneurons (asterisk in A) are relatively spared in the neurodegenerative process.

necrosis. Juvenile HD is characterized by a larger length of the CAG triplet repeats than in adult-onset cases (Duyao et al., 1993; Stine et al., 1993; Ikeuchi et al., 1995; Komure et al., 1995). Thus, the neuropathological differences between juvenile- and adult-onset cases may be correlated, at least in part, with triplet repeat length (Snell et al., 1993; Ikeuchi et al., 1995; Komure et al., 1995).

EXPERIMENTAL MODELS OF HD

Excitotoxin-Based Models

Studies of the "excitotoxic" actions of various analogs of the neurotransmitter glutamate in rodents revealed that overactivation of ionotropic glutamate receptors can selectively destroy striatal medium spiny neurons and induce motor dysfunction (McGeer and McGeer, 1976; Coyle and Schwarcz, 1976; Olney and Ishimaru, 1998). The recent parcellation of different glutamate receptors as *N*-Methyl-D-aspartate (NMDA) and non-NMDA glutamate receptors allowed a further clarification of the striatal vulnerability to glutamate receptor agonists. Although non-NMDA agonists, when infused into the striatum, reproduce many neuropathological features of HD (Coyle and Schwarcz, 1976; McGeer et al., 1978), NMDA receptor agonists such as quinolinic acid (QA) cause lesions that are more selective for striatal projection (medium spiny) neurons with relative sparing of large somatostatin, neuropeptide Y, and cholinergic nerve cells (Beal, et al., 1986, 1993b; Young et al., 1988). QA-treated medium spiny neurons show morphological changes in their processes and increases in calbindin D_{28K} (a Ca^{2+} binding protein) immunoreactivity similar to those seen in HD brains (Huang et al., 1995). These findings suggest a pattern of NMDA receptor-mediated selective excitotoxicity, which resembles the vulnerability of medium spiny neurons in HD and is a plausible pathogenic mechanism, given the dense and direct innervation of these neurons by cortical glutamatergic inputs. In fact, ablation of glutamatergic corticostriatal projections prior to treatment with glutamate agonists attenuates striatal excitotoxicity (McGeer et al., 1978). In addition, a recent magnetic resonance spectroscopy study has suggested that glutamate is increased in the striatum of patients with early HD (Taylor-Robinson et al., 1996).

Despite the kinds of data just described, NMDA receptor-mediated excitotoxicity cannot be a unifying pathogenetic hypothesis for HD. The preferential death of striatal medium spiny neurons is incongruent with the wide distribution of NMDA receptors in the brain (Monaghan and Cotman, 1985; Cotman et al., 1987; Monaghan et al., 1989; Young and Fagg, 1990). NMDA receptors are equally dense in neocortex, striatum, and hippocampus, whereas pathology in HD is relatively selective for the striatum, with the hippocampus remaining essentially unaffected. Furthermore, within cortex, degenerative changes in HD are more prominent in deep layers, whereas NMDA receptors are more abundant in superficial layers. In summary, although excitotoxic models show a striking resemblance to cellular pathologies encountered in the neostriatum in HD, they fall short of explaining the selective vulnerability of this structure.

Mitochondrial Toxins

Metabolic toxicity involves the subacute disruption of energy generation within neurons by mitochondrial toxins, such as malonate and 3-nitropropionic acid (3NPA) and produces a pattern of pathology very similar to that of HD (Beal et al., 1993a, b, c; Brouillet et al., 1993). In humans, ingestion of 3NPA via contaminated food produces selective lesions of the basal ganglia (Ludolph et al., 1991). In rats, systemic injection of 3NPA causes selective loss of striatal medium spiny neurons. The neurotoxic effects of mitochondrial toxins can be prevented by prior decortication or, in vitro, by excitatory amino acid antagonists. Malonate-induced injury can be blocked by coenzyme

Q10 and nicotinamide, an effect consistent with a mechanism involving energy depletion (Beal et al., 1994). Chronic 3NPA treatment in baboons causes selective striatal lesions and replicates the cognitive and motoric deficits of HD (Persichetti et al., 1995). These results suggest that an impairment of mitochondrial energy metabolism can selectively increase the vulnerability of medium spiny neurons to excitotoxicity (Albin and Greenamyre, 1992; Greene and Greenamyre, 1995).

Evidence of energy defects in neurons has been obtained in studies of HD patients. For example, magnetic resonance spectroscopy shows altered energy metabolism in HD patients (Jenkins et al., 1993, 1996). Other studies have revealed neurochemical defects in several mitochondrial enzyme complexes in postmortem brain tissues from HD patients (Gu et al., 1996). An attractive feature of the metabolic models is that disruption of energy generation within cells at different points can cause distinct patterns of pathology. For example, 3NPA and malonate, both inhibitors of complex II in the mitochondrial electron transport chain, cause selective striatal degeneration. By contrast, cyanide, which is an electron-transport-chain inhibitor selective for complex IV, causes a more widespread damage involving the putamen, globus pallidus, substantia nigra, subthalamic nucleus, and cerebellum (Uitti et al., 1985; Rosenow et al., 1995). Despite differences in the location/extent of neuronal injury, metabolic inhibitors tend to give rise to subcortical/cerebellar rather than cortical/hippocampal pathology, i.e., a pattern characteristic of polyglutamine diseases.

Excitotoxic and metabolic models of HD do not represent disparate mechanisms of neuronal injury. For example, free radicals have been implicated in both excitotoxic and metabolic models (Coyle and Puttfarcken, 1993; Dugan et al., 1995; Schulz et al., 1995a, b). Lesions produced by intrastriatal injections of glutamate agonists in rats are attenuated by pretreatment of the animals with free-radical spin-trap compounds. One free radical that may be involved in excitotoxicity is nitric oxide (NO), synthesized in the course of the conversion of arginine to citrulline via the action of nitric oxide synthase (NOS) (Bredt and Snyder, 1992; Dawson et al., 1993). In cell cultures, NO is a mediator of NMDA toxicity (Dawson et al., 1993). Treatment with NOS inhibitors ameliorates excitotoxic injury in several models (Schulz et al., 1995b) and mice with deletions of neuronal NOS are resistant to both excitotoxicity (Dawson et al., 1996) and metabolic toxicity (Schulz et al., 1996).

Transgenic Mouse Models

The cloning of the huntingin gene provided the opportunity to produce genetic mouse models of HD. The first such mouse was created by injecting a fragment of genomic human DNA that encoded the first exon of huntingtin with 115-150 CAG repeats (Exon 1 HD mouse) (Mangiarini et al., 1996). Transgenic mice exhibited significant weight loss and a neurologic phenotype including tremor, myoclonic movements, seizures, and early death (Mangiarini et al., 1996). There was no significant loss of neurons in the brains of these animals. Ubiquitin and huntingtin-positive intranuclear inclusions were present in many neurons and some glial cells before the onset of phenotype (Davies et al., 1997). Although the localization of normal huntingtin is mostly cytoplasmic (Gutekunst et al., 1995; Sharp et al., 1995) (*see* below), expanded huntingtin aggregates in the nucleus. This pattern may relate to the tendency of long stretches of polyglutamines to aggregate into "polar zippers" in vitro (Perutz, 1994; Scherzinger et al., 1997). Recently, huntingtin-immunoreactive intranuclear inclusions were detected in a variety of peripheral tissues of the Exon-1 transgenic mice including heart and skeletal muscle and in the Langerhans cells of pancreas. The presence of the inclusions in Langerhans cells may be related to the hyperglycemia seen in these animals (i.e., may cause a type of diabetes).

Transgenic mice expressing a N-terminal piece of human HD cDNA including 82 glutamines (N171-82Q) develop a progressive neurological phenotype including early death and display intranuclear inclusions in many areas of the brain (Schilling et al., 1999). In addition, the N171-82Q mice have high frequency of cytoplasmic huntingtin aggregates, which have been observed in HD patients (DiFiglia et al., 1997; Gutekunst et al., 1999). These aggregates and inclusions may contain other important proteins such as proteasome components and transcription factors. The N171-82Q mice do not show evidence of brain pathology by TUNEL (cell death) and GFAP immunostaining (astrocytosis).

A full-length huntingtin cDNA transgenic mouse was also created with 16, 48, or 89 CAGs using the CMV promoter (Reddy et al., 1998). Transgenic mice with 89 glutamines display a behavioral phenotype including an initial hyperactivity (circling), followed by hypoactivity. Transgenic mice with 48Q and 89Q show cell loss and increased TUNEL labeling in the striatum, hippocampus, and cerebellum, but very few intranuclear inclusions in the affected regions.

Recently, a YAC (yeast artificial chromosome) transgenic mouse model for HD was developed encoding 18, 48, or 72 glutamines (Hodgson et al., 1999). YAC transgenic mice with 72 glutamines show a behavioral phenotype and progressive electrophysiological abnormalities. These mice also show nuclear aggregation of mutant huntingtin and neurodegeneration that is specific for the neostriatum.

Transgenic mouse models expressing short forms of huntingtin and higher number of glutamines (i.e., biochemical signatures associated with greater toxicity in vitro), have a tendency for a more severe neurological phenotype, including premature death, but there is no obvious death of neurons. In contrast, full-length huntingtin-transgenic mice have a more delayed and subtle behavioral phenotype and possibly more neurodegeneration; these mice usually do not die prematurely. The difference in the phenotypes among the various models might be explained by differences in the length of the transgene construct, difference in the distribution of transgene expression, or difference in the genetic backgrounds of these animals.

EVIDENCE FOR APOPTOTIC DEATH OF STRIATAL AND CORTICAL NEURONS IN HD

Studies of HD Patients

Although HD has been studied extensively with classical neuropathological approaches (Rodda, 1981; Vonsattel et al., 1985; Hedreen et al., 1991; Albin and Greenamyre, 1992; Jackson et al., 1995; Sharp and Ross, 1996), direct methods for detecting dying neurons have only recently been applied to postmortem tissues from HD patients. In a recent study of ours (Portera-Cailliau et al., 1995), we used techniques that are especially sensitive to apoptotic cell death in 13 cases of HD representing various stages of the disease (Vonsattel grades 0–4) and 8 age-matched controls (subjects with no neuropsychiatric disease or with neurological illnesses other than HD). Terminal-transferase-mediated (TdT) deoxyuridine triphosphate (d-UTP)-biotin nick-end labeling (TUNEL) and DNA electrophoresis were used to detect DNA strand breaks *in situ* and in homogenized samples from neostriatum. TUNEL was intense in the striatum in most grade 3 cases (intermediate disease) and significantly less in grade 2 (early disease); labeling was greatly diminished in advanced disease (grade 4). Striatal labeling was especially intense in the putamen, followed by globus pallidus and the caudate nucleus. TUNEL in the putamen showed a striking dorsoventral pattern, with dorsal regions displaying much more intense TUNEL than the midputamen and with only minimal labeling evident in ventral putamen.

The types of TUNEL-stained brain cells differ according to disease stage. Oligodendrocytes appear to predominate early on in the course of the illness, although data on grade 0 cases are limited (Portera-Cailliau et al., 1995). In grades 2 and 3, many more neurons than glial cells are labeled. In grade 4, in which there are few surviving medium spiny neurons in the neostriatum, the vast majority of TUNEL-positive nuclei belong to glia. Many TUNEL-positive oligodendrocytes are located within the bundles of striato-pallidal fibers (pencils of Wilson). Astrocytes are essentially negative for TUNEL throughout the course of the illness. TUNEL staining is also present in neurons in most layers of middle frontal cortex from grade 3 subjects that show robust TUNEL labeling in putamen.

DNA electrophoresis has not revealed classical "laddering" of DNA into internucleosomal fragments in HD. DNA isolated from the dorsal putamen of grade 3–4 HD cases (i.e., samples with maximal representation of dying cells) showed variable migration profiles, including intact "single band" patterns (Portera-Cailliau et al., 1995). Some grade 3 cases did show evidence of nonrandom DNA fragmentation. DNA samples from well-preserved striatal tissues of healthy controls usually do not show any evidence of degradation, whereas positive controls from areas with anoxic/ischemic injury (stroke) show classical laddering, comparable to that of DNA from animal models of programmed cell death.

Although neuronal death in HD occurs continuously over many years, there appears to be an accelerated rate of neuronal cell death as HD progresses, especially in the dorsal putamen and the caudate nucleus, a pattern consistent with the early degeneration of these two neostriatal regions in HD (Dunlap, 1927; McCaughey, 1961; Vonsattel et al., 1985). TUNEL staining in cortex and globus pallidus may represent retrograde and anterograde cell degeneration, respectively, or could reflect the independent effects of abnormal huntingtin in those regions. The possibility of anterograde/postsynaptic death of neurons in external globus pallidus is consistent with the well-recognized death of striatal neurons projecting to this region in early HD (Reiner et al., 1988; Albin et al., 1992).

In addition to dying neurons, TUNEL also detected dying oligodendrocytes in the striatum at different grades of HD severity. There are two possible explanations for this observation. One possibility is that oligodendrocytes die as part of the Wallerian degeneration of striato-pallidal axons subsequent to death of medium spiny neurons; alternatively, oligodendrocytes may die as a direct result of the genetic alteration in huntingtin, in exactly the same mechanism as medium spiny neurons.

Several features of the cell death that occur in HD are consistent with an apoptotic mechanism. In contrast to necrosis, in which whole patches of cells usually die rapidly and synchronously after an exogenous insult, TUNEL staining is seen in isolated medium spiny neurons in brains with HD. Both the slow time course of progression and the multifocal single-cell death patterns in HD are consistent with an apoptotic mechanism triggered within affected cells. Identical patterns of TUNEL labeling have been described in retinal degenerations of genetic origin (Portera-Cailliau et al., 1994). The fact that TUNEL-stained cells were interspersed among many unlabeled cells may explain why many cases of HD did not show a "ladder" type of DNA fragmentation, i.e., DNA fragments were present but were diluted in the larger mass of intact DNA. Generally, it is difficult to detect DNA fragmentation in tissues with limited TUNEL staining. For example, in the developing retina, where TUNEL can detect DNA fragmentation in isolated dying cells, DNA laddering is not seen by agarose gel electrophoresis (Portera-Cailliau et al., 1994).

Apoptosis has often been associated with programmed cell death (PCD) (Gavrieli et al., 1992; Obeid et al., 1993; Peitsch et al., 1993; Rabacchi et al., 1994), a link suggesting that a genetic program of cell death may be activated in HD. However, because it is nearly impossible to gather evidence of prevention of cell death by macromolecular synthesis inhibitors in the setting of disease, we cannot conclude that aberrant activation of PCD is a central mechanism in HD. Neither can we definitively rule out a necrotic mechanism, possibly triggered by pathogenic events intrinsic to individual neurons that renders them vulnerable to excitotoxic injury.

The pathway of injury from the expanded CAG repeat in huntingtin to apoptotic death is far from clear. Recent insights include evidence from patterns of expression of huntingtin in the forebrain, the localization of huntingtin within neurons, and the discovery of neuronal proteins interacting with huntingtin.

Huntingtin-expression data suggest that, although this gene is widely expressed throughout the forebrain, it is not as abundant in neurons at risk in HD (i.e., medium spiny neurons), as is in striatal interneurons, which are spared in the illness; in addition, huntingtin is universally expressed in corticostriatal neurons (Fusco et al., 1999). This pattern has been suggested to indicate that CAG-expanded huntingtin may not directly kill medium spiny neurons, but instead render corticostriatal neurons "destructive" for striatal projection neurons (Fusco et al., 1999).

Regarding the localization of mutant huntingtin, despite earlier data showing an "exclusive" localization in the cytoplasm, recent evidence suggests that a fraction of abnormal huntingtin is localized to the nucleus of certain populations of neurons (Davies et al., 1997). As discussed earlier (*see* section on transgenic mice), this fraction may participate in the formation of the neuronal intranuclear inclusions observed in HD (especially juvenile-onset) (DiFiglia et al., 1997; Gourfinkel-An et al., 1998). These inclusions stain with antibodies against the N-terminal domain of huntingtin and are also ubiquitinated. The distribution of huntingtin-containing inclusions is much broader than the neuropathology in HD and there may also be a general mismatch between neurodegeneration and localization of these inclusions in HD transgenic mice. More recent experiments using ataxin 1 transgenic mice suggest that the presence of mutant polyglutamine-containing proteins in nuclei, rather than the nuclear inclusions per se, is conducive to degeneration and death of neurons (Klement et al., 1998).

Finally, a number of proteins have been identified that interact with huntingtin. Two novel proteins include huntingtin-associated protein 1 (HAP1) (Li et al., 1995) and huntingtin-interacting protein 1 (HIP1) (Wanker et al., 1997). HIP1 may have significant homologies to cytoskeletal proteins, an observation that suggests a possible role of huntingtin in the formation and maintenance of the neuronal cytoskeleton. Among known proteins found to interact with huntingtin, GAPDH and calmodulin may provide some clues on a possible pathway of injury and degeneration. GAPDH participates in the formation of ATP and interactions with mutant huntingtin may cause energy depletion in neuronal cells. Calmodulin is a calcium-binding protein that controls levels of free intracellular calcium and interactions of this enzyme with mutant huntingtin may cause across-the-board activation of lytic enzymes, including proteases and endonucleases. Although these are reasonable scenarios of neuronal injury and death linking the genetic cause of HD with downstream events, there is very little direct evidence that such mechanisms operate in the illness.

Studies of Animal Models

Recent studies have elucidated the mode of neuronal death in excitotoxin and mitochondrial toxin animal models of HD. Excitotoxic lesions have been classically associated with necrotic cell death. For example, data from many studies have excluded internucleosomal DNA fragmentation in this setting (for example, *see* Ignatowicz et al., 1991), although some investigators have disputed this pattern (Kure et al., 1991). We have revisited the issue of cell-death mechanisms in excitotoxic models by performing a detailed analysis by TUNEL, EM, and DNA electrophoresis, of neostriatal neurons in vivo 4 h to 20 d after exposure to QA (Portera-Cailliau et al., 1995). TUNEL stains a large number of nuclei in the QA-injected caudate-putamen starting at 12 h and disappearing by 7 d after the lesion (peak at 3 d). Nearly all medium spiny neurons are labeled by the injection site, whereas, peripheral to the injection, labeled medium spiny neurons are interspersed with unlabeled healthy neurons. There is an inverse correlation of staining intensity with degree of progression of cell death; for example, at 5–7 d postlesion, staining is limited to some nuclei in the periphery of the injection site, i.e., an area with the mildest initial impact of the excitotoxin. As evidenced by the presence of degenerating axons in the striatopallidal axonal bundles on the QA-injected side, TUNEL-labeled neurons in caudate-putamen truly undergo degeneration. TUNEL is selective to discrete forebrain areas; in cases with large QA injections extending beyond the caudate-putamen, neurons in the endopiriform nucleus and thalamus, but no pallidal neurons, occasionally demonstrate TUNEL staining (Portera-Cailliau et al., 1995). Only neurons are vulnerable to the effect of QA; double-labeling histochemistry with TUNEL and markers specific for astrocytes or oligodendrocytes failed to show any colocalization (Portera-Cailliau et al., 1995).

Ultrastructural analysis of the QA-injected rat striatum 6 h to 3 d following the injection showed a variable pattern of degeneration, evident at 6–10 h postinjection. Some affected neurons contained large clumps of chromatin within nuclei as well as numerous vacuoles and swollen organelles in the cytoplasm, although perikayal synapses were characteristically intact, including identifiable postsynaptic densities. A subset of neurons was severely affected, showing evidence of membrane disintegration with no recognizable synapses and severely distorted organelles; these neurons were

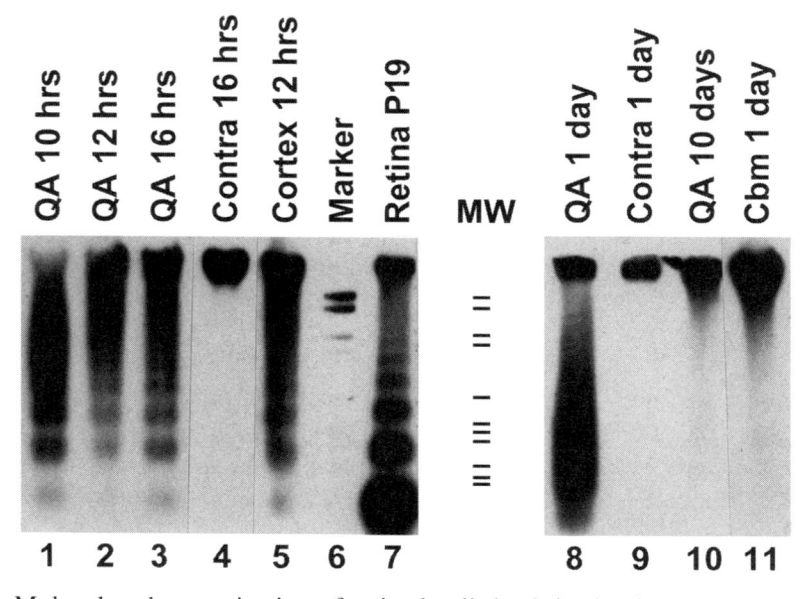

Fig. 2. Molecular characterization of striatal cell death in the QA model based on DNA electrophoretic patterns. DNA was isolated from lesioned striatum at various time points postinjection, including 10, 12, and 16 h (lanes 1–3), 1 d (lane 8), and 10 d (lane 10). Control DNA was isolated from the contralateral striatum at 16 h (lane 4) and 1 d (lane 9) postinjection as well as from the ipsilateral cerebellum 1 d postinjection (lane 11). Control apoptotic ladder of DNA (lane 7) represents cell death in the retina of a mouse model of letinitis pigmentosa (Portera-Cailliau et al., 1994) at postnatal d 19. DNA was also isolated from a portion of cortex surrounding the needle tract (lane 5).

especially abundant near the injection site and became evident as early as 6 h postinjection. At the periphery of the injection, we occasionally encountered cells with chromatin fragmentation and a relative preservation of organelles, i.e., morphologies closer to apoptosis.

DNA isolated from the QA-lesioned rat striatum at times corresponding to the beginning of TUNEL labeling (i.e., 10, 12, or 16 h postinjection) migrated as a classical ladder of fragments representing multiples of 180–200 base pairs (Fig. 2). However, DNA isolated from the lesioned striatum 1 d after the QA injection showed a random DNA degradation resulting in a smear of labeled DNA. DNA isolated from the striatum 10 d after the QA injection (when the cell death process is complete, with no more TUNEL labeling) did not show evidence of fragmentation (Portera-Cailliau et al., 1995) (Fig. 2).

The above findings appear to contrast traditional views on excitotoxic cell death, i.e., the contention that excitotoxicity is not associated with internucleosomal DNA fragmentation (Ignatowicz et al., 1991) and that it leads to a necrotic type of neuronal death. We have shown that, if one looks in the first few hours of the evolution of an excitotoxic lesion, DNA undergoes internucleosomal digestion. Of course, at later time-points (including 1 d after the excitotoxin injections, i.e., a time-point studied extensively in the literature), DNA appears in a state of random digestion. Ultrastructurally, QA-treated neurons display a necrotic morphology at all stages, including the early ones where DNA is cleaved in an endonuclease pattern. One possible explanation is that internucleosomal DNA fragmentation is a common early step in all types of cell death and, depending on whether proteases also play a significant role in the digestion of the nucleus, the final outcome will be apoptosis or necrosis. In necrosis, the membrane integrity of neurons may be universally compromised at some point after endonuclease activation so that proteases released from ruptured lysosomes cause histone degradation. The result is that the entire DNA is accessible to endonuclease digestion, causing a DNA smear on agarose gel electrophoresis. In apoptosis, the preservation of membrane integrity does not allow proteases to play a major role in the cell-death process and the nucleosomal histone backbones continue to protect the DNA segments coiled around them (Model 1, Fig. 3). A second possibility is that necrosis and apoptosis are two idealized extremes in a spectrum of cellular phenomena associated with neuronal death with different ultrastructural morphologies and DNA fragmentation patterns (Model 2, Fig. 3). On the one end of the spectrum ("pure apoptosis"; top diagram in Model 2), proteases never become activated or become activated very late in the process. On the other end of the spectrum, proteases become activated immediately ("pure necrosis"; bottom diagram in Model 2). Excitotoxic models may stand somewhere in the middle of the spectrum, with a combination of apoptotic and necrotic processes within individual degenerating cells; proteases become promptly (but not immediately) activated and any early apoptotic morphology is masked by the eventual necrotic lysis of cells. The relative predominance of necrotic vs apoptotic morphologies in excitotoxic models appears to depend on the degree of injury (i.e., the dose of excitotoxin to which cells are exposed). This is suggested by the presence of apoptotic features in some dying neurons at the periphery of the excitotoxin injection site present, both in striatal injections and in injections at other sites in the basal forebrain (Portera-Cailliau et al., 1995; Wilcox et al., 1995).

The combination of apoptotic and necrotic features is not unique to purely excitotoxic models. For example, in a case of stroke we studied 1–2 d after the onset of an infarct event, DNA gels showed a nonrandom pattern of fragmentation, despite evident swollen neurons on histological preparations (Portera-Cailliau et al., 1995).

Studies of the neuronal death process in the 3NPA model have also provided evidence for a role for alterations of apoptotic and anti-apoptotic signaling mechanisms in HD. Prostate apoptosis response-4 (Par-4) is a protein recently linked to neuronal death in Alzheimer's disease (Guo et al., 1998) and Parkinson's disease (Duan et al., 1999). Levels of Par-4 increase in striatum, and to a lesser extent in cortex and hippocampus, after systemic administration of 3NPA to adult rats (Duan et al., 2000). The increase in Par-4 levels occurs within 6 h of 3NPA administration and is followed by caspase activation and delayed neuronal death. Suppression of Par-4 expression in cultured striatal neurons protects them against death induced by 3NPA, a pattern demonstrating a necessary role for Par-4 in the cell death process, and suggesting that Par-4 may play an important role in the degeneration of striatal neurons in HD. Additional evidence supporting the involvement of apoptotic cascades in the 3NPA model comes from studies of mice lacking the p50 subunit of the transcription factor NF-κB (Yu et al., in press). Activation of NF-κB can prevent neuronal apoptosis in many different experimental models of neurodegenerative disorders; NF-κB induces the expression of gene products that suppress free radical production, and stabilizes mitochondrial function and

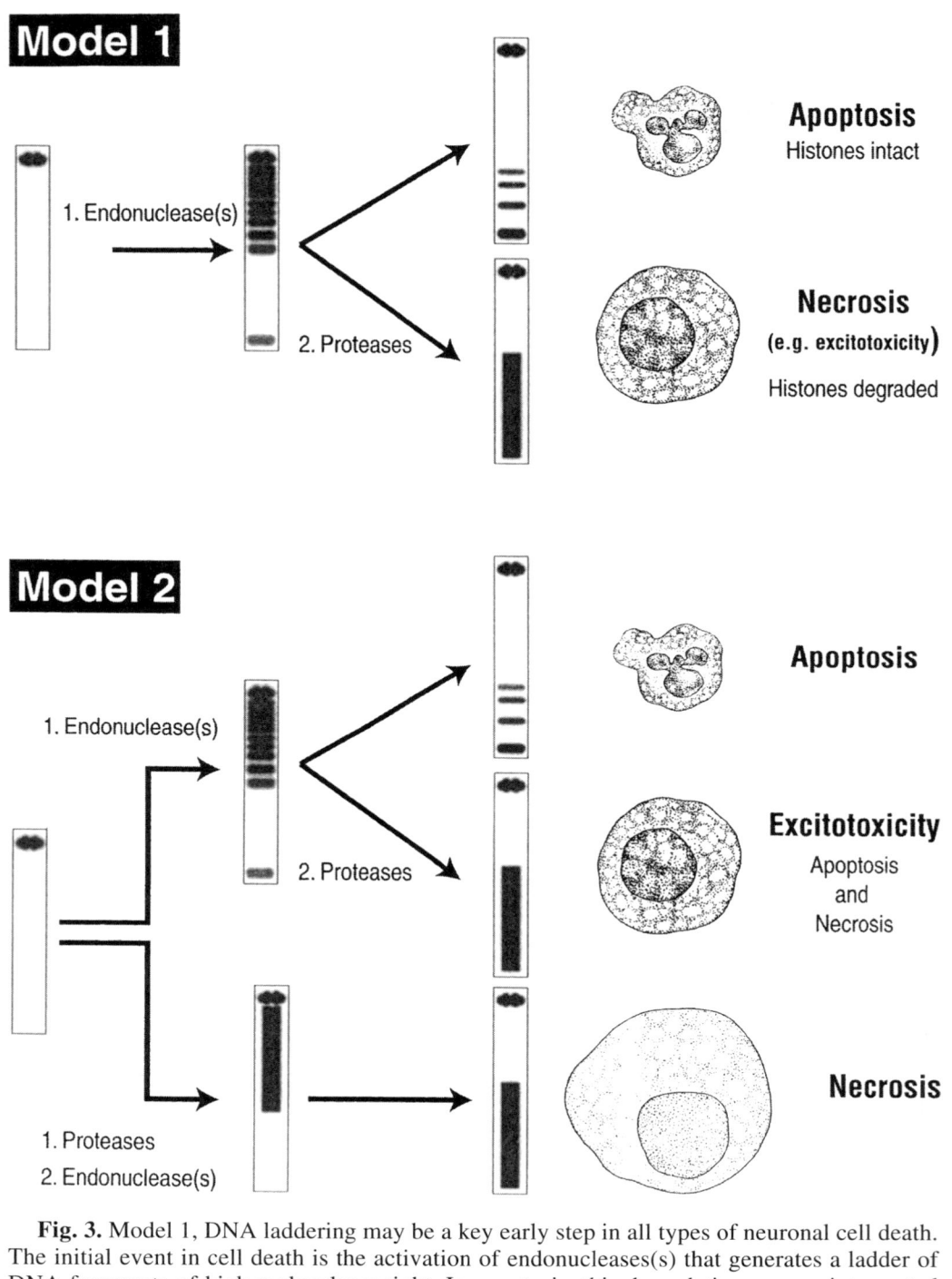

Fig. 3. Model 1, DNA laddering may be a key early step in all types of neuronal cell death. The initial event in cell death is the activation of endonucleases(s) that generates a ladder of DNA fragments of high molecular weight. In apoptosis, this degradation goes uninterrupted until most of the DNA is fragmented into mononucleosomes, which are DNA fragments protected from further endonuclease digestion by intact histone octamers. In necrosis, release of proteases from disrupted lysosomes occurs some time after endonuclease activation, leading to degradation of the protective histones and full exposure of the DNA to endonuclease action.

cellular calcium homeostasis (Mattson et al., 2000). Striatal damage is significantly increased after 3NPA administration in mice lacking p50 compared to wild-type mice (Yu et al., in press. The enhanced death occurs by apoptosis, as demonstrated by increased caspase activation and nuclear DNA laddering.

PREVENTIVE AND THERAPEUTIC STRATEGIES

The ultimate strategy for eradicating HD and other trinucleotide repeat disorders would be to correct the genetic defect. This might be accomplished by either germline therapy (eliminating the defective gene from the fertilized egg) or somatic gene therapy. Because it is unlikely that such genetic approaches will be available in the near future, pharmacological and dietary/behavioral approaches should be pursued. Our current understanding of the molecular and cellular mechanisms underlying the neuro-degenerative process, as described above, suggest several potential classes of drugs that might prove effective in suppressing neuronal death. Glutamate receptor antago-nists have been effective in experimental models of HD in which rodents are administered metabolic toxins (Schulz et al., 1996), and it will be of considerable interest to determine whether blocking glutamate receptors will slow the neurodegenerative process and motor deficits in transgenic mice expressing expanded huntingtin or other trinucleotide repeat proteins. Antioxidant therapy is a long-standing approach for prevention and treatment of essentially all neurodegenerative disorders, including polyglutamine repeat disorders. Antioxidants that suppress membrane lipid peroxidation, such as vitamin E and estrogens, have proven effective in suppressing neuronal degen-eration in experimental models of AD, and data further suggest benefit in patients (Mattson, 1998). Based on the evidence that activation of apoptotic cascades causes neuronal death in HD, inhibitors of key steps in such cascades may prove beneficial. Caspase inhibitors, agents that stabilize mitochondrial function (e.g., cyclosporin A), and inhibitors of pro-apoptotic proteins that act upstream of mitochondrial dysfunction (inhibitors of Bax, p53, or Par-4) are examples that merit further preclinical investiga-tion (Guo et al., 1998; Ona et al., 1999; Nakai et al., 2000). Finally, a preventive strategy that has proven effective in reducing risk for and severity of other age-related disorders (e.g., cardiovascular disease and diabetes) may also be effective in suppressing age-related neurodegenerative disorders, including HD. Dietary restric-tion (reduced caloric intake with maintained nutrition) can extend life-span in rodents and was shown to reduce neuronal damage and improve behavioral outcome in experi-mental models of AD, Parkinson's disease (PD), and HD (Bruce-Keller et al., 1999; Duan et al., 1999; Zhu et al., 1999).

Epidemiological data further suggest that individuals with a low caloric intake are at reduced risk for AD and PD (Logroscino et al., 1996; Mayeux et al., 1999). Collec-tively, the available experimental, clinical, and epidemiological data suggest that

---

(continued) This exposure results in random DNA degradation seen as a "smear" on agarose gel electrophoresis. Model 2, Apoptosis and necrosis may be two extremes of cell death with different ultrastructural morphologies and DNA fragmentation patterns. In excitotoxic injury, both types of cell death occur simultaneously within individual affected neurons, as evi-denced by the DNA fragmentation patterns, but ultrastructurally the apoptotic morphology is masked by the overriding necrotic lysis of neurons. For more details, see text.

multiple preventive and therapeutic strategies for trinucleotide repeat disorders should be pursued.

CONCLUSIONS

The last several years have been marked by unprecedented progress in our understanding of HD and other polyglutamine diseases. The genetic cause of HD has been identified; enormous progress has been made in the molecular properties of expanded huntingtin; the pattern of neuronal death in HD has been better characterized; and transgenic animal models that recapitulate aspects of the illness more faithfully may soon replace the more traditional excitotoxic and metabolic toxic paradigms. Despite these remarkable developments, there is a big gap of knowledge between the expansion of CAG in huntingtin and the massive, as well as selective, death of neostriatal neurons and glial cells. Some clues begin to emerge from the combination of both molecular and systems analysis, but the evident complexity of the problem will require much more work across a number of disciplines in the years to come.

ACKNOWLEDGMENTS

The authors wish to thank our many colleagues at JHMI and other institutions for their contributions to some of the original work cited in this review and for helpful discussions. This article is modified and updated from Koliatsos et al. (2000). Aspects of this work were supported by grants from the U.S. Public Health Service (AG 05146, AG 14248, NS 10580) as well as the American Health Assistance Foundation.

REFERENCES

Albin, R. L. and Greenamyre, J. T. (1992) Alternative excitotoxic hypotheses. *Neurology* **42,** 733–738.

Albin, R. L., Reiner, A., Anderson, K. D., Dure, L. S., IV, Handelin, B., Balfour, R., et al. (1992) Preferential loss of striato-external pallidal projection neurons in presymptomatic Huntington's disease. *Ann. Neurol.* **31,** 425–430.

Bates, G. (1996) Expanded glutamines and neurodegeneration: a gain of insight. *Bioessays* **18,** 175–178.

Beal, M. F., Brouillet, E., Jenkins, B., Henshaw, R., Rosen, B., and Hyman, B. T. (1993a) Age-dependent striatal excitotoxic lesions produced by the endogenous mitochondrial inhibitor malonate. *J. Neurochem.* **61,** 1147–1150.

Beal, M. F., Brouillet, E., Jenkins, B. G., Ferrante, R. J., Kowall, N. W., Miller, J. M., et al. (1993b) Neurochemical and histologic characterization of striatal excitotoxic lesions produced by the mitochondrial toxin 3-nitropropionic acid. *J. Neurosci.* **13,** 4181–4192.

Beal, M. F., Henshaw, D. R., Jenkins, B. G., Rosen, B. R., and Schulz, J. B. (1994) Coenzyme Q_{10} and nicotinamide block striatal lesions produced by the mitochondrial toxin malonate. *Ann. Neurol.* **36,** 882–888.

Beal, M. F., Hyman, B. T., and Koroshetz, W. (1993c) Do defects in mitochondrial energy metabolism underlie the pathology of neurodegenerative diseases? *Trends Neurosci.* **16,** 125–131.

Beal, M. F., Kowall, N. W., Ellison, D. W., Mazurek, M. F., Swartz, K. J., and Martin, J. B. (1986) Replication of the neurochemical characteristics of Huntington's disease by quinolinic acid. *Nature* **321,** 168–171.

Bredt, D. S. and Snyder, S. H. (1992) Nitric oxide, a novel neuronal messenger. *Neuron* **8,** 3–11.

Brouillet, E., Jenkins, B. G., Hyman, B. T., Ferrante, R. J., Kowall, N. W., Srivastava, R., et al. (1993) Age-dependent vulnerability of the striatum to the mitochondrial toxin 3-nitropropionic acid. *J. Neurochem.* **60,** 356–369.

Bruce-Keller, A. J., Umberger, G., McFall, R., and Mattson, M. P. (1999) Food restriction reduces brain damage and improves behavioral outcome following excitotoxic and metabolic insults. *Ann. Neurol.* **45,** 8–15.

Byers, R. K. and Dodge, J. A. (1967) Huntington's chorea in children. *Neurology* **17,** 587–596.

Byers, R. K., Gilles, F. H., and Gung, C. (1973) Huntington's disease in children: neuropathologic study of four cases. *Neurology* **23,** 561–561.

Cotman, C. W., Monaghan, D. T., Ottersen, O. P., and Storm-Mathisen, J. (1987) Anatomical organization of excitatory amino acid receptors and their pathways. *Trends Neurosci.* **10,** 273–280.

Coyle, J. T. and Puttfarcken, P. (1993) Oxidative stress, glutamate and neurodegenerative disorders. *Science* **262,** 689–695.

Coyle, J. T. and Schwarcz, R. (1976) Lesion of striatal neurons with kainic acid provides a model for Huntington's chorea. *Nature* **263,** 244–246.

Davies, S. W., Turmaine, M., Cozens, B. A., DiFiglia, M., Sharp, A. H., Ross, C. A., et al. (1997) Formation of neuronal intranuclear inclusions underlies the neurological dysfunction in mice transgenic for the HD mutation. *Cell* **90,** 537–548.

Dawson, V. L., Dawson, T. M., Bartley, D. A., Uhl, G. R., and Snyder, S. H. (1993) Mechanisms of nitric oxide-mediated neurotoxicity in primary brain cultures. *J. Neurosci.* **13,** 2651–2661.

Dawson, V. L., Kizushi, V. M., Huang, P. L., Snyder, S. H., and Dawson, T. M. (1996) Resistance to neurotoxicity in cortical cultures from neuronal nitric oxide synthase-deficient mice. *J. Neurosci.* **16,** 2479–2487.

DiFiglia, M., Sapp, E., Chase, K. O., Davies, S. W., Bates, G. P., Vonsattel, J. P., and Aronin, N. (1997) Aggregation of huntingtin in neuronal intranuclear inclusions and dystrophic neurites in brain. *Science* **277,** 1990–1993.

Duan, W. and Mattson, M. P. (1999) Dietary restriction and 2-deoxyglucose administration improve behavioral outcome and reduce degeneration of dopaminergic neurons in models of Parkinson's disease. *J. Neurosci. Res.* **57,** 195–206.

Duan, W., Gash, D. M., Zhang, Z., and Mattson, M. P. (1999) Participation of Par-4 in degeneration of dopaminergic neurons in models of Parkinson's disease. *Ann. Neurol.* **6,** 587–597.

Duan, W., Guo, Z., and Mattson, M. P. (2000) Participation of Par-4 in the degeneration of striatal neurons induced by metabolic compromise with 3-nitropropionic acid. *Exp. Neurol.* **165,** 1–11.

Dugan, L. L., Sensi, S. L., Canzoniero, L. M. T., Handran, S. D., Rothman, S. M., Lin, T. S., et al. (1995) Mitochondrial production of reactive oxygen species in cortical neurons following exposure to N-methyl-D-aspartate. *J. Neurosci.* **15,** 6377–6388.

Dunlap, C. B. (1927) Pathologic changes in Huntington's chorea. With special reference to the corpus striatum. *Arch. Neurol. Psychiatry* **18,** 867–943.

Duyao, M., Ambrose, C., Myers, R., Novelletto, A., Persichetti, F., Frontali, M., et al. (1993) Trinucleotide repeat length instability and age of onset in Huntington's disease. *Nat. Genet.* **4,** 387–392.

Fusco, F. R., Chen, Q., Lamoreaux, W. J., Figueredo-Cardenas, G., Jiao, Y., Coffman, J. A., et al. (1999) Cellular localization of huntingtin in striatal and cortical neruons in rats: lack of correlation with neuronal vulnerability in Huntington's disease. *J. Neurosci.* **19,** 1189–1202.

Gavrieli, Y., Sherman, Y., and Ben-Sasson, S. A. (1992) Identification of programmed cell death in situ via specific labeling of nuclear DNA fragmentation. *J. Cell Biol.* **119,** 493–501.

Gourfinkel-An, I., Cancel, G., Duyckaerts, C., Faucheux, B., Hauw, J. J., Trottier, Y., et al. (1998) Neuronal distribution of intranuclear inclusions in Huntington's disease with adult onset. *Neuroreport* **9,** 1823–1826.

Greene, J. C. and Greenamyre, J. T. (1995) Manipulation of membrane potential modulates malonate-induced striatal excitotoxicity *in vivo. Soc. Neurosci. Abstr.* **21,** 1039–1039 (Abstract).

Gu, M., Gash, M. T., Mann, V. M., Javoy-Agid, F., Cooper, J. M., and Schapira, A. H. V. (1996) Mitochondrial defect in Huntington's disease caudate nucleus. *Ann. Neurol.* **39,** 385–389.

Guo, Q., Fu, W., Xie, J., Luo, H., Sells, S. F., Geddes, J. W., et al. (1998) Par-4 is a mediator of neuronal degeneration associated with the pathogenesis of Alzheimer disease. *Nature Med.* **4,** 957–962.

Gutekunst, C. A., Li, S. H., Yi, H., Mulroy, J. S., Kuemmerle, S., Jones, R., et al. (1999) Nuclear and neuropil aggregates in Huntington's disease: relationship to neuropathology. *J. Neurosci.* **19,** 2522–2534.

Gutekunst, C.-A., Levey, A. I., Heilman, C. J., Whaley, W. L., Yi, H., Nash, N. R., et al. (1995) Identification and localization of huntingtin in brain and human lymphoblastoid cell lines with anti-fusion protein antibodies. *Proc. Natl. Acad. Sci. USA* **92,** 8710–8714.

Hedreen, J. C., Peyser, C. E., Folstein, S. E., and Ross, C. A. (1991) Neuronal loss in layers V and VI of cerebral cortex in Huntington's disease. *Neurosci. Lett.* **133,** 257–261.

Hodgson, J. G., Agopyan, N., Gutekunst, C. A., Leavitt, B. R., LePiane, F., Singaraja, R., et al. (1999) A YAC mouse model for Huntington's disease with full-length mutant huntingtin, cytoplasmic toxicity, and selective striatal neurodegeneration. *Neuron* **23,** 181–192.

Housman, D. (1995) Gain of glutamines, gain of function? *Nat. Genet.* **10,** 3–4.

Huang, Q., Zhou, D., Sapp, E., Aizawa, H., Ge, P., Bird, E. D., et al. (1995) Quinolinic acid-induced increases in calbindin D_{28k} immunoreactivity in rat striatal neurons in vivo and in vitro mimic the pattern seen in Huntington's disease. *Neuroscience* **65,** 397–407.

Huntington's Disease Collaborative Research Group. (1993) A novel gene containing a trinucleotide repeat that is expanded and unstable on Huntington's disease chromosomes. *Cell* **72,** 971–983.

Ignatowicz, E., Vezzani, A.-M., Rizzi, M., and D'Incalci, M. (1991) Nerve cell death induced *in vivo* by kainic acid and quinolinic acid does not involve apoptosis. *Neuroreport* **2,** 651–654.

Ikeuchi, T., Koide, R., Tanaka, H., Onodera, O., Igarashi, S., Takahashi, H., et al. (1995) Dentatorubral-pallidoluysian atrophy: clinical features are closely related to unstable expansions of trinucleotide (CAG) repeat. *Ann. Neurol.* **37,** 769–775.

Jackson, M., Gentleman, S., Lennox, G., Ward, L., Gray, T., Randall, K., et al. (1995) The cortical neuritic pathology of Huntington's disease. *Neuropathol. Appl. Neurobiol.* **21,** 18–26.

Jenkins, B. E., Brouillet, E., Chen, Y.-C., Storey, E., Schulz, J. B., Kirschner, P., et al. (1996) Non-invasive neurochemical analysis of focal excitotoxic lesions in models of neurodegenerative illness using spectroscopic imaging. *J. Cereb. Blood Flow Metab.* **16,** 450–461.

Jenkins, B. G., Koroshetz, W. J., Beal, M. F., and Rosen, B. R. (1993) Evidence for impairment of energy metabolism in vivo in Huntington's disease using localized ^{1}H NMR spectroscopy. *Neurology* **43,** 2689–2695.

Jervis, G. A. (1963) Huntington's chorea in childhood. *Arch. Neurol.* **9,** 244–257.

Klement, I. A., Skinner, P. J., Kaytor, M. D., Yi, H., Hersch, S. M., Brent Clark, H., et al. (1998) Ataxin-1 nuclear localization and aggregation: role in polyglutamine-induced disease in *SCA1* transgenic mice. *Cell* **95,** 41–53.

Koliatsos, V. E., Portera-Cailliau, C., Schilling, G., Borchelt, D. B., Becher, M. W., and Ross, C. A. (2000) Apoptosis in Huntington disease and animal models, in *Programmed Cell Death, Vol. 2: Roles in Disease Pathogenesis and Prevention* (Mattson, M. P., Estus, S., and Rangnekar, V., eds.), JAI Press, *Adv. Cell Aging Gerontol.*

Komure, O., Sano, A., Nishino, N., Yamauchi, N., Ueno, S., Kondoh, K., et al. (1995) DNA analysis in hereditary dentatorubral-pallidoluysian atrophy: correlation between CAG repeat length and phenotypic variation and the molecular basis of anticipation. *Neurology* **45,** 143–149.

Kure, S., Tominaga, T., Yoshimoto, T., Tada, K., and Narisawa, K. (1991) Glutamate triggers internucleosomal DNA cleavage in neuronal cells. *Biochem. Biophys. Res. Commun.* **179,** 39–45.

LaSpada, A. R., Paulson, H. L., and Fischbeck, K. H. (1994) Trinucleotide repeat expansion in neurological disease. *Ann. Neurol.* **36,** 814–822.

Li, X.-J., Li, S.-H., Sharp, A. H., Nucifora, F. C., Jr., Schilling, G., Lanahan, A., et al. (1995) A huntingtin-associated protein enriched in brain with implications for pathology. *Nature* **378,** 398–402.

Logroscino, G., Marder, K., Cote, L., Tang, M. X., Shea, S., and Mayeux, R. (1996) Dietary lipids and antioxidants in Parkinson's disease: a population-based, case-control study. *Ann. Neurol.* **39,** 89–94.

Ludolph, A. C., He, F., Spencer, P. S., Hammerstad, J., and Sabri, M. (1991) 3-Nitropropionic acid-exogenous animal neurotoxin and possible human striatal toxin. *Can. J. Neurol. Sci.* **18,** 492–498.

MacDonald, M. E. and Gusella, J. F. (1997) Huntington's disease: Translating a CAG repeat into a pathogenic mechanism. *Curr. Op. Neurobiol.* **6,** 638–643.

Mangiarini, L., Sathasivam, K., Seller, M., Cozens, B., Harper, A., Hetherington, C., et al. (1996) Exon 1 of the *HD* gene with an expanded CAG repeat is sufficient to cause a progressive neurological phenotype in transgenic mice. *Cell* **87,** 493–506.

Mattson, M. P. (1998) Modification of ion homeostasis by lipid peroxidation: roles in neuronal degeneration and adaptive plasticity. *Trends Neurosci.* **21,** 53–57.

Mattson, M. P., Culmsee, C., Yu, Z., and Camandola, S. (2000) Roles of NF-κB in neuronal survival and plasticity. *J. Neurochem.* **74,** 443–456.

Mayeux, R., Costa, R., Bell, K., Merchant, C., Tung, M. X., and Jacobs, D. (1999) Reduced risk of Alzheimer's disease among individuals with low calorie intake. *Neurology* **59,** S296–S297.

McCaughey, W. T. E. (1961) The pathologic spectrum of Huntington's chorea. *J. Nerv. Ment. Dis.* **133,** 91–103.

McGeer, E. G. and McGeer, P. L. (1976) Duplication of biochemical changes of Huntington's chorea by intrastriatal injections of glutamic and kainic acids. *Nature* **263,** 517–519.

McGeer, E. G., McGeer, P. L., and Singh, K. (1978) Kainate-induced degeneration of neostriatal neurons: dependency upon corticostriatal tract. *Brain Res.* **139,** 381–383.

Monaghan, D. T. and Cotman, C. W. (1985) Distribution of N-methyl-D-aspartate-sensitive L-[^{3}H]glutamate- binding sites in rat brain. *J. Neurosci.* **5,** 2909–2919.

Monaghan, D. T., Bridges, R. J., and Cotman, C. (1989) The excitatory amino acid receptors: their classes, pharmacology, and distinct properties in the function of the central nervous system. *Annu. Rev. Pharmacol. Toxicol.* **29,** 365–402.

Nakai, M., Qin, Z. H., Chen, J. F., Wang, Y., and Chase, T. N. (2000) Kainic acid-induced apoptosis in rat striatum is associated with nuclear factor-kappaB activation. *J. Neurochem.* **74,** 647–658.

Nasir, J., Goldberg, Y. P., and Hayden, M. R. (1996) Huntington disease: new insights into the relationship between CAG expansion and disease. *Hum. Mol. Genet.* **5,** 1431–1435.

Nelson, J. S. (1995) Diseases of the basal ganglia, in *Pediatric Neuropathology* (Ducket, S., ed.), Williams & Wilkins, Baltimore, pp. 212–214.

Obeid, L. M., Linardic, C. M., Karolak, L. A., and Hannun, Y. A. (1993) Programmed cell death induced by ceramide. *Science* **259,** 1769–1771.

Olney, J. W. and Ishimaru, M. J. (1998) Excitotoxic cell death, in *Cell Death and Diseases of the Nervous System* (Koliatsos, V. E. and Ratan, R. R., eds.), Humana Press, Totowa, NJ, pp. 197–220.

Ona, V. O., Li, M., Vonsattel, J. P., Andrews, L. J., Khan, S. Q., Chung, W. M., et al. (1999) Inhibition of caspase-1 slows disease progression in a mouse model of Huntington's disease. *Nature* **399,** 263–267.

Oppenheimer, D. R. and Esiri, M. M. (1992) Diseases of the basal ganglia, cerebellum and motor neurons, in *Greenfield's Neuropathology* (Adams, J. H. and Duchen, L. W., eds.), Oxford University Press, New York, pp. 988–1045.

Peitsch, M. C., Polzar, B., Stephan, H., Crompton, T., MacDonald, H. R., Mannherz, H. G., and Tschopp, J. (1993) Characterization of the endogenous deoxyribonuclease involved in nuclear DNA degradation during apoptosis (programmed cell death). *EMBO J.* **12,** 371–377.

Persichetti, F., Ambrose, C. M., Ge, P., McNeil, S. M., Srinidhi, J., Anderson, M. A., et al. (1995) Normal and expanded Huntington's disease gene alleles produce distinguishable proteins due to translation across the CAG repeat. *Mol. Med.* **1,** 374–383.

Perutz, M. (1994) Polar zippers: their role in human disease. *Protein Sci.* **3,** 1629–1637.

Portera-Cailliau, C., Hedreen, J. C., Price, D. L., and Koliatsos, V. E. (1995) Evidence for apoptotic cell death in Huntington disease and excitotoxic animal models. *J. Neurosci.* **15,** 3775–3787.

Portera-Cailliau, C., Sung, C.-H., Nathans, J., and Adler, R. (1994) Apoptotic photoreceptor cell death in mouse models of retinitis pigmentosa. *Proc. Natl. Acad. Sci. USA* **91,** 974–978.

Rabacchi, S. A., Bonfanti, L., Liu, X.-H., and Maffei, L. (1994) Apoptotic cell death induced by optic nerve lesion in the neonatal rat. *J. Neurosci.* **14,** 5292–5301.

Reddy, P. H., Williams, M., Charles, V., Garrett, L., Pike-Buchanan, L., Whetsell, W. O., Jr., et al. (1998) Behavioral abnormalities and selective neuronal loss in HD transgenic mice expressing mutated full-length HD cDNA. *Nat. Genet.* **20,** 198–202.

Reiner, A., Albin, R. L., Anderson, K. D., D'Amato, C. J., Penney, J. B., and Young, A. B. (1988) Differential loss of striatal projection neurons in Huntington disease. *Proc. Natl. Acad. Sci. USA* **85,** 5733–5737.

Rodda, R. A. (1981) Cerebellar atrophy in Huntington's disease. *J. Neurol. Sci.* **50,** 147–157.

Rosenow, F., Herholz, K., Lanfermann, H., Weuthen, G., Ebner, R., Kessler, J., et al. (1995) Neurological sequelae of cyanide intoxication- the patterns of clinical, magnetic resonance imaging, and positron emission tomography findings. *Ann. Neurol.* **38,** 825–828.

Ross, C. A., Margolis, R. L., Becher, M. W., Wood, J. D., Engelender, S., and Sharp, A. H. (1997) Pathogenesis of polyglutamine neurodegenerative diseases: towards a unifying mechanism, in *Genetic Instabilities and Hereditary Neurological Diseases* (Wells, R. D. and Warren, S. T., eds.), Academic, New York.

Ross, C. A. (1995) When more is less: pathogenesis of glutamine repeat neurodegenerative diseases. *Neuron* **15,** 493–496.

Scherzinger, E., Lurz, R., Turmaine, M., Mangiarini, L., Hollenbach, B., Hasenbank, R., et al. (1997) Huntingtin-encoded polyglutamine expansions form amyloid-like protein aggregates in vitro and in vivo. *Cell* **90,** 549–558.

Schilling, G., Becher, M. W., Sharp, A. H., Jinnah, H. A., Duan, K., Kotzuk, J. A., et al. (1999) Intranuclear inclusions and neuritic pathology in transgenic mice expressing a mutant N-terminal fragment of huntingtin. *Hum. Mol. Genet.* **8,** 397–407.

Schulz, J. B., Henshaw, D. R., Siwek, D., Jenkins, B. G., Ferrante, R. J., Cipolloni, P. B., et al. (1995a) Involvement of free radicals in excitotoxicity in vivo. *J. Neurochem.* **64,** 2239–2247.

Schulz, J. B., Huang, P. L., Matthews, R. T., Passov, D., Fishman, M. C., and Beal, M. F. (1996) Striatal malonate lesions are attenuated in neuronal nitric oxide synthase knockout mice. *J. Neurochem.* **67,** 430–433.

Schulz, J. B., Matthews, R. T., Jenkins, B. G., Ferrante, R. J., Siwek, D., Henshaw, D. R., et al. (1995b) Blockade of neuronal nitric oxide synthase protects against excitotoxicity *in vivo*. *J. Neurosci.* **15,** 8419–8429.

Schulz, J. B., Matthews, R. T., Henshaw, D. R., and Beal, M. F. (1996) Neuroprotective strategies for treatment of lesions produced by mitochondrial toxins: implications for neurodegenerative diseases. *Neuroscience* **71,** 1043–1048.

Sharp, A. H. and Ross, C. A. (1996) Neurobiology of Huntington's disease. *Neurobiol. Dis.* **3**, 3–15.

Sharp, A. H., Loev, S. J., Schilling, G., Li, S.-H., Li, X.-J., Bao, J., et al. (1995) Widespread expression of the Huntington's disease gene (IT-15) protein product. *Neuron* **14**, 1065–1074.

Snell, R. G., MacMillan, J. C., Cheadle, J. P., Fenton, I., Lazarou, L. P., Davies, P., et al. (1993) Relationship between trinucleotide repeat expansion and phenotypic variation in Huntington's disease. *Nat. Genet.* **4**, 393–397.

Stine, O. C., Pleasant, N., Franz, M. L., Abbott, M. H., Folstein, S. E., and Ross, C. A. (1993) Correlation between the onset age of Huntington's disease and length of the trinucleotide repeat in IT-15. *Hum. Mol. Genet.* **2**, 1547–1549.

Taylor-Robinson, S. D., Weeks, R. A., Bryant, D. J., Sargentoni, J., Marcus, C. D., Harding, A. E., and Brooks, D. J. (1996) Proton magnetic resonance spectroscopy in Huntington's disease: evidence in favor of the glutamate excitotoxic theory? *Mov. Disord.* **11**, 167–173.

Uitti, R. J., Rajput, A. H., Ashengurst, E. M., and Rozdilsky, B. (1985) Cyanide-induced parkinsonism: a clinicopathologic report. *Neurology* **35**, 921–925.

Vonsattel, J.-P., Myers, R. H., Stevens, T. J., Ferrante, R. J., Bird, E. D., and Richardson, E. P., Jr. (1985) Neuropathological classification of Huntington's disease. *J. Neuropathol. Exp. Neurol.* **44**, 559–577.

Wanker, E. E., Rovira, C., Scherzinger, E., Hasenbank, R., Walter, S., Tait, D., et al. (1997) HIP-I: a huntingtin interacting protein isolated by the yeast two-hybrid system. *Hum. Mol. Genet.* **6**, 487–495.

Warren, S. and Nelson, D. L. (1993) Trinucleotide repeat expansions in neurological disease. *Curr. Opin. Neurobiol.* **3**, 752–759.

Wilcox, B. J., Applegate, M. D., Portera-Cailliau, C., and Koliatsos, V. E. (1995) Nerve growth factor prevents apoptotic cell death in injured central cholinergic neurons. *J. Comp. Neurol.* **359**, 573–585.

Young, A. B. and Fagg, G. E. (1990) Excitatory amino acid receptors in the brain: membrane binding and receptor autoradiographic approaches. *Trends Pharmacol. Sci.* **11**, 126–133.

Young, A. B., Greenamyre, J. T., Hollingsworth, Z., Albin, R., D'Amato, C., Shoulson, I., and Penney, J. B. (1988) NMDA receptor losses in putamen from patients with Huntington's disease. *Science* **241**, 981–983.

Yu, Z., Zhou, D., and Mattson, M. P. (2000) Increased vulnerability of striatal neurons to 3-nitropropionic acid in mice lacking the p50 subunit of NF-κB. *J. Mol. Neurosci.*, in press.

Zhu, H., Guo, Q., and Mattson, M. P. (1999) Dietary restriction protects hippocampal neurons against the death-promoting action of a presenilin-1 mutation. *Brain Res.* **842**, 224–229.

6

Cellular and Molecular Mechanisms Underlying Synaptic Degeneration and Neuronal Death in Alzheimer's Disease

Mark P. Mattson, Ward A. Pedersen, and Carsten Culmsee

INTRODUCTION

Cognitive dysfunction and behavioral abnormalities resulting from synaptic degeneration and death of neurons in limbic and cortical brain regions are defining features of Alzheimer's disease (AD). The impact of AD on our aging society is dramatically increasing, and projections indicate that up to 20 million Americans will have AD by the year 2020. When considered together with the other chapters in this volume, it is therefore apparent that age-related neurodegenerative disorders will very soon become the major health problem in industrialized countries, superseding cardiovascular disease and cancer. A synthesis of the data described in the different chapters reveals that several features of the neurodegenerative process are shared among the disorders including: increased levels of oxidative stress, metabolic compromise, mitochondrial dysfunction, activation of apoptotic biochemical cascades, perturbed calcium regulation, and abnormalities in protein processing/degradation. The present chapter integrates the available data obtained from analyses of AD patients and experimental cell culture and animal models of AD so as to provide a framework that will guide future research in this field, as well as the development of preventative and therapeutic approaches. Because increasing age is the major risk factor for AD, it is not surprising that age-related alterations, including increased levels of oxidative stress and metabolic aberrancies, are believed to contribute greatly to the neurodegenerative process (Hoyer et al., 1991; Jagust et al., 1991; Butterfield and Stadtman, 1997; Mattson and Pedersen, 1998). Examples of important unanswered questions in the field of AD research therefore include:

1. What are the factors that promote oxidative stress and mitochondrial dysfunction in sporadic AD?
2. How do specific genetic factors promote neurodegeneration?
3. What are the mechanisms that allow for preservation of neuronal circuits and maintained cognitive function in those individuals who age without developing AD?
4. Does altered peripheral energy metabolism (e.g., increased insulin resistance) initiate and/or predispose to neurodegenerative cascades?

From: *Pathogenesis of Neurodegenerative Disorders* Edited by: M. P. Mattson © Humana Press Inc., Totowa, NJ

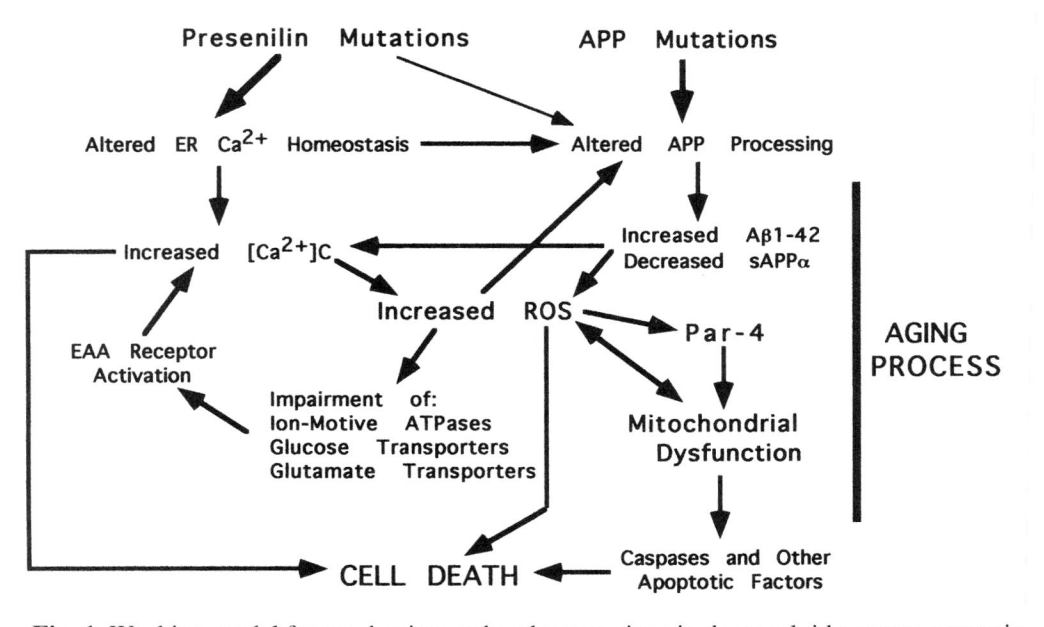

Fig. 1. Working model for mechanisms whereby mutations in the amyloid precursor protein and presenilin-1 promote neuronal degeneration in AD. *See text* for discussion.

5. Are risk factors for other age-related diseases such as cardiovascular disease, diabetes and cancer also risk factors for AD?
6. Can diet and lifestyle influence risk for AD?

Synapses and neurons in brain regions that subserve learning and memory functions, including the hippocampus, entorhinal cortex, basal forebrain, and neocortical association cortices degenerate in AD (DeKosky et al., 1996). The histological alterations that define AD include extensive extracellular deposits of amyloid β-peptide (Aβ), and degenerating neurons that contain abnormal hyperphosphorylated filaments comprised mainly of the microtubule-associated protein tau (Mattson, 1997; Mattson et al., 1997a). The amyloid precursor protein (APP), encoded by a gene on chromosome 21, is the source of Aβ. Aberrant proteolytic processing of APP is strongly implicated as an important alteration that contributes to the neurodegenerative cascade and may promote neurodegeneration by increasing production of neurotoxic forms of Aβ and/or by decreasing production of neuroprotective forms of secreted APP (Mattson, 1997). Indeed, mutations in APP are causally linked to a small percentage of cases of early-onset inherited AD, and experimental data have shown that the APP mutations result in increased production of Aβ, particularly the longer Aβ1-42 peptide. In addition to APP, two proteins called presenilin-1 (PS1; chromosome 14) and presenilin-2 (PS2; chromosome 1) can harbor mutations that cause early-onset AD (Hardy, 1997; Mattson et al., 1998a). Two general consequences of mutations in PS1 and PS2, elucidated in studies of transfected cells and transgenic mice harboring the mutations, are aberrant APP processing and disturbed cellular calcium regulation resulting in increased neuronal vulnerability to excitotoxic and metabolic insults, and a form of programmed cell death called apoptosis (Fig. 1).

BIOCHEMICAL AND CELLULAR ALTERATIONS IN AD: DATA FROM PATIENTS

Studies of postmortem brain tissue from AD patients, in comparison with tissue from aged-matched neurologically normal control patients, have provided evidence for increased levels of cellular oxidative stress in vulnerable regions of AD brain (Bruce et al., 1997; Moccoci et al., 1994; Smith et al., 1991). Immunohistochemical analyses of brain sections from AD patients reveal increased protein oxidation, protein nitration, and lipid peroxidation in neurofibrillary tangles and neuritic plaques (Good et al., 1996; Smith et al., 1997). Moreover, analyses of cerebrospinal fluid have demonstrated increased levels of lipid peroxidation products such as 4-hydroxynonenal in AD patients (Lovell et al., 1997). Further evidence for increased levels of cellular oxidative stress in AD are data showing alterations in levels of antioxidant enzymes such as catalase, Cu/Zn-superoxide dismutase, and Mn-superoxide dismutase (Bruce et al., 1997). Additional biochemical analyses have shown that mitochondrial function is compromised in brain cells of AD patients. For example, levels of α-ketoglutarate dehydrogenase complex activity are greatly reduced in AD brain tissue (Gibson et al., 1998). Interestingly, such mitochondrial abnormalities have also been reported to occur in fibroblasts from AD patients (Sheu et al., 1994), suggesting that AD may involve a widespread underlying metabolic disturbance.

Evidence for perturbations in cellular Ca^{2+} regulation have been obtained in studies of AD patients. For example, levels of activation of Ca^{2+}-dependent proteases including calpain II are increased in neurons exhibiting neurofibrillary changes (Grynspan et al., 1997). Calcium-calmodulin-dependent protein kinase II is associated with neurofibrillary tangles in AD, suggesting a role for this kinase in the neurodegenerative process (Xiao et al., 1996). Radioligand binding studies have demonstrated increased levels of ryanodine binding in vulnerable neuronal populations in hippocampus of AD patients in the early stages of the disease, and reduced levels in end-stage patients (Kelliher et al., 1999). In addition, reductions of IP_3 binding sites in entorhinal cortex and hippocampus occur in AD, and such decreases are strongly correlated with neurofibrillary pathology (Kurumatani et al., 1998). These findings suggest that neurons with relatively large pools of endoplasmic reticulum calcium release sites may be prone to degeneration in AD (Mattson et al., 2000a). Further evidence supporting a role for Ca^{2+} overload in neurodegeneration in AD comes from studies showing that neurons containing high levels of the Ca^{2+}-binding proteins calretinin and calbindin are resistant to neurofibrillary degeneration (Mattson et al., 1991; Sampson et al., 1997).

Comparisons of expression levels of various genes in brain tissue from AD patients vs controls have provided evidence for a variety of alterations in cellular signaling pathways. Alterations in levels of several different growth factors in the brains of AD patients have been reported. For example, expression of brain-derived neurotrophic factor (BDNF) and its high-affinity receptor trkB were selectively decreased in frontal cortex and hippocampus of AD patients (Ferrer et al., 1999). BDNF and several other neurotrophic factors (e.g., nerve growth factor [NGF] and basic fibroblast growth factor [BFGF]) can protect cortical and hippocampal neurons against a variety of oxidative and metabolic insults relevant to the pathogenesis of AD (Mattson and Lindvall, 1997), suggesting that decreased levels of neurotrophic factors and/or their receptors

could contribute to the neurodegenerative process in AD. Levels of cell-cycle-associ-
ated proteins such as cyclin D and cdk4 are increased in vulnerable regions of AD brain
(Busser et al., 1998), suggesting that cells may attempt to re-enter the cell cycle, perhaps as a response to injury. Cytokine-signaling cascades serve important functions in
tissue-injury responses, and alterations in cytokine levels in the brains of AD patients
are certainly consistent with the presence of such injury responses. For example, levels
of tumor necrosis factor-α (TNF-α) are altered in brain tissue and cerebrospinal fluid
(CSF) of AD patients (Maes et al., 1999). In addition, levels of interleukin-1α (IL-1α)
are increased in association with the neurodegenerative process in AD (Griffin et al.,
1995), suggesting an important role for immune cells, particularly microglia, in AD
pathogenesis.

EXPERIMENTAL MODELS OF AD

Animal and cell-culture studies have proven invaluable in advancing knowledge of
the cellular and molecular mechanisms that result in neuronal dysfunction and degeneration in AD. Studies of cultured neuron-like cell lines and primary rodent and human
neurons have provided evidence that alterations in proteolytic processing of APP may
play a major role in the increased levels of oxidative stress in neurons in AD (Mattson,
1997). On the other hand, studies have also shown that oxidative stress and metabolic
impairment may be necessary for the altered APP processing that occurs in AD
(Gabuzda et al., 1994; Zhang et al., 1997). Because of its extensive deposition in the
brain parenchyma and its association with degenerating neurons in most cases of AD,
Aβ is strongly implicated in the neurodegenerative process. Our studies and those of
others have elucidated the mechanism whereby Aβ might promote neuronal degeneration. Aβ, when in an aggregating form, can induce membrane lipid peroxidation in
cultured rat hippocampal and cortical neurons (Mark et al., 1995; 1997a) and cortical
synaptosomes (Keller et al., 1997). The oxidative stress induced by Aβ can render
neurons vulnerable to excitotoxicity and apoptosis (Mattson et al., 1992; Mark et al.,
1995, 1997a; Kruman et al., 1997). Exposure of hippocampal neurons to Aβ also
induces time- and dose-dependent decreases in catalase activity and increases in Cu/Zn-
and Mn-superoxide dismutase (SOD) activities (Bruce et al., 1997), suggesting a role
for Aβ in the altered antioxidant enzyme profile in AD brain. Lipid peroxidation promotes neuronal death, at least in part, by impairing the function of membrane ion-
motive ATPases (Na$^+$/K$^+$-ATPase and Ca^{2+}-ATPase) and glucose transporters (Mark
et al., 1995, 1997b). ATP levels are decreased following exposure of neurons to Aβ or
the lipid-peroxidation product 4-hydroxynonenal (Mark et al., 1995, 1997b), which
promotes membrane depolarization, energy depletion, and disruption of cellular Ca^{2+}
homeostasis. Membrane lipid peroxidation, as induced by Fe^{2+} or Aβ, also impairs
glutamate transport in astrocytes and synaptosomes (Keller et al., 1997; Blanc et al.,
1998), which would be expected to further promote excitotoxic injury. Antioxidants
such as vitamin E, uric acid, propyl gallate, glutathione, and estrogens can protect neurons against Aβ toxicity, strongly suggesting a pivotal role for oxidative stress in the
mechanism of Aβ-induced neuronal death in AD (Goodman and Mattson, 1994;
Goodman et al., 1996; Keller and Mattson, 1997; Mark et al., 1997a; Keller et al.,
1998a; Mattson et al., 1997c). Aβ may also have adverse effects on the cerebral vasculature that may promote metabolic compromise and neuronal degeneration. For example,

Table 1
Examples of Data Supporting Roles for Oxidative Stress, Excitotoxicity, and Metabolic Compromise in Neurofibrillary Degeneration in AD

System studied	Observation	References
AD patient brain tissue	Altered antioxidant enzymes in NFT	Bruce et al., 1997
AD patient brain tissue	HNE immunoreactivity in NFT	Sayre et al., 1997
AD patient brain tissue	Nitrotyrosine immunoreactivity in NFT	Good et al., 1996
Cultured brain neurons	Glutamate induces NFT-like changes	Mattson, 1990
Cultured brain neurons	Aβ induces NFT-like changes	Cheng and Mattson, 1992
Cultured brain neurons	HNE prevents tau dephosphorylation	Mattson et al., 1997
Adult rats	Kainate induces NFT-like changes	Elliot et al., 1993
Adult rats	Stress/glucocorticoids promote NFT-like changes	Stein-Behrens et al., 1994
APP mutant mice	NFT-like immunoreactivity in some neurons	Sturchler-Pierrat et al., 1997

we have found that exposure of vascular endothelial cells to Aβ results in impairment of glucose transport and barrier functions in these cells (Blanc et al., 1997). Antioxidants attenuated the adverse effects of Aβ on vascular endothelial cells, again implicating oxidative stress in Aβ's untoward actions.

Several different lines of transgenic mice have been generated that express AD-linked APP mutations. Several of the mutant lines exhibit increased production of Aβ1-42 and extensive age-dependent deposition of Aβ in the cerebral cortex and hippocampus (Games et al., 1995; Hsiao et al., 1996). In addition, neuritic and synaptic alterations have been documented in APP mutant mice (Larson et al., 1999), but whether neuronal death occurs remains unclear. There is evidence for increased levels of oxidative stress in association with Aβ deposits in APP mutant mice (Pappolla et al., 1998), supporting previous cell-culture data showing that Aβ induces oxidative stress in neurons (Goodman and Mattson, 1994; Hensley et al., 1994; Mark et al., 1995, 1997a). Interestingly, it was recently reported that APP mutant transgenic mice exhibit abnormal sensitivity to stress, which is associated with severe hypoglycemia (Pedersen et al., 1999). The latter findings are of considerable interest because many AD patients exhibit dysregulation of the hypothalamic–pituitary–adrenal axis (Molchan et al., 1990) and increased insulin resistance (Messier and Gagnon, 1996; Vanhanen and Soininen, 1998).

Additional experimental findings implicating oxidative stress, metabolic compromise, and overactivation of glutamate receptors, in the pathogenesis of AD, include studies documenting alterations in the neuronal cytoskeleton in cell culture and animal models (Table 1). Exposure of cultured rat hippocampal neurons to glutamate, Aβ, and glucose deprivation results in antigenic changes in tau similar to those present in the neurofibrillary tangles in AD (Mattson, 1990; Cheng and Mattson, 1992). Calcium influx and increased oxidative stress may be important mediators of such cytoskeletal alterations. Excitotoxin administration to adult rats can induce antigenic changes in tau in hippocampal neurons similar to those present in neurofibrillary tangles (Elliot et al.,

1993). Physiological stress can exacerbate these AD-like alterations in the cytoskeleton of hippocampal neurons by a glucocorticoid-mediated mechanism (Stein-Behrens et al., 1994). Neurofibrillary degenerative changes can be suppressed by treatment of neurons with neurotrophic factors and antioxidants, supporting the involvement of oxidative stress (Table 1 and M.P.M., unpublished data). Indeed, recent studies have shown that the lipid-peroxidation product 4-hydroxynonenal can prevent dephosphorylation of tau and promote crosslinking of tau (Mattson et al., 1997a).

Mitochondrial dysfunction occurs in cultured neurons exposed to AD-relevant insults including Aβ, Fe^{2+}, and glucose deprivation (Mattson et al., 1993; Keller et al., 1997). The alterations in mitochondria include those commonly observed in cells undergoing apoptosis including membrane depolarization, increased levels of mitochondrial oxyradicals, and membrane permeability transition. Such mitochondrial dysfunction in these models is secondary to increased oxidative stress because administration of antioxidants or overexpression of Mn-SOD results in preservation of mitochondrial function (Mattson et al., 1997b; Keller et al., 1998a). The mitochondrial dysfunction, in turn, contributes to the disruption of calcium homeostasis believed to occur in neurons in AD. In addition to causing mitochondrial membrane depolarization, oxidative stress, and calcium dysregulation (Keller et al., 1997, 1998a; Kruman et al., 1999), Aβ can cause damage to mitochondrial DNA (Bozner et al., 1997), although it is not known if the latter alteration precedes or follows mitochondrial dysfunction.

Many cases of early-onset autosomal dominant AD are caused by mutations in PS1 (Hardy, 1997; Mattson and Guo, 1998). PS1 is an integral membrane protein localized predominantly in the endoplasmic reticulum (ER). When mutant PS1 is expressed in cultured neural cells, and transgenic and knockin mice, it results in aberrant processing of APP (Scheuner et al., 1996; Guo et al., 1999b) and to increased vulnerability of neurons to apoptosis and excitotoxicity (Guo et al., 1996, 1997, 1998a; 1999a). The primary alteration in neurons expressing mutant PS1 is perturbed ER Ca^{2+} homeostasis resulting in enhanced Ca^{2+} release when neurons are "challenged" with physiological (e.g., glutamate) or pathophysiological (e.g., exposure to Aβ) insults (Guo et al., 1996, 1998a, 1999a). Experiments performed in PS1 mutant knockin mice show that a PS1 mutation increases vulnerability of hippocampal neurons to excitotoxicity (Guo et al., 1999a). Perturbed Ca^{2+} homeostasis appears to contribute to increased levels of cellular oxidative stress and mitochondrial dysfunction in neurons expressing mutant PS1. Accordingly, manipulations that block Ca^{2+} release from ER (dantrolene and xestospongin), block influx through plasma membrane voltage-dependent channels (nifedipine), or buffer cytoplasmic calcium (overexpression of calbindin D-28k) protect neurons expressing mutant PS-1 against Aβ-induced death (Guo et al., 1997, 1998a). Increased oxidative stress and mitochondrial dysfunction contribute to the cell death-enhancing actions of PS1 mutations because the following manipulations counteract the adverse actions of PS1 mutations: 1) antioxidants including vitamin E, uric acid, and 17β-estradiol (Guo et al., 1997; Mattson et al., 1997b); 2) overexpression of Mn-SOD (Guo et al., 1999c); 3) treatment of neurons with cyclosporin A, an inhibitor of mitochondrial permeability transition pore formation (Keller et al., 1998b).

Although the major focus of altered APP processing has been on increased Aβ production, it is clear that APP mutations also result in decreased levels of the secreted form of APP (sAPPα) (Lannfelt et al., 1995; Furukawa et al., 1996b). sAPPα exerts

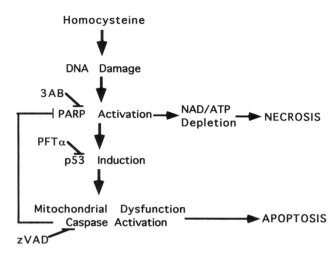

Fig. 2. Model for the involvement of homocysteine in the pathogenesis of AD.

potent neuroprotective actions on hippocampal neurons in cell culture (Mattson et al., 1993; Furukawa et al., 1996a, 1996b) and in vivo (Smith-Swintosky et al., 1994). Indeed, pretreatment of neurons with picomolar to nanomolar concentrations of sAPPα increases their resistance to oxidative injury induced by Fe^{2+} and Aβ (Furukawa et al., 1996b). In addition, we recently found that sAPPα induces an increase in the basal level of glucose transport, and attenuates oxidative impairment of glucose transport in cortical synaptosomes (Mattson et al., 1999). A signal transduction pathway that mediates the neuroprotective effects of sAPPα has been elucidated and involves cyclic GMP production, activation of potassium channels (Furukawa et al., 1996a), and activation of the transcription factor NF-κB (Barger and Mattson, 1996). Studies have shown that NF-κB can protect neurons against apoptosis and excitotoxicity in several different culture and in vivo models (Barger et al., 1995; Mattson et al., 1997b; Yu et al., 1999). Interestingly, PS1 mutations have been shown to decrease production of sAPPα (Ancolio et al., 1997), and sAPPα treatment can counteract the pro-apoptotic action of PS1 mutations in cultured neural cells by a mechanism involving NF-κB activation (Guo et al., 1998c). Collectively, these findings suggest that decreased production of sAPPα may contribute to the pathogenic actions of mutations of APP and PS1.

Risk for AD is increased in individuals with an elevated plasma homocysteine concentration (see Miller, 1999 for review). Homocysteine, a metabolite of the essential amino acid methionine, can either be remethylated to methionine by enzymes that require folate or cobalamin (vitamin B12) or catabolized by cystathionine β-synthase, a pyridoxine (vitamin B6)-dependent enzyme, to form cysteine. It is now well-established that individuals with high plasma homocysteine levels are at increased risk for heart disease and stroke, and that folic-acid supplementation can reduce homocysteine levels and risk (Giles et al., 1995; Refsum et al., 1998). We have found that homocysteine induces apoptosis in cultured hippocampal neurons by a mechanism involving DNA damage, activation of poly-ADP-ribose polymerase (PARP) and p53 induction (Kruman et al., 2000; Fig. 2). Homocysteine markedly increases neuronal vulnerability to excitotoxic and oxidative insults relevant to the pathogenesis of AD.

"SYNAPTIC APOPTOSIS" AND THE PATHOGENESIS OF AD

A unique feature of the nervous system that may contribute to selective vulnerability of neurons to age-related neurodegenerative conditions is that interneuronal signaling is localized to sites called synapses. A consequence of excitatory synaptic transmission is that synapses (presynaptic terminals and postsynaptic dendritic spines) are exposed to very high levels of Ca^{2+} influx, and oxidative and metabolic stress. Indeed, glutamate receptors and Ca^{2+} channels are concentrated in synaptic compartments, and the membrane depolarization and Ca^{2+} influx resulting from activation of these ion channels results in oxidative stress and a high energy (ATP) demand. Several observations suggest that synapses are the sites where the neurodegenerative process begins in AD. Studies of AD patients have shown that the extent of synapse loss is tightly correlated with cognitive deficits in AD patients (DeKosky et al., 1996). Overactivation of glutamate receptors, which are localized to postsynaptic regions of neuronal dendrites, plays an important role in the neuronal death process in several different animal and cell-culture models of AD (Mattson et al., 1992; Elliot et al., 1993). Third, recent studies have shown that apoptotic biochemical cascades which are activated in vulnerable neuronal populations in AD can also be activated locally in synaptic compartments following exposure to insults relevant to AD (e.g., Aβ) (Mattson et al., 1998b, 1998c) (Fig. 3). Exposure of synaptosomes or intact synaptically connected neurons to Aβ and related oxidative insults results in caspase activation, loss of plasma membrane phospholipid asymmetry, increased Par-4 levels, mitochondrial membrane depolarization, mitochondrial oxyradical production, mitochondrial calcium uptake, and release into the cytosol of factors capable of inducing nuclear chromatin condensation and fragmentation (Mattson et al., 1998b, 1998c; Duan et al., 1999).

Studies of AD brain tissue, and animal and cell-culture models of AD, suggest that prostate apoptosis response-4 (Par-4) may serve as a critical link in the chain of events that leads neuronal degeneration in AD. Par-4 is a leucine zipper- and death-domain-containing protein originally identified for its role in apoptosis of prostate cells (Sells et al., 1997), and more recently implicated as a pivotal effector of neuronal apoptosis (Guo et al., 1998). Par-4 mRNA and protein levels are increased in vulnerable regions, and to a lesser extent in nonvulnerable regions, of AD brain (Guo et al., 1998b). Moreover, approx 40% of neurofibrillary tangle-bearing neurons are also Par-4 immunoreactive, suggesting a direct and possibly causal relationship between increased Par-4 levels and neuronal degeneration in AD. Experimental data also suggest a central role for Par-4 in the cell-death process in AD. For example, Par-4 levels increase within 1–2 h of exposure cultured hippocampal neurons to Aβ, and treatment of the neurons with a Par-4 antisense oligonucleotide prevents neuronal apoptotis (Guo et al., 1998b). Increased Par-4 levels occur prior to, and are required for, mitochondrial dysfunction and caspase activation following exposure of neurons to Aβ. Overexpression of the Par-4 leucine zipper domain acts in a dominant-negative manner to prevent apoptosis, suggesting a necessary role for Par-4 interactions with another protein in its pro-apoptotic action (Guo et al., 1998b). Several proteins have been shown to interact with Par-4. One such interacting protein is PKCζ, and this interaction may inhibit the enzymatic activity of PKCζ (Diaz-Meco et al., 1996). Interaction of Par-4 with PKCζ may promote apoptosis by preventing activation of NF-κB, since we have recently found

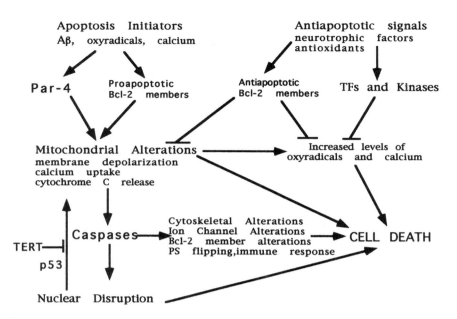

Fig. 3. Working model of apoptotic biochemical cascades and their roles in synaptic degeneration and neuronal death. See text for discussion.

that Par-4 induction suppresses activation of NF-κB in PC12 cells (Camandola and Mattson, 2000). Par-4 may also interact with the anti-apoptotic Bcl-2, and this interaction may reduce association of Bcl-2 with mitochondria and thereby enhance membrane permeability transition (M. P. Mattson, unpublished data).

Recent studies of cortical synaptosomes and cultured primary hippocampal neurons have provided evidence that Par-4 may act locally within synaptic terminals to induce apoptosis. Exposure of synaptosomes to Aβ and other apoptotic insults results in a rapid increase in levels of Par-4 protein (occurring within 1–2 h), which precedes mitochondrial dysfunction and caspase activation (Duan et al., 1999). When the increase in Par-4 levels is prevented by treatment of synaptosomes with Par-4 antisense oligonucleotides, mitochondrial function is preserved and caspase activation is suppressed. Studies of brain tissue from AD patients support a role for caspase activation in synaptic degeneration, as caspase activation is increased in degenerating neurons and neurites (Chan et al., 1999). Such caspase activation in synaptic terminals results in selective degradation of the AMPA subtype of ionotropic glutamate receptors, which appears to play a role in enhancing apoptosis (Glazner et al., 2000).

Mutations in APP and PS1 have been shown to have direct adverse consequences for synaptic function, and likely promote degeneration of synapses. As described above, APP mutations result in increased production of neurotoxic forms of Aβ that have been shown to directly impair synaptic ion-motive ATPases and glucose and glutamate transporters. APP mutations may also decrease levels of sAPPα, which would also compromise synaptic function because sAPPα plays a role in synaptic plasticity (Ishida et al., 1997) and can protect synaptic terminals against oxidative and metabolic insults (Mattson et al., 1999). Analyses of cortical synaptic terminals of PS1 mutant transgenic mice have shown that PS1 mutations perturb synaptic calcium regulation and promote mitochondrial dysfunction (Begley et al., 1999). Moreover, electrophysiological analyses of hippocampal slices have shown that PS1 mutations result in enhanced potentiation of synaptic transmission (Zaman et al., 2000).

PERIPHERAL ALTERATIONS IN IMMUNE FUNCTION AND ENERGY METABOLISM: ROLES IN THE PATHOGENESIS OF AD

Beyond the evidence that inflammation-like processes occur in association with amyloid plaques in the brains of AD patients (see Mattson, 2001a for review), emerging findings suggest that rather widespread alterations in immune function occur in AD. Studies of the immune system in AD patients, as compared to age-matched control patients, have revealed evidence for several abnormalities in AD. For example, natural killer (NK) cell cytotoxicity is increased (Solerte et al., 1998), oxidative damage to DNA is increased (Mecocci et al., 1998), and cellular calcium homeostasis is perturbed (Eckert et al., 1996) in lymphocytes from AD patients. Interestingly, lymphocytes from AD patients also exhibit increased vulnerability to apoptosis (Eckert et al., 1998).

As described above, studies of PS1 mutant knockin mice have shown that an important consequence of the mutations is that calcium regulation is altered in neurons such that calcium release from ryanodine-sensitive ER calcium stores is increased (Guo et al., 1999a, b, c). The perturbed calcium regulation endangers neurons such that they are more vulnerable to Aβ, excitotoxicity, and apoptosis. Recent studies have shown that splenocytes from PS1 mutant mice exhibit altered proliferative responses to mitogens, perturbed calcium homeostasis, and increased vulnerability to apoptosis (Morgan et al., 2000). The latter alterations were age-dependent and occurred prior to any evidence for neurodegenerative changes in the brain. These findings suggest the possibility that alterations in the immune system precede, and possibly contribute to, the neurodegenerative process in AD.

Many studies have documented alterations in glucose metabolism in AD patients (Hoyer et al., 1991; Jagust et al., 1991). Studies of fibroblasts and platelets from AD and PD patients add further weight to the evidence for widespread metabolic alterations in these two disorders (Sheu et al., 1994; Sorbi et al., 1995). In addition, several studies have documented "diabetes-like" alterations in AD patients including increased insulin resistance and abnormal glucose tolerance (Messier and Gagnon, 1996; Vanhanen and Soininen, 1998). Moreover, alterations in the hypothalamic–pituitary–adrenal axis that controls glucocorticoid production have been widely reported (Molchan et al., 1990). Despite these findings, there is as yet no definitive evidence that a generalized metabolic disturbance precedes and contributes to the neurodegenerative process. However, we have recently found that glucose regulation is altered in transgenic mice expressing the APP "Swedish" mutation. Specifically, we have found that the APP mutant mice exhibit altered responses to food deprivation and restraint stress such that they become severely hypoglycemic (Pedersen et al., 1999) (Fig. 4). Basal levels of glucose and insulin were not different in the wild-type and APP mutant transgenic mice. The APP mutant mice are hypersensitive to fasting such that they die within days to weeks when subjected to an alternate day feeding regimen (Pedersen et al., 1999). This perturbation in stress responses and energy metabolism appears to be related to amyloid deposition in the brain and/or hypothalamus.

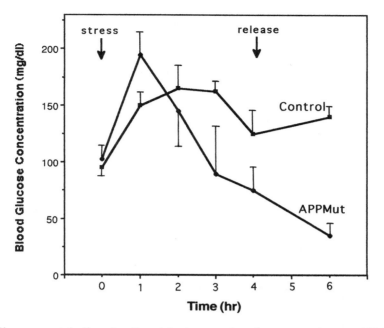

Fig. 4. Glucose metabolism is altered in transgenic mice expressing an AD-linked APP mutation. Glucose levels were measured in blood samples from nontransgenic control mice and APP mutant transgenic mice immediately prior to, and at the indicated time-points following restraint stress. Mice were released from restraint immediately after taking the blood sample at the 4 h time-point. Modified from Pedersen et al. (2000).

MECHANISMS THAT ALLOW PRESERVATION
OF NEURONAL CIRCUITS
AND MAINTAINED COGNITIVE FUNCTION IN SUCCESSFUL AGING

Some individuals flourish in their twilight years, maintaining a high level of cognitive function even as they reach and surpass the century mark. What is it about their genetic make-up and/or environment that permits the neuronal circuits in their brains to live long and prosper? A look at their brains reveals several clues. Golgi-stained sections from cognitively normal 80–100-yr-old individuals manifest a remarkable preservation of dendritic arbors in pyramidal neurons in the cerebral cortex and hippocampus (Buell and Coleman, 1979). In fact, there is evidence that the extent of dendritic arbors and perhaps synapses/neuron actually increases in neurons in the brains of cognitively normal elderly individuals (Buell and Coleman, 1979; Scheff and Price, 1993). Although rare, some persons reach their tenth decade of life with essentially no evidence of cerebral amyloid deposition (Dickson et al., 1992). However, some changes that appear to occur in the vast majority of persons of such advanced age do occur in cognitively normal persons over the age of 80 including accumulation of lipofuscin, some amyloid deposition, and scattered neurofibrillary tangles. While neurofibrillary tangles are present in cognitively normal elderly individuals, their numbers are clearly reduced compared to age-matched cognitively impaired persons. At the biochemical level, markers of oxidative stress (e.g., accumulation of lipid-peroxidation products, advanced glycation end-products, and oxidized proteins) are increased (Stadtman,

1992; Vlassara et al., 1994). Collectively, these data suggest that two general mechanisms account for successful brain aging, one being an attenuation of inevitable age-related biochemical changes, and the other being the ability of neuronal circuits to compensate for such changes by expanding their synaptic connections.

Genetic factors are clear determinants of longevity and successful brain aging in humans. Population-based epidemiological analyses and studies of twins have demonstrated a highly significant heritable component to longevity (Iachine et al., 1998). Whether there exist a few key genes that determine lifespan and successful aging remains to be determined. However, the latter appears to be the case in invertebrates such as *Caenorhabditis elegans* and *Drosophila*, wherein genes involved in insulin signaling pathways, energy metabolism, and oxyradical metabolism appear to play particularly important roles (Dorman et al., 1995; Tower, 1996; Mihaylova et al., 1999). Many of the genes that affect lifespan are also important determinants of age-related disease in many different organ systems including the brain. One example is the gene for apolipoprotein E. Persons with an E4 allele have reduced lifespan and increased risk for atherosclerosis and AD, whereas persons with an E2 and E3 alleles are likely to live longer and not develop AD (Kervinen et al., 1994; Schachter et al., 1994; Siest et al., 1995). It has been proposed that such "longevity assurance genes" are a major factor in successful aging (Hodes et al., 1996).

At this point, the molecular and cellular mechanisms that promote increased healthspan of the brain are not known. However, consideration of the kinds of data described in the preceding sections of this chapter, and in other chapters in this volume, provides considerable fuel for thought (and hopefully bench-side action). Signal transduction pathways are clearly operative in neuronal circuits that are designed to promote cell survival and plasticity of neuritic structure and function. Prominent among such pathways are those involving activity-dependent neurotrophic factors (see Mattson and Lindvall, 1997 for review). Experimental data described in the latter review article have clearly shown that activation of such neurotrophic factor signaling pathways can protect neurons and promote plasticity in animal and cell culture models of AD, PD, Huntington's disease (HD), and stroke. In general, the different neurotrophic factors may protect neurons against age-related degeneration by preventing/suppressing neurodegenerative cascades and stimulating survival-promoting mechanisms. For example, BDNF stimulates production of antioxidant enzymes, which may account, in part, for its ability to protect neurons against oxidative and metabolic insults relevant to the pathogenesis of AD and PD (Cheng and Mattson, 1994; Frim et al., 1994). Insulin-like growth factors and basic FGF can also protect neurons against excitotoxic, metabolic, and oxidative insults in various experimental models of AD (Zhang et al., 1993; Mark et al., 1997c; Guo et al., 1999b). In addition to suppressing oxidative stress, these neurotrophic factors enhance neuronal calcium homeostasis by modulating the expression and/or function of glutamate receptors, ion-motive ATPases, and calcium-binding proteins. Neurotrophic factors may also induce production of anti-apoptotic proteins such as Bcl-2 (Allsopp et al., 1995; Furukawa et al., 1997). Finally, recent findings suggest that neurotrophic factors can also act directly on synapses in a manner that enhances their ability to cope with age-related oxidative and metabolic stress (Guo and Mattson, 2000).

Environmental factors that may contribute to successful brain aging are suggested by several lines of investigation. Increased education is associated with reduced risk for AD (Katzman, 1993), and it was recently reported that individuals with a high linguistic ability in early life are at reduced risk for AD (Snowdon et al., 1996). This relationship suggests that increased activity in neuronal circuits involved in cognition increases the resistance of those neurons to age-related degeneration. The underlying mechanism may involve activity-dependent production of neurotrophic factors (Mattson and Lindvall, 1997). Traumatic brain injury is a well-established risk for AD (Nemetz et al., 1999) and the underlying mechanism presumably involves direct damage to neurons. Similarly, cerebrovascular disease predisposes to AD (Scheinberg, 1988) by a mechanism that presumably involves reduced energy availability and oxidative stress (*see* Chapter 10 by Mattson and Culmsee on cerebral ischemia). Exposure to toxic agents such as heavy metals have been proposed to contribute to some cases of sporadic AD, but the evidence is not convincing (Mattson, 2001b).

PREVENTATIVE STRATEGIES FOR AD

We believe that there is now quite compelling evidence that dietary restriction (DR; reduced calorie intake with maintenance of micronutrient levels) during adult life will reduce risk for AD. Despite the convincing evidence that high food intake is a risk factor for age-related disorders such as cardiovascular disease and diabetes, the possibility that high food intake might also increase risk for neurodegenerative disorders is largely unexplored. There are several reasons why it might be expected that reduced life-long food intake might ward off age-related neurodegenerative disorders such as AD and PD. First, DR dramatically extends lifespan and reduces development of age-related disease in rodents (see Sohal and Weindruch, 1996 for review) and monkeys (Lane et al., 1996; Ramsey et al., 1997). Second, DR reduces levels of cellular oxidative stress in several different organ systems including the brain (Dubey et al., 1996). Third, DR attenuates age-related deficits in learning and memory ability and motor function in rodents (Ingram et al., 1987; Stewart et al., 1989). Fourth, epidemiological data suggest that the incidences of AD and PD are lower in countries with low per capita food consumption (e.g., China and Japan) compared to countries with high per capita food consumption (e.g., US and Canada) (Grant, 1997).

We have recently tested the hypothesis that DR will protect neurons against degeneration in experimental models of AD and other age-related neurodegenerative disorders. One DR regimen that extends lifespan of rats and mice by 30–40% involves an alternate day feeding regimen; over time the animals consume approx 20–30% fewer calories than do animals fed daily. Administration of the excitotoxin kainate to adult rats or mice results in selective damage to the hippocampus and associated deficits in learning and memory. Rats maintained on a DR feeding regimen for 2–4 mo exhibited increased resistance of hippocampal neurons to kainate-induced degeneration (Bruce-Keller et al., 1999). The reduced damage to hippocampal neurons was correlated with a striking preservation of learning and memory in a water-maze spatial learning task. Maintenance of PS1 mutant knockin mice on a DR regimen for 3 mo results in increased resistance of hippocampal CA1 and CA3 neurons to kainate-induced injury compared to mice fed ad libitum (Zhu et al., 1999). Levels of kainate-induced oxidative stress in the hippocampus were lower the DR mice compared to mice fed ad libitum, indicating that suppression of oxidative stress may be one mechanism underlying the neuroprotective effect of DR. Thus, the neurodegeneration-promoting effect of a mutation that causes AD can be counteracted by DR.

DR has also proven beneficial in increasing resistance of neurons to degeneration in experimental models of PD, HD, and stroke. Administration of the toxin MPTP to mice and monkeys results in PD-like pathology and behavioral symptoms. This involves metabolism of MPTP to 1-methyl-4-phenylpyridinium, which is then transported selectively into dopaminergic terminals, concentrates in their mitochondria, and induces oxidative stress and impairment of complex I activity. Maintenance of mice on DR increases resistance of dopaminergic neurons in their substantia nigra to MPTP toxicity and improves motor function (Duan and Mattson, 1999). HD, a disorder involving selective degeneration of neurons in the striatum, which results in the inability to control body movements properly, can be mimicked (in part) by administration of the mitochondrial toxin 3-nitropropionic acid to rats and mice. Maintenance of rats on a DR regimen for several months prior to administration of 3-nitropropionic acid decreases degeneration of striatal neurons and improves motor function (Bruce-Keller et al., 1999).

A stroke model in rats involves transient occlusion of the middle cerebral artery, which results in damage to the cerebral cortex, and striatum supplied by that artery, and associated motor dysfunction. Rats maintained on DR exhibit reduced brain damage and improved behavioral outcome following transient occlusion of the middle cerebral artery (Yu and Mattson, 1999).

The experimental findings just described demonstrate clear neuroprotective effects of DR. Recent epidemiological findings support the hypothesis that DR can reduce risk for AD and other age-related neurodegenerative disorders. In a prospective study of a large cohort of people living in New York City, it was found that those with the lowest daily calorie intakes had the lowest risk for AD (Mayeux et al., 1999). In a population-based case-control study, it was found that patients with PD had a higher calorie intake and a higher energy-adjusted animal fat intake (Logroscino et al., 1996). The striking beneficial effects of DR in experimental models of neurodegenerative disorders, when considered in light of such epidemiological data, suggest that DR may prove beneficial in reducing the incidence and/or severity of many different human neurodegenerative disorders including AD.

Two behavioral modifications that may also reduce risk for AD are increased mental activity and increased physical activity. As described above, there is a strong negative correlation between education level and risk for AD. As with DR, increased mental activity can be considered as a mild stress that may increase resistance of neurons to age-related degeneration via a preconditioning mechanism. As evidence, rodents maintained in an intellectually "enriched" environment exhibit improved performance in spatial learning tasks and increased levels of BDNF (Falkenberg et al., 1992). Moreover, it was recently shown that rats maintained in an enriched environment exhibit increased resistance of neurons to excitotoxic and apoptotic injury (Young et al., 1999). An enriched environment may also increase proliferation of neural stem cells (Nilsson et al., 1999), and may thereby "replenish" lost neurons or provide a mechanism for enhanced recovery following injury. Interestingly, it was recently found that physical

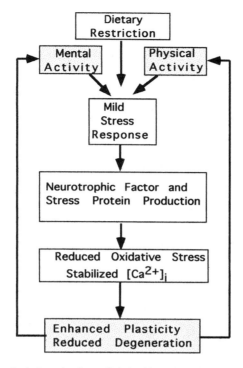

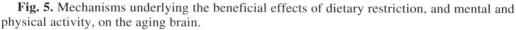

Fig. 5. Mechanisms underlying the beneficial effects of dietary restriction, and mental and physical activity, on the aging brain.

activity can increase neurogenesis (van Praag et al., 1999) and increase levels of neurotrophic factors (Neeper et al., 1996) in the brains of rodents. When taken together with the data concerning the mechanism whereby DR benefits the brain in experimental models of neurodegenerative disorders, the data suggest that a similar "preconditioning" mechanism can be activated by DR, an enriched environment, and physical exercise (Fig. 5).

INTERVENTION AND RESTORATION STRATEGIES

Long-standing therapeutic strategies that are designed to slow down or halt the neurodegenerative process include:

1. Administration of drugs that inhibit amyloid deposition or enhance clearance of amyloid from the brain. Examples include agents such as Congo red that possess anti-amyloid activity (Burgevin et al., 1994), and amyloid-based vaccines that stimulate the immune system to remove Aβ from the brain (Schenk et al., 1999).
2. Chronic consumption of antioxidants/anti-inflammatory agents. Examples include: vitamin E, which has already proven effective in experimental models (Goodman and Mattson, 1994; Mark et al., 1995) and a clinical trial (Sano et al., 1997); estrogens (Goodman et al., 1996; Tang et al., 1996); and nonsteroidal anti-inflammatory drugs (Mattson, 2000a).
3. Neurotrophic factor-based therapies, either administration of exogenous neurotrophic factors or drugs that induce production of neurotrophic factors in brain cells (Mattson and Lindvall, 1997).

There are numerous emerging approaches that hold promise for stopping the neurodegenerative process in AD and/or restoring damaged neuronal circuits. Two such strategies that are particularly exciting are based on the biology of neural stem cells and the "anti-aging" enzyme telomerase. The brain contains populations of neural stem cells in the subventricular zone and dentate gyrus of the hippocampus that are capable of proliferating and then differentiating into neurons and glial cells (*see* Scheffler et al., 1999; Svendsen and Smith, 1999 for review). The neural stem cells can be removed from the brains of developing or adult rodents and propagated in culture in the presence of neurotrophic factors (*see* Weiss et al., 1996; Palmer et al., 1999; Vescovi and Snyder, 1999 for review). By manipulating the culture environment (e.g., removing trophic factors and providing a growth substrate with extracellular matrix proteins), neural stem cells can be induced to differentiate into neurons, astrocytes, and oligodendrocytes. Moreover, stem cells can be transplanted into the brain, wherein they can differentiate into neurons and glia. Transplanted stem cells may even integrate into neuronal circuits in the adult brain (Snyder et al., 1997). Reports are emerging that, at least in some experimental models of neurodegenerative conditions, transplanted stem cells can promote functional recovery (McDonald et al., 1999; Studer et al., 1998). Neural-stem-cell-based approaches will most likely be applied to AD patients in the early stages of dysfunction with the hope of replacing damaged neuronal circuits and thereby stabilizing and/or restoring learning and memory abilities.

An enzyme complex called telomerase, which adds a six-base DNA repeat sequence (TTAGGG) to chromosome ends, has recently received considerable attention because of its apparent "anti-aging" function (*see* Harley, 1991; Rhyu, 1995; Greider, 1998; Liu, 1999 for review). Telomerase consists of an RNA template, a reverse transcriptase catalytic protein called TERT, and several telomere-associated proteins including TRF-1 (telomerase repeat-binding factor-1), TRF-2, and TEP-1 (telomerase-associated protein-1). Telomerase activity and levels of TERT decrease during cellular differentiation and senescence, and the lifespan of cultured mitotic cells can be extended by overexpression of TERT (Bodnar et al., 1998), suggesting a pivotal role for telomerase in preventing cellular senescence. Recent studies of rodents have shown that, whereas TERT and telomerase activity are undetectable in neurons in the adult brain, they are present in neurons throughout the brain during embryonic and early postnatal development We have found that TERT and telomerase activity are present in many tissues, including the brain during embryonic and early postnatal development (Fu et al., 2000). We (Fu et al., 1999), recently provided evidence that TERT and telomerase activity serve an anti-apoptotic function in neurons. Suppression of TERT expression in cultured pheochromocytoma cells and embryonic hippocampal and cortical neurons (using antisense technology and drugs that inhibit telomerase) results in increased vulnerability of the cells to death induced by oxidative and apoptotic insults including exposure to Aβ (Fu et al., 1999, 2000; Zhu et al., 2000). Overexpression of TERT results in increased resistance of neural cells to apoptosis. TERT may suppress an apoptotic signal(s), possibly resulting from damaged DNA, that results in mitochondrial dysfunction and caspase activation (Fu et al., 2000). These findings suggest the possibility that methods that induce TERT expression and telomerase activity in neurons in the adult brain may increase neuronal resistance to age-related neurodegeneration.

CONCLUSIONS

Studies of brain tissue and cerebrospinal fluid (CSF) from AD patients has provided evidence for increased levels of oxidative stress, mitochondrial dysfunction, and impaired glucose uptake in vulnerable neuronal populations. Experimental cell-culture and animal models of AD indicate that increased levels of oxidative stress, impaired energy metabolism, and perturbed cellular calcium homeostasis are critical events underlying synaptic degeneration and neuronal death in AD. Abnormal production and aggregation of Aβ may play a key role in promoting oxidative stress, resulting in impaired function of membrane ion-motive ATPases and glucose and glutamate transporters. In this way oxidative and metabolic compromise render neurons vulnerable to excitotoxicity and apoptosis. Studies of inherited forms of AD caused by mutations in APP and presenilins strongly support central roles for perturbed cellular calcium homeostasis and aberrant proteolytic processing of APP in the pathogenesis of AD. Novel mediators of neuronal cell death in AD and related disorders are being identified and include Par-4 and caspases. On the other hand, metabolic and signaling pathways that may increase neuronal resistance to aging and AD are being identified and include dietary restriction and neurotrophic factors. Although cognitive dysfunction dominates the clinical presentation, recent findings from studies of AD patients and mouse models of AD suggest rather widespread alterations of energy metabolism (e.g., increased insulin resistance and dysregulation of glucose metabolism), stress responses (dysregulation of hypothalamic–pituitary–adrenal system), and immune function (altered mitogenic and apoptotic responses in lymphocytes). Exciting new findings concerning mechanisms of neuronal degeneration and neuroprotection are leading to the development of new approaches aimed at preventing AD and treating AD patients.

REFERENCES

Aksenova, M. V., Aksenov, M. Y., Carney, J. M., and Butterfield, D. A. (1998) Protein oxidation and enzyme activity decline in old brown Norway rats are reduced by dietary restriction. *Mech. Ageing Dev.* **100,** 157–168.

Allsopp, T. E., Kiselev, S., Wyatt, S., and Davies, A. M. (1995) Role of Bcl-2 in the brain-derived neurotrophic factor survival response. *Eur. J. Neurosci.* **7,** 1266–1272.

Ancolio, K., Marambaud, P., Dauch, P., and Checler, F. (1997) Alpha-secretase-derived product of beta-amyloid precursor protein is decreased by presenilin 1 mutations linked to familial Alzheimer's disease. *J. Neurochem.* **69,** 2494–2499.

Barger, S. W., Horster, D., Furukawa, K., Goodman, Y., Krieglstein, J., and Mattson, M. P. (1995) Tumor necrosis factors α and β protect neurons against amyloid β-peptide toxicity: evidence for involvement of a κB-binding factor and attenuation of peroxide and Ca^{2+} accumulation. *Proc. Natl. Acad. Sci. USA* **92,** 9328–9332.

Barger, S. W. and Mattson, M. P. (1996) Induction of neuroprotective κB-dependent transcription by secreted forms of the Alzheimer's β-amyloid precursor. *Mol. Brain Res.* **40,** 116–126.

Begley, J. G., Duan, W., Duff, K., and Mattson, M. P. (1999) Altered calcium homeostasis and mitochondrial dysfunction in cortical synaptic compartments of presenilin-1 mutant mice. *J. Neurochem.* **72,** 1030–1039.

Blanc, E. M., Keller, J. N., Fernandez, S., and Mattson, M. P. (1998) 4-Hydroxynonenal, a lipid peroxidation product, inhibits glutamate transport in astrocytes. *Glia* **22,** 149–160.

Blanc, E. M., Toborek, M., Mark, R. J., Hennig, B., and Mattson, M. P. (1997) Amyloid β-peptide disrupts barrier and transport functions and induces apoptosis in vascular endothelial cells. *J. Neurochem.* **68,** 1870–1881.

Bodnar, A. G., Ouellette, M., Frolkis, M., Holt, S. E., Chiu, C. P., Morin, G. B., et al. (1998) Extension of life-span by introduction of telomerase into normal human cells. *Science* **279,** 349–352.

Bozner, P., Grishko, V., LeDoux, S. P., Wilson, G. L., Chyan, Y. C., and Pappolla, M. A. (1997) The amyloid beta protein induces oxidative damage of mitochondrial DNA. *J. Neuropathol. Exp. Neurol.* **56,** 1356–1362.

Bruce, A. J., Bose, S., Fu, W., Butt, C. M., Mirault, M. E., Taniguchi, N., and Mattson, M. P. (1997) Amyloid β-peptide alters the profile of antioxidant enzymes in hippocampal cultures in a manner similar to that observed in Alzheimer's disease. *Pathogenesis* **1,** 15–30.

Bruce-Keller, A. J., Umberger, G., McFall, R., and Mattson, M. P. (1999) Food restriction reduces brain damage and improves behavioral outcome following excitotoxic and metabolic insults. *Ann. Neurol.* **45,** 8–15.

Buell, S. J. and Coleman, P. D. (1979) Dendritic growth in the aged human brain and failure of growth in senile dementia. *Science* **206,** 854–856.

Burgevin, M. C., Passat, M., Daniel, N., Capet, M., and Doble, A. (1994) Congo red protects against toxicity of beta-amyloid peptides on rat hippocampal neurones. *Neuroreport* **5,** 2429–2432.

Busser, J., Geldmacher, D. S., and Herrup, K. (1998) Ectopic cell cycle proteins predict the sites of neuronal cell death in Alzheimer's disease brain. *J. Neurosci.* **18,** 2801–2807.

Butterfield, D. A. and Stadtman E. R. (1997) Protein oxidation processes in aging brain, in *The Aging Brain* (Mattson, M. P. and Geddes, J. W., eds.), JAI Press, Greenwich, CT. *Adv. Cell Aging Gerontol.* **2,** 161–191.

Camandola, S. and Mattson, M. P. (2000) Pro-apoptotic action of Par-4 involves inhibition of NF-κB activity and suppression of Bcl-2 expression. *J. Neurochem. Res.* **61,** 134–139.

Cefalu, W. T., Wagner, J. D., Wang, Z. Q., Bell-Farrow, A. D., Collins, J., Haskell, D., et al. (1997) A study of caloric restriction and cardiovascular aging in cynomolgus monkeys (Macaca fascicularis): a potential model for aging research. *J. Gerontol. A Biol. Sci. Med. Sci.* **52,** B10–B19.

Chan, S. L., Griffin, W. S. T., and Mattson, M. P. (1999a) Evidence for caspase-mediated cleavage of AMPA receptor subunits in neuronal apoptosis and in Alzheimer's disease. *J. Neurosci. Res.* **57,** 315–323.

Chan, S. L., Tammariello, S. P., Estus, S., and Mattson, M. P. (1999b) Par-4 mediates trophic factor withdrawal-induced apoptosis of hippocampal neurons: actions prior to mitochondrial dysfunction and caspase activation. *J. Neurochem.* **73,** 502–512.

Cheng, B. and Mattson, M. P. (1994) NT-3 and BDNF protect CNS neurons against metabolic/ excitotoxic insults. *Brain Res.* **640,** 56–67.

Cheng, B. and Mattson, M. P. (1992) Glucose deprivation elicits neurofibrillary tangle-like antigenic changes in hippocampal neurons: prevention by NGF and bFGF. *Exp. Neurol.* **117,** 114–123.

DeKosky, S. T., Scheff, S. W., and Styren, S. D. (1996) Structural correlates of cognition in dementia: quantification and assessment of synapse change. *Neurodegeneration* **5,** 417–421.

Diaz-Meco, M. T., Municio, M. M., Frutos, S., Sanchez, P., Lozano, J., Sanz, L., and Moscat, J. (1996) The product of par-4, a gene induced during apoptosis, interacts selectively with the atypical isoforms of protein kinase C. *Cell* **86,** 777–786.

Dickson, D. W., Crystal, H. A., Mattiace, L. A., Masur, D. M., Blau, A., Davies, P., et al. (1992) Identification of normal and pathological aging in prospectively studied nondemented elderly humans. *Neurobiol. Aging* **13,** 1–11.

Dorman, J. B., Albinder, B., Shroyer, T., and Kenyon, C. (1995) The age-1 and daf-2 genes function in a common pathway to control the lifespan of Caenorhabditis elegans. *Genetics* **141,** 1399–1406.

Duan, W. and Mattson, M. P. (1999) Dietary restriction and 2-deoxyglucose administration improve behavioral outcome and reduce degeneration of dopaminergic neurons in models of Parkinson's disease. *J. Neurosci. Res.* **57,** 195–206.

Duan, W., Rangnekar, V. M., and Mattson, M. P. (1999) Prostate apoptosis response-4 production in synaptic compartments following apoptotic and excitotoxic insults: evidence for a pivotal role in mitochondrial dysfunction and neuronal degeneration. *J. Neurochem.* **72,** 2312–2322.

Dubey, A., Forster, M. J., Lal, H., and Sohal, R. S. (1996) Effect of age and caloric intake on protein oxidation in different brain regions and on behavioral functions of mouse. *Arch. Biochem. Biophys.* **333,** 189–197.

Eckert, A., Cotman, C. W., Zerfass, R. Hennerici, M., and Muller, W. E. (1998) Lymphocytes as cell model to study apoptosis in Alzheimer's disease: vulnerability to programmed cell death appears to be altered. *J. Neural Transm. Suppl.* **54,** 259–267.

Eckert, A., Forstl, H., Zerfass, R., Hartmann, H., and Muller, W. E. (1996) Lymphocytes and neutrophils as peripheral models to study the effect of β-amyloid on cellular calcium signalling in Alzheimer's disease. *Life Sci.* **59,** 499–510.

Elliott, E., Mattson, M. P., Vanderklish, P., Lynch, G., Chang, I., and Sapolsky, R. M. (1993) Corticosterone exacerbates kainate-induced alterations in hippocampal tau immunoreactivity and spectrin proteolysis in vivo. *J. Neurochem.* **61,** 57–67.

Falkenberg, T., Mohammed, A. K., Henriksson, B., Persson, H., Winblad, B., and Lindefors, N. (1992) Increased expression of brain-derived neurotrophic factor mRNA in rat hippocampus is associated with improved spatial memory and enriched environment. *Neurosci. Lett.* **138,** 153–156.

Ferrer, I., Marin, C., Rey, M. J., Ribalta, T., Goutan, E., Blanco, R., et al. (1999) BDNF and full-length and truncated TrkB expression in Alzheimer disease. Implications in therapeutic strategies. *J. Neuropathol. Exp. Neurol.* **58,** 729–739.

Frim, D. M., Uhler, T. A., Galpern, W. R., Beal, M. F., Breakefield, X. O., and Isacson, O. (1994) Implanted fibroblasts genetically engineered to produce brain-derived neurotrophic factor prevent 1-methyl-4-phenylpyridinium toxicity to dopaminergic neurons in the rat. *Proc. Natl. Acad. Sci. USA* **91,** 5104–5108.

Fu, W., Begley, J. G., Killen, M. W., and Mattson, M. P. (1999) Anti-apoptotic role of telomerase in pheochromocytoma cells. *J. Biol. Chem.* **274,** 7264–7271.

Fu, W., Killen, M., Pandita T., and Mattson, M. P. (2000) The catalytic subunit of telomerase is expressed in developing brain neurons and serves a cell survival-promoting function. *J. Mol. Neurosci.* **14,** 3–15.

Furukawa, K., Barger, S. W., Blalock, E., and Mattson, M. P. (1996a) Activation of K^+ channels and suppression of neuronal activity by secreted β-amyloid precursor protein. *Nature* **379,** 74–78.

Furukawa, K., Estus, S., Fu, W., Mark, R. J., and Mattson, M. P. (1997) Neuroprotective action of cycloheximide involves induction of bcl-2 and antioxidant pathways. *J. Cell Biol.* **136,** 1137–1149.

Furukawa, K., Sopher, B., Rydel, R. E., Begley, J. G., Martin, G. M., and Mattson, M. P. (1996b) Increased activity-regulating and neuroprotective efficacy of α-secretase-derived secreted APP is conferred by a C-terminal heparin-binding domain. *J. Neurochem.* **67,** 1882–1896.

Gabuzda, D., Busciglio, J., Chen, L., Matsudaira, P., and Yankner, B. A. (1994) Inhibition of energy metabolism alters the processing of amyloid precursor protein and induces a potentially amyloidogenic derivative. *J. Biol. Chem.* **269,** 13,623–13,628.

Games, D., Adams, D., Alessandrini, R., Barbour, R., Berthelette, P., Blackwell, C., et al. (1995) Alzheimer-type neuropathology in transgenic mice overexpressing V717F beta-amyloid precursor protein. *Nature* **373,** 523–527.

Gibson, G. E., Sheu, K. F., and Blass, J. P. (1998) Abnormalities of mitochondrial enzymes in Alzheimer disease. *J. Neural. Transm.* **105,** 855–870.

Giles, W. H., Kittner, S. J., Anda, R. F., Croft, J. B., and Casper, M. L. (1995) Serum folate and risk for ischemic stroke. First National Health and Nutrition Examination Survey epidemiologic follow-up study. *Stroke* **26,** 1166–1170.

Glazner, G. W., Chan, S. L., Lu, C., and Mattson, M. P. (2000) Caspase-mediated degradation of AMPA receptor subunits: a mechanism for preventing excitotoxic necrosis and ensuring apoptosis. *J. Neurosci.* **20,** 3641–3649.

Good, P. F., Werner, P., Hsu, A., Olanow, C. W., and Perly, D. P. (1996) Evidence of neuronal oxidative damage in Alzheimer's disease. *Am. J. Pathol.* **149,** 21–28.

Goodman, Y., Bruce, A. J., Cheng, B., and Mattson, M. P. (1996) Estrogens attenuate and corticosterone exacerbates excitotoxicity, oxidative injury and amyloid β-peptide toxicity in hippocampal neurons. *J. Neurochem.* **66,** 1836–1844.

Goodman, Y. and Mattson, M. P. (1994) Secreted forms of β-amyloid precursor protein protect hippocampal neurons against amyloid β-peptide-induced oxidative injury. *Exp. Neurol.* **128,** 1–12.

Grant, W. (1997) Dietary links to Alzheimer's disease. *Alz. Dis. Rev.* **2,** 42–55.

Greider, C. W. (1998) Telomerase activity, cell proliferation, and cancer. *Proc. Natl. Acad. Sci. USA* **95,** 90–92.

Griffin, W. S., Sheng, J. G., Roberts, G. W., and Mrak, R. E. (1995) Interleukin-1 expression in different plaque types in Alzheimer's disease: significance in plaque evolution. *J. Neuropathol. Exp. Neurol.* **54,** 276–281.

Grynspan, F., Griffin, W. R., Cataldo, A., Katayama, S., and Nixon, R. A. (1997) Active site-directed antibodies identify calpain II as an early-appearing and pervasive component of neurofibrillary pathology in Alzheimer's disease. *Brain Res.* **763,** 145–158.

Guo, Q., Christakos, S., Robinson, N., and Mattson, M. P. (1998a) Calbindin blocks the pro-apoptotic actions of mutant presenilin-1: reduced oxidative stress and preserved mitochondrial function. *Proc. Natl. Acad. Sci. USA* **95,** 3227–3232.

Guo, Q., Fu, W., Sopher, B. L., Holtsberg, F. W., Steiner, S. M., and Mattson, M. P. (1999c) Superoxide mediates the apoptosis-enhancing action of presenilin-1 mutations. *J. Neurosci. Res.* **56,** 457–470.

Guo, Q., Fu, W., Sopher, B. L., Miller, M. W., Ware, C. B., Martin, G. M., and Mattson, M. P. (1999a) Increased vulnerability of hippocampal neurons to excitotoxic necrosis in presenilin-1 mutant knockin mice. *Nature Med.* **5,** 101–107.

Guo, Q., Furukawa, K., Sopher, B. L., Pham, D. G., Martin, G. M., and Mattson, M. P. (1996) Alzheimer's PS-1 mutation perturbs calcium homeostasis and sensitizes PC12 cells to death induced by amyloid β-peptide. *Neuroreport* **8,** 379–383.

Guo, Z. H. and Mattson, M. P. (2000) Neurotrophic factors protect cortical synaptic terminals against amyloid and oxidative stress-induced impairment of glucose transport, glutamate transport and mitochondrial function. *Cereb. Cortex* **10,** 50–57.

Guo, Q., Robinson, N., and Mattson, M. P. (1998c) Secreted APPα counteracts the pro-apoptotic action of mutant presenilin-1 by activation of NF-κB and stabilization of calcium homeostasis. *J. Biol. Chem.* **273,** 12,341–12,351.

Guo, Q., Fu, W., Xie, J., Luo, H., Sells, S. F., Geddes, J. W., et al. (1998b) Par-4 is a mediator of neuronal degeneration associated with the pathogenesis of Alzheimer's disease. *Nature Med.* **4,** 957–962.

Guo, Q., Sebastian, L., Sopher, B. L., Miller, M. W., Glazner, G. W., Ware, C. B., et al. (1999b) Neurotrophic factors [activity-dependent neurotrophic factor (ADNF) and basic fibroblast growth factor (bFGF)] interrupt excitotoxic neurodegenerative cascades promoted by a presenilin-1 mutation. *Proc. Natl. Acad. Sci. USA* **96,** 4125–4130.

Guo, G., Sopher, B. L., Pham, D. G., Furukawa, K., Robinson, N., Martin, G. M., and Mattson, M. P. (1997) Alzheimer's presenilin mutation sensitizes neural cells to apoptosis induced by trophic factor withdrawal and amyloid β-peptide: involvement of calcium and oxyradicals. *J. Neurosci.* **17**, 4212–4222.

Hardy, J. (1997) Amyloid, the presenilins and Alzheimer's disease. *Trends Neurosci.* **20**, 154–159.

Harley, C. B. (1991) Telomere loss: mitotic clock or genetic time bomb? *Mutat. Res.* **256**, 271–282.

Hensley, K., Carney, J. M., Mattson, M. P., Aksenova, M., Harris, M., Wu, J. F., et al. (1994) A model for β-amyloid aggregation and neurotoxicity based on free radical generation by the peptide: relevance to Alzheimer's disease. *Proc. Natl. Acad. Sci. USA* **91**, 3270–3274.

Hodes, R. J., McCormick, A. M., and Pruzan, M. (1996) Longevity assurance genes: how do they influence aging and life span? *J. Am. Geriatr. Soc.* **44**, 988–991.

Hoyer, S., Nitsch, R., and Oesterreich, K. (1991) Predominant abnormality in cerebral glucose utilization in late-onset dementia of the Alzheimer type: a cross-sectional comparison against advanced late-onset and incipient early-onset cases. *J. Neural Transm.* **3**, 1–14.

Hsiao, K., Chapman, P., Nilsen, S., Eckman, C., Harigaya, S., Younkin, S., et al. (1996) Correlative memory deficits, Aβ elevation, and amyloid plaques in transgenic mice. *Science* **274**, 99–102.

Iachine, I. A., Holm, N. V., Harris, J. R., Begun, A. Z., Iachina, M. K., Laitinen, M., et al. (1998) How heritable is individual susceptibility to death? The results of an analysis of survival data on Danish, Swedish and Finnish twins. *Twin Res.* **1**, 196–205.

Ingram, D. K., Weindruch, R., Spangler, E. L., Freeman, J. R., and Walford, R. L. (1987) Dietary restriction benefits learning and motor performance of aged mice. *J. Gerontol.* **42**, 78–81.

Ishida, A., Furukawa, K., Keller, J. N., and Mattson, M. P. (1997) Secreted form of β-amyloid precursor protein shifts the frequency dependence for induction of LTD, and enhances LTP in hippocampal slices. *Neuroreport* **8**, 2133–2137.

Jagust, W. J., Seab, J. P., Huesman, R. H., Valk, P. E., Mathis, C. A., Reed, B. R., et al. (1991) Diminished glucose transport in Alzheimer's disease: dynamic PET studies. *J. Cereb. Blood Flow Metab.* **11**, 323–330.

Katzman, R. (1993) Education and the prevalence of dementia and Alzheimer's disease. *Neurology* **43**, 13–20.

Keller, J. N., Guo, Q., Holtsberg, F. W., Bruce-Keller, A. J., and Mattson, M. P. (1998b) Increased sensitivity to mitochondrial toxin-induced apoptosis in neural cells expressing mutant presenilin-1 is linked to perturbed calcium homeostasis and enhanced oxyradical production. *J. Neurosci.* **18**, 4439–4450.

Keller, J. N., Kindy, M. S., Holtsberg, F. W., St Clair, D. K., Yen, H. C., Germeyer, A., et al. (1998a) Mitochondrial MnSOD prevents neural apoptosis and reduces ischemic brain injury: suppression of peroxynitrite production, lipid peroxidation and mitochondrial dysfunction. *J. Neurosci.* **18**, 687–697.

Keller, J. N. and Mattson, M. P. (1997) 17β-estradiol attenuates oxidative impairment of synaptic Na⁺/K⁺-ATPase activity, glucose transport and glutamate transport induced by amyloid β-peptide and iron. *J. Neurosci. Res.* **50**, 522–530.

Keller, J. N., Pang, Z., Geddes, J. W., Begley, J. G., Germeyer, A., Waeg, G., and Mattson, M. P. (1997) Impairment of glucose and glutamate transport and induction of mitochondrial oxidative stress and dysfunction in synaptosomes by amyloid β-peptide: role of the lipid peroxidation product 4-hydroxynonenal. *J. Neurochem.* **69**, 273–284.

Kelliher, M., Fastbom, J., Cowburn, R. F., Bonkale, W., Ohm, T. G., Ravid, R., et al. (1999) Alterations in the ryanodine receptor calcium release channel correlate with Alzheimer's disease neurofibrillary and beta-amyloid pathologies. *Neuroscience* **92**, 499–513.

Kervinen, K., Savolainen, M. J., Salokannel, J., Hynninen, A., Heikkinen, J., Ehnholm, C., et al. (1994) Apolipoprotein E and B polymorphisms-longevity factors assessed in nonagenarians. *Atherosclerosis* **105**, 89–95.

Kruman, I., Bruce-Keller, A. J., Bredesen, D. E., Waeg, G., and Mattson, M. P. (1997) Evidence that 4-hydroxynonenal mediates oxidative stress-induced neuronal apoptosis. *J. Neurosci.* **17,** 5097-5108.

Kruman, I., Chan, S. L., Culmsee, C., Kruman, Y., Penix, L., and Mattson, M. P. (2000) Homocysteine elicits a DNA damage response in neurons that promotes apoptosis and hypersensitivity to excitotoxicity. *J. Neurosci.* **20,** 6920–6926.

Kruman, I., Pang, Z., Geddes, J. W., and Mattson, M. P. (1999) Pivotal role of mitochondrial calcium uptake in neural cell apoptosis and necrosis. *J. Neurochem.* **72,** 529–540.

Kurumatani, T., Fastbom, J., Bonkale, W. L., Bogdanovic, N., Winblad, B., Ohm, T. G., and Cowburn, R. F. (1998) Loss of inositol 1,4,5-trisphosphate receptor sites and decreased PKC levels correlate with staging of Alzheimer's disease neurofibrillary pathology. *Brain Res.* **796,** 209–221.

Lane, M. A., Baer, D. J., Rumpher, W. V., Weindruch, R., Ingram, D. K., Tilmont, E. M., et al. (1996) Calorie restriction lowers body temperature in rhesus monkeys, consistent with a postulated anti-aging mechanism in rodents. *Proc. Natl. Acad. Sci. USA* **93,** 4159–4164.

Lannfelt, L., Basun, H., Wahlund, L. O., Rowe, B. A., and Wagner, S. L. (1995) Decreased α-secretase-cleaved amyloid precursor protein as a diagnostic marker for Alzheimer's disease. *Nature Med.* **1,** 829–832.

Larson, J., Lynch, G., Games, D., and Seubert, P. (1999) Alterations in synaptic transmission and long-term potentiation in hippocampal slices from young and aged PDAPP mice. *Brain Res.* **840,** 23–35.

Liu, J. P. (1999) Studies of the molecular mechanisms in the regulation of telomerase activity. *FASEB J.* **13,** 2091–2104.

Logroscino, G., Marder, K., Cote, L., Tang, M. X., Shea, S., and Mayeux, R. (1996) Dietary lipids and antioxidants in Parkinson's disease: a population-based, case-control study. *Ann. Neurol.* **39,** 89–94.

Lovell, M. A., Ehmann, W. D., Mattson, M. P., and Markesbery, W. R. (1997) Elevated 4-hydroxynonenal levels in ventricular fluid in Alzheimer's disease. *Neurobiol. Aging* **18,** 457–461.

Maes, M., DeVos, N., Wauters, A., Demedts, P., Maurits, V. W., Neels, H., et al. (1999) Inflammatory markers in younger vs elderly normal volunteers and in patients with Alzheimer's disease. *J. Psychiatr. Res.* **33,** 397–405.

Mark, R. J., Hensley, K., Butterfield, D. A., and Mattson, M. P. (1995) Amyloid β-peptide impairs ion-motive ATPase activities: evidence for a role in loss of neuronal Ca^{2+} homeostasis and cell death. *J. Neurosci.* **15,** 6239–6249.

Mark, R. J., Keller, J. N., Kruman, I., and Mattson, M. P. (1997c) Basic FGF attenuates amyloid beta-peptide-induced oxidative stress, mitochondrial dysfunction, and impairment of Na+/K+-ATPase activity in hippocampal neurons. *Brain Res.* **756,** 205–214.

Mark, R. J., Lovell, M. A., Markesbery, W. R., Uchida, K., and Mattson, M. P. (1997a) A role for 4-hydroxynonenal in disruption of ion homeostasis and neuronal death induced by amyloid β-peptide. *J. Neurochem.* **68,** 255–264.

Mark, R. J., Pang, Z., Geddes, J. W., and Mattson, M. P. (1997b) Amyloid β-peptide impairs glucose uptake in hippocampal and cortical neurons: involvement of membrane lipid peroxidation. *J. Neurosci.* **17,** 1046–1054.

Mattson, M. P. (1990) Antigenic changes similar to those seen in neurofibrillary tangles are elicited by glutamate and calcium influx in cultured hippocampal neurons. *Neuron* **4,** 105–117.

Mattson, M. P. (1997) Cellular actions of β-amyloid precursor protein, and its soluble and fibrillogenic peptide derivatives. *Physiol. Rev.* **77,** 1081–1132.

Mattson, M. P. (1998) Modification of ion homeostasis by lipid peroxidation: roles in neuronal degeneration and adaptive plasticity. *Trends Neurosci.* **21,** 53–57.

Mattson, M. P. (2001a) Roles of cytokines and inflammation in Alzheimer's disease, in *Functional Neurobiology of Aging* (Hoff, P. R. and Mobbs, C. V., eds.), Academic, San Diego, CA, pp. 349–352.

Mattson, M. P. (2001b) Metals and the pathophysiology of Alzheimer' disease, in *Functional Neu-robiology of Aging* (Hoff, P. R. and Mobbs, C. V., eds.), Academic, San Diego, CA, pp. 361–365.

Mattson, M. P., Chan, S. L., LaFerla, F. M., Leissring, M., Shepel, P. N., and Geiger, J. D. (2000) Calcium signaling in the ER: its role in neuronal plasticity and neurodegenerative disorders. *Trends Neurosci.* **23,** 222–229.

Mattson, M. P., Cheng, B., Davis, D., Bryant, K., Lieberburg, I., and Rydel, R. E. (1992) β-Amyloid peptides destabilize calcium homeostasis and render human cortical neurons vulnerable to excitotoxicity. *J. Neurosci.* **12,** 376–389.

Mattson, M. P., Cheng, B., Culwell, A., Esch, F., Lieberburg, I., and Rydel, R. E. (1993) Evidence for excitoprotective and intraneuronal calcium-regulating roles for secreted forms of β-amyloid precursor protein. *Neuron* **10,** 243–254.

Mattson, M. P., Fu, W., Waeg, G., and Uchida, K. (1997a) 4-Hydroxynonenal, a product of lipid peroxidation, inhibits dephosphorylation of the microtubule-associated protein tau. *NeuroReport* **8,** 2275–2281.

Mattson, M. P., Goodman, Y., Luo, H., Fu, W., and Furukawa, K. (1997b) Activation of NF-κB protects hippocampal neurons against oxidative stress-induced apoptosis: evidence for induction of Mn-SOD and suppression of peroxynitrite production and protein tyrosine nitration. *J. Neurosci. Res.* **49,** 681–697.

Mattson, M. P. and Guo, Q. (1998) The Presenilins. *The Neuroscientist* **5,** 112–124.

Mattson, M. P., Guo, Q., Furukawa, K., and Pedersen, W. A. (1998a) Presenilins, the endoplasmic reticulum, and neuronal apoptosis in Alzheimer's disease. *J. Neurochem.* **70,** 1–14.

Mattson, M. P., Guo, Z. H., and Geiger, J. D. (1999) Secreted form of amyloid precursor protein attenuates oxidative impairment of glucose and glutamate transport in synaptosomes by a cyclic GMP-mediated mechanism. *J. Neurochem.* **73,** 532–537.

Mattson, M. P., Keller, J. N., and Begley, J. G. (1998b) Evidence for synaptic apoptosis. *Exp. Neurol.* **153,** 35–48.

Mattson, M. P. and Lindvall, O. (1997) Neurotrophic factors and the aging brain, in *The Aging Brain: Advances in Cell Aging and Gerontology, vol. 2* (Mattson, M. P. and Geddes, J. W., eds.), JAI Press, Greenwich, CT, pp. 299–345.

Mattson, M. P., Partin, J., and Begley, J. G. (1998c) Amyloid b-peptide induces apoptosis-related events in synapses and dendrites. *Brain Res.* **807,** 167–176.

Mattson, M. P. and Pedersen, W. A. (1998) Effects of amyloid precursor protein derivatives and oxidative stress on basal forebrain cholinergic systems in Alzheimer's disease. *Int. J. Dev. Neurosci.* **16,** 737–753.

Mattson, M. P., Robinson, N., and Guo, Q. (1997c) Estrogens stabilize mitochondrial function and protect neural cells against the pro-apoptotic action of mutant presenilin-1. *NeuroReport* **8,** 3817–3821.

Mattson, M. P., Rychlik, B., Chu, C., and Christakos, S. (1991) Evidence for calcium-reducing and excito-protective roles for the calcium-binding protein calbindin-D28k in cultured hippocampal neurons. *Neuron* **6,** 41–51.

Mayeux, R., Costa, R., Bell, K., Merchant, C., Tung, M. X., and Jacobs, D. (1999) Reduced risk of Alzheimer's disease among individuals with low calorie intake. *Neurology* **59,** S296–297.

McDonald, J. W., Liu, X. Z., Qu, Y., Liu, S., Mickey, S. K., Turetsky, D., et al. (1999) Transplanted embryonic stem cells survive, differentiate and promote recovery in injured rat spinal cord. *Nature Med.* **5,** 1410–1412.

Mecocci, P., Polidori, M. C., Ingegni, T., Cherubini, A., Chionne, F., Cecchetti, R., and Senin, U. (1998) Oxidative damage to DNA in lymphocytes from AD patients. *Neurology* **51,** 1014–1017.

Messier, C. and Gagnon, M. (1996) Glucose regulation and cognitive functions: relation to Alzheimer's disease and diabetes. *Behav. Brain Res.* **75**, 1–11.

Mihaylova, V. T., Borland, C. Z., Manjarrez, L., Stern, M. J., and Sun. H. (1999) The PTEN tumor suppressor homolog in Caenorhabditis elegans regulates longevity and dauer formation in an insulin receptor-like signaling pathway. *Proc. Natl. Acad. Sci. USA* **96**, 7427–7432.

Miller, J. W. (1999) Homocysteine and Alzheimer's disease. *Nutr. Rev.* **57**, 126–129.

Moccoci, P., MacGarvey, M. S., and Beal, M. F. (1994) Oxidative damage to mitochondrial DNA is increased in Alzheimer's disease. *Ann. Neurol.* **36**, 747–751.

Molchan, S. E., Hill, J. L., Mellow, A. M., Lawlor, B. A., Martinez, R., and Sunderland, T. (1990) The dexamethasone suppression test in Alzheimer's disease and major depression: relationship to dementia severity, depression, and CSF monoamines. *Int. Psychogeriatr.* **2**, 99–122.

Morgan, G., Chan, S. L., Osborne, B. A., and Mattson, M. P. (2000) Defects of immune regulation in the presenilin-1 mutant knockin mouse. Submitted.

Neeper, S. A., Gomez-Pinilla, F., Choi, J., and Cotman, C. W. (1996) Physical activity increases mRNA for brain-derived neurotrophic factor and nerve growth factor in rat brain. *Brain Res.* **726**, 49–56.

Nemetz, P. N., Leibson, C., Naessens, J. M., Beard, M., Kokmen, E., Annegers, J. F., and Kurland, L. T. (1999) Traumatic brain injury and time to onset of Alzheimer's disease: a population-based study. *Am. J. Epidemiol.* **149**, 32–40.

Nilsson, M., Perfilieva, E., Johansson, U., Orwar, O., and Eriksson, P. S. (1999) Enriched environment increases neurogenesis in the adult rat dentate gyrus and improves spatial memory. *J. Neurobiol.* **39**, 569–578.

Palmer, T. D., Markakis, E. A., Willhoite, A. R., Sfar, F., and Gage, F. H. (1999) Fibroblast growth factor-2 activates a latent neurogenic program in neural stem cells from diverse regions of the adult CNS. *J. Neurosci.* **19**, 8487–8497.

Pappolla, M. A., Chyan, Y. J., Omar, R. A., Hsiao, K., Perry, G., Smith, M. A., and Bozner, P. (1998) Evidence of oxidative stress and in vivo neurotoxicity of beta-amyloid in a transgenic mouse model of Alzheimer's disease: a chronic oxidative paradigm for testing antioxidant therapies in vivo. *Am. J. Pathol.* **152**, 871–877.

Pedersen, W. A., Culmsee, C., Ziegler, D., Herman, J. P., and Mattson, M. P. (1999) Aberrant stress response associated with severe hypoglycemia in a transgenic mouse model of Alzheimer's disease. *J. Mol. Neurosci.* **13**, 159–165.

Ramsey, J. J., Roecker, E. B., Weindruch, R., and Kemnitz, J. W. (1997) Energy expenditure of adult male rhesus monkeys during the first 30 mo of dietary restriction. *Am. J. Physiol.* **272**, E901–907.

Refsum, H., Ueland, P. M., Nygard, O., and Vollset, S. E. (1998) Homocysteine and cardiovascular disease. *Annu. Rev. Med.* **49**, 31–62.

Rhyu, M. S. (1995) Telomeres, telomerase and immortality. *J. Nat. Canc. Inst.* **87**, 884–894.

Sampson, V. L., Morrison, J. H., and Vickers, J. C. (1997) The cellular basis for the relative resistance of parvalbumin and calretinin immunoreactive neocortical neurons to the pathology of Alzheimer's disease. *Exp. Neurol.* **145**, 295–302.

Sano, M., Ernesto, C., Thomas, R. G., Klauber, M. R., Schafer, K., Grundman, M., et al. (1997) A controlled trial of selegiline, alpha-tocopherol, or both as treatment for Alzheimer's disease. The Alzheimer's Disease Cooperative Study. *N. Engl. J. Med.* **336**, 1216–1222.

Sayre, L. M., Zelasko, D. A., Harris, P. L., Perry, G., Salomon, R. G., and Smith, M. A. (1997) 4-Hydroxynonenal-derived advanced lipid peroxidation end products are increased in Alzheimer's disease. *J. Neurochem.* **68**, 2092–2097.

Schachter, F., Faure-Delanef, L., Guenot, F., Rouger, H., Froguel, P., Lesueur-Ginot, L., and Cohen, D. (1994) Genetic associations with human longevity at the APOE and ACE loci. *Nat. Genet.* **6**, 29–32.

Scheff, S. W. and Price, D. A. (1993) Synapse loss in the temporal lobe in Alzheimer's disease. *Ann. Neurol.* **33**, 190–199.

Scheffler, B., Horn, M., Blumcke, I., Laywell, E. D., Coomes, D., Kukekov, V. G., and Steindler, D. A. (1999) Marrow-mindedness: a perspective on neuropoiesis. *Trends Neurosci.* **22**, 348–357.

Scheinberg, P. (1988) Dementia due to vascular disease—a multifactorial disorder. *Stroke* **19**, 1291–1299.

Schenk, D., Barbour, R., Dunn, W., Gordon, G., Grajeda, H., Guido, T., et al. (1999) Immunization with amyloid-beta attenuates Alzheimer-disease-like pathology in the PDAPP mouse. *Nature* **400**, 173–177.

Scheuner, D., Eckman, C., Jensen, M., Song, X., Citron, M., Suzuki, N., et al. (1996) The amyloid β protein deposited in the senile plaques of Alzheimer's disease is increased in vivo by the presenilin 1 and 2 and APP mutations linked to familial Alzheimer's disease. *Nature Med.* **2**, 864–870.

Sells, S. F., Han, S.-S., Muthukkumar, S., Maddiwar, N., Johnstone, R., Boghaert, E., Gillis, D., et al. (1997) Expression and function of the leucine zipper protein Par-4 in apoptosis. *Mol. Cell. Biol.* **17**, 3823–3832.

Sheu, K. F., Cooper, A. J., Koike, K., Koike, M., Lindsay, J. G., and Blass, J. P. (1994) Abnormality of the alpha-ketoglutarate dehydrogenase complex in fibroblasts from familial Alzheimer's disease. *Ann. Neurol.* **35**, 312–318.

Siest, G., Pillot, T., Regis-Bailly, A., Leininger-Muller, B., Steinmetz, J., Galteau, M. M., Visvikis, S. (1995) Apolipoprotein E: an important gene and protein to follow in laboratory medicine. *Clin. Chem.* **41**, 1068–1086.

Smith, C. D., Carney, J. M., Starke-Reed, P. E., Oliver, C. N., Stadtman, E. R., Floyd, R. A., and Markesbery, W. R. (1991) Excess brain protein oxidation and enzyme dysfunction in normal aging and in Alzheimer disease. *Proc. Natl. Acad. Sci. USA* **88**, 10,540–10,543.

Smith, M. A., Harris, P. L. R., Sayre, L. M., Beckman, J. S., and Perry, G. (1997) Widespread peroxynitrite-mediated damage in Alzheimer's disease. *J. Neurosci.* **17**, 2653–2657.

Smith-Swintosky, V. L., Pettigrew, L. C., Craddock, S. D., Culwell, A. R., Rydel, R. E., and Mattson, M. P. (1994) Secreted forms of β-amyloid precursor protein protect against ischemic brain injury. *J. Neurochem.* **63**, 781–784.

Snowdon, D. A., Kemper, S. J., Mortimer, J. A., Greiner, L. H., Wekstein, D. R., and Markesbery W. R. (1996) Linguistic ability in early life and cognitive function and Alzheimer's disease in late life. Findings from the Nun Study. *JAMA* **275**, 528–532.

Snyder, E. Y., Yoon, C., Flax, J. D., and Macklis, J. D. (1997) Multipotent neural precursors can differentiate toward replacement of neurons undergoing targeted apoptotic degeneration in adult mouse neocortex. *Proc. Natl. Acad. Sci. USA* **94**, 11,663–11,668.

Sohal, R. S. and Weindruch, R. (1996) Oxidative stress, caloric restriction, and aging. *Science* **273**, 59–63.

Solerte, S. B., Fioravanti, M., Pascale, A., Ferrari, E., Govoni, S., and Battaini, F. (1998) Increased natural killer cell cytotoxicity in Alzheimer's disease may involve protein kinase C dysregulation. *Neurobiol. Aging* **19**, 191–199.

Sorbi, S., Piacentini, S., Latorraca, S., Piersanti, P., and Amaducci, L. (1995) Alterations in metabolic properties in fibroblasts in Alzheimer disease. *Alzheimer Dis. Assoc. Disord.* **9**, 73–77.

Stadtman, E. R. (1992) Protein oxidation and aging. *Science* **257**, 1220–1224.

Stein-Behrens, B., Mattson, M. P., Chang, I., Yeh, M., and Sapolsky, R. M. (1994) Stress exacerbates neuron loss and cytoskeletal pathology in the hippocampus. *J. Neurosci.* **1**, 5373–5380.

Stewart, J., Mitchell, J., and Kalant, N. (1989) The effects of life-long food restriction on spatial memory in young and aged Fischer 344 rats measured in the eight-arm radial and the Morris water mazes. *Neurobiol. Aging* **10**, 669–675.

Studer, L., Tabar, V., and McKay, R. D. (1998) Transplantation of expanded mesencephalic precursors leads to recovery in parkinsonian rats. *Nat. Neurosci.* **1,** 290–295.

Sturchler-Pierrat, C., Abramowski, D., Duke, M., Wiederhold, K. H., Mistl, C., Rothacher, S., et al. (1997) Two amyloid precursor protein transgenic mouse models with Alzheimer disease-like pathology. *Proc. Natl. Acad. Sci. USA* **94,** 13,287–13,292.

Svendsen, C. N. and Smith, A. G. (1999) New prospects for human stem-cell therapy in the nervous system. *Trends Neurosci.* **22,** 357–364.

Tang, M. X., Jacobs, D., Stern, Y., Marder, K., Schofield, P., Gurland, B., Andrews, H., and Mayeux, R. (1996) Effect of oestrogen during menopause on risk and age at onset of Alzheimer's disease. *Lancet* **348,** 429–432.

Tower, J. (1996) Aging mechanisms in fruit flies. *Bioessays* **18,** 799–807.

Vanhanen, M. and Soininen, H. (1998) Glucose intolerance, cognitive impairment and Alzheimer's disease. *Curr. Opin. Neurol.* **11,** 673–677.

van Praag, H., Kempermann, G., and Gage, F. H. (1999) Running increases cell proliferation and neurogenesis in the adult mouse dentate gyrus. *Nat. Neurosci.* **2,** 266–270.

Vescovi, A. L. and Snyder, E. Y. (1999) Establishment and properties of neural stem cell clones: plasticity in vitro and in vivo. *Brain Pathol.* **9,** 569–598.

Vlassara, H., Bucala, R., and Striker, L. (1994) Pathogenic effects of advanced glycosylation: biochemical, biologic, and clinical implications for diabetes and aging. *Lab. Invest.* **70,** 138–151.

Weiss, S., Reynolds, B. A., Vescovi, A. L., Morshead, C., Craig, C. G., and van der Kooy, D. (1996) Is there a neural stem cell in the mammalian forebrain. *Trends Neurosci.* **19,** 387–393.

Xiao, J., Perry, G., Troncoso, J., and Monteiro, M. J. (1996) Alpha-calcium-calmodulin-dependent kinase II is associated with paired helical filaments of Alzheimer's disease. *J. Neuropathol. Exp. Neurol.* **55,** 954–963.

Young, D., Lawlor, P. A., Leone, P., Dragunow, M., and During, M. J. (1999) Environmental enrichment inhibits spontaneous apoptosis, prevents seizures and is neuroprotective. *Nature Med.* **5,** 448–453.

Yu, Z. F. and Mattson, M. P. (1999) Dietary restriction and 2-deoxyglucose administration reduce focal ischemic brain damage and improve behavioral outcome: evidence for a preconditioning mechanism. *J. Neurosci. Res.* **57,** 830–839.

Yu, Z. F., Zhou, D., Bruce-Keller, A. J., Kindy, M. S., and Mattson, M. P. (1999) Lack of the p50 subunit of NF-κB increases the vulnerability of hippocampal neurons to excitotoxic injury. *J. Neurosci.* **19,** 8856–8865.

Zaman, S. H., Parent, A., Laskey, A., Lee, M. K., Borchelt, D. R., Sisodia, S. S., and Malinow, R. (2000) Enhanced synaptic potentiation in transgenic mice expressing presenilin 1 familial Alzheimer's disease mutation is normalized with a Benzodiazepine. *Neurobiol. Dis.* **7,** 54–63.

Zhang, L., Zhao, B., Yew, D. T., Kusiak, J. W., and Roth, G. S. (1997) Processing of Alzheimer's amyloid precursor protein during H2O2-induced apoptosis in human neuronal cells. *Biochem. Biophys. Res. Commun.* **235,** 845–848.

Zhang, Y., Tatsuno, T., Carney, J., and Mattson, M. P. (1993) Basic FGF, NGF, and IGFs protect hippocampal neurons against iron-induced degeneration. *J. Cerebral Blood Flow Metab.* **13,** 378–388.

Zhu, H., Fu, W., and Mattson, M. P. (2000) The catalytic subunit of telomerase protects neurons against amyloid β-peptide-induced apoptosis. *J. Neurochem.* **75,** 117–124.

Zhu, H., Guo, Q., and Mattson, M. P. (1999) Dietary restriction protects hippocampal neurons against the death-promoting action of a presenilin-1 mutation. *Brain Res.* **842,** 224–229.

7

Oxidative Stress in Down Syndrome

A Paradigm for the Pathogenesis of Neurodegenerative Disorders

Rocco C. Iannello and Ismail Kola

INTRODUCTION

Late-onset neurological disorders, including Alzheimer's disease (AD), Parkinson's disease (PD), and motor neuron disease, are characterized as complex traits involving many genes and pathways. Consequently, each has unique pathophysiological indicators and clinical outcomes. Gene defects specific for each neurological disorder have been identified, although, in most cases, these only account for a small proportion of the total number of the population afflicted. Despite this complexity, neurodegenerative disorders show some commonality in several parameters. These include elevated levels of reactive oxygen species, increased neural inflammatory responses, initiation of apoptotic and/or necrotic processes, and a reduction in mitochondrial function concomitant with a generalized depletion in cellular energy. From these observations, one can visualize a pattern of convergence initiating from a specific predisposing gene to the activation of genes that may be involved in common pathways, such as those involving inflammatory responses, reactive oxygen species (ROS), cell cycle, and apoptosis through to clinical outcomes in which some neuropathological indices are shared (Fig. 1). If a detailed knowledge of the pertinent pathways and genes were known and mapped, then this type of modeling would be highly informative. Unfortunately, with the exception of some familial cases, the gene components and, to a lesser extent, the pathways underlying many neurological disorders have not yet been elucidated. In this review, we will discuss how Down syndrome can serve as a model to gain insight into the molecular pathways, which may not only account for the specific neurological perturbations seen in Down syndrome, but which may also have a broader significance in neurodegeneration.

DOWN SYNDROME

Down syndrome results from either a total or partial trisomy of chromosome 21 and is highly prevalent throughout the world. Although characterized by very specific morphological features (Patterson, 1987), individuals with Down syndrome also exhibit a broad and varied range of clinical pathologies including mental retardation, microcephaly, bone and skeletal abnormalities, congenital malformations, and increased

From: *Pathogenesis of Neurodegenerative Disorders* Edited by: M. P. Mattson © Humana Press Inc., Totowa, NJ

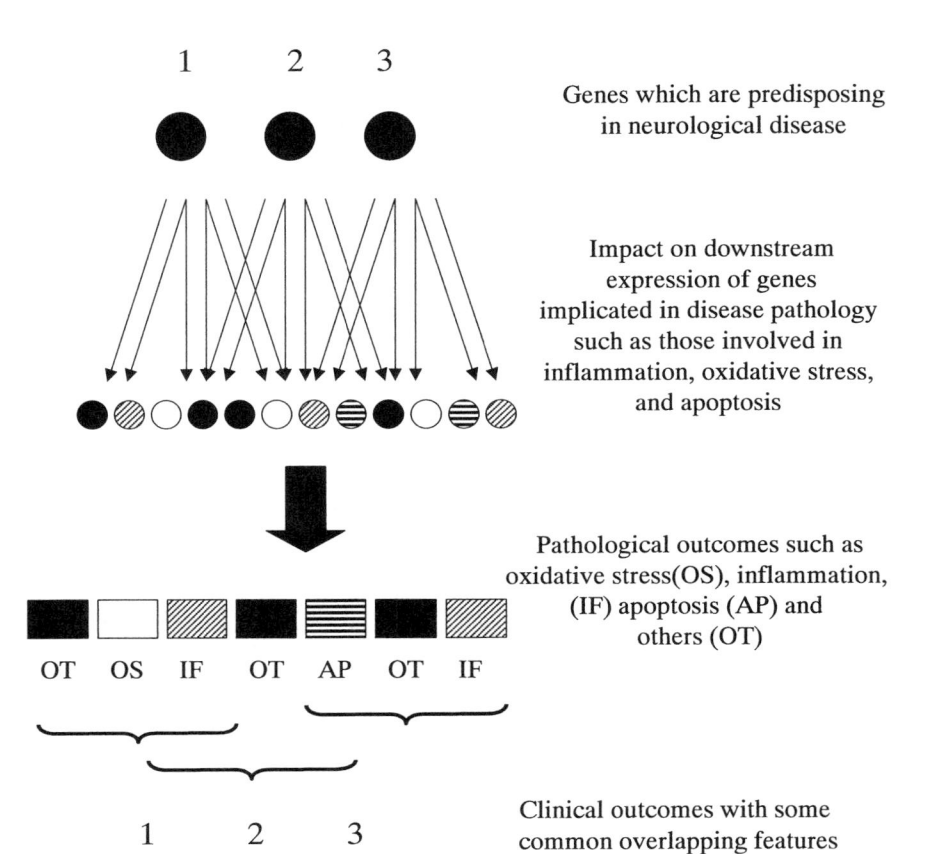

Fig. 1. Genes predisposing individuals to different neurological disorders activate numerous molecular events. Specific and overlapping subsets of downstream genes are modulated, which then act to generate pathophysiological responses which can either be unique for that disorder or common amongst several.

incidence of leukemia and diabetes (Epstein, 1986; Kola, 1997; Kola and Hertzog, 1997; Pueschel, 1990). In addition to a high incidence of mental retardation, there are also several neuroanatomical abnormalities associated with the Down syndrome phenotype. These include decreased brain weight, diminished gyral size, and a simplified convolutional pattern (Crome and Stern 1972). At the microscopic level, Down syndrome brains present decreased neuronal number/density (Colon, 1972; Galaburda and Kemper, 1979; Ross et al., 1984), trans-synaptic degeneration (Marin-Padilla, 1976), and alterations in dendritic spine structure (Marin-Padilla, 1972). Several neurotransmitters are also depleted in the Down syndrome brain (Coyle et al., 1986), and electrophysiological alterations have been described in dorsal root ganglion cells in culture (Caviedes et al., 1990, 1991; Scott et al., 1981). One of the most obvious clinical features, however, is premature aging, with many individuals often developing an early onset of Alzheimer's-like dementia (Kesslak et al., 1994; Korenberg, 1995). The determination of the biological roles of chromosome 21-linked genes, will clearly allow progress to be made toward elucidating mechanisms that contribute to the different pathologies in Down syndrome (Kola and Hertzog, 1997).

It is hypothesized that one consequence of acquiring additional copies of some chromosome 21-specific genes is the perturbation of ROS homeostasis. In this review, we will describe the experimental evidence that supports this and how this may relate to the neural abnormalities associated with the Down's phenotype. ROS are molecules formed spontaneously as part of cellular processes involving oxygen and serve as important signaling molecules and regulators of gene expression when present in low concentrations (Khan and Wilson, 1995; Winyard and Blake, 1997). At high concentrations, however, they have the potential to cause cellular damage either directly or by serving as substrates for the formation of other ROS (Kehrer, 1993). A number of genes located on chromosome 21 have been shown or suspected of playing a role in regulating the redox state of the cell (Brodsky et al., 1995; Fuentes et al., 1995; Kraus et al., 1998; Kelner, 1997; Wermuth, 1982). In this review, we will focus on two extensively studied chromosome 21 genes, *SOD-1* and *APP*. While SOD-1 functions in maintaining the redox state, the biological function of APP is less understood. However, there is increasing evidence to suggest a strong correlative association between amyloid β-peptide (Aβ) deposition and ROS (Mattson, 1997).

CU/ZN-SUPEROXIDE DISMUTASE-1

The gene coding for the antioxidant enzyme, Cu/Zn-superoxide dismutase-1 (SOD-1), is localized to 21q22.1 and plays a key role in the metabolism of ROS (Fridovich, 1986) by catalyzing the dismutation of superoxide anions to hydrogen peroxide. Hydrogen peroxide is subsequently neutralized to water and oxygen through the actions of glutathione peroxidase (GPX) and/or catalase (de Haan et al., 1995, 1997; Yu, 1994).

It has been demonstrated that the levels of SOD-1 in individuals with Down syndrome are elevated in a variety of cell types and organs (Anneren and Epstein, 1987; Brooksbank and Balazs, 1984; de Haan et al., 1995, 1997; Feaster et al., 1977). Therefore, it has been speculated that the steady-state levels of ROS in these tissues are altered leading to metabolic impairment and loss of cellular function.

Of particular interest is the observation that in most diploid tissues, GPX and/or catalase activities increase to compensate elevated levels of SOD-1 as a function of age (Cristiano et al., 1995). However, very little adaptation by either GPX or catalase was observed in aging mice brain (Cristiano et al., 1995; de Haan et al., 1992, 1996). This has important implications, particularly with respect to the neuropathologies, because it suggests that the brain would be particularly prone to oxidative stress. Indeed, a number of studies have already demonstrated that Down syndrome neurons generate increased levels of ROS. In one study, primary cultures of Down syndrome cortical neurons differentiate normally in culture but subsequently degenerate and undergo apoptosis, whereas control neurons remain viable (Busciglio and Yankner, 1995). Further support for the involvement of increased SOD-1 activity in neurological perturbations is derived from Gahtan et. al (1998). In their study, *SOD-1* transgenic mice were shown to have an increase in tetanic stimulation-evoked formation of hydrogen peroxide, resulting in cognitive deficits and impaired hippocampal long-term potentiation.

The role of SOD-1 in chronic neurodegeneration can be highlighted with respect to its involvement in familial amyotrophic lateral sclerosis (FALS). Approximately 15–25% of FALS patients have mutations in Cu/Zn-SOD. Transgenic mice carrying human

SOD-1 mutations have been shown to develop selective motor-neuron impairment, which resembles the pathology seen in FALS patients (Gurney et al., 1994). Data obtained from studies using these transgenic models suggest that the mutant enzymes in SOD-1-linked FALS enhance the production of hydroxyl radical and hydrogen peroxide, thereby resulting in oxidative damage of vital cellular targets (Bruijn et al., 1997; Bogdanov et al., 1998). Avraham et al. (1991) and Yarom et al. (1988) also reported that transgenic mice overexpressing SOD-1 undergo premature aging with respect to neuromuscular junction morphology. In these mice, the degeneration and withdrawal of terminal axons in tongue muscle, along with the degeneration of endplate structures of the leg muscle, were similar to the changes observed in muscles of aging rats and mice (Cardasis, 1983; Fahim and Robbins, 1982), and tongue muscles of individuals with Down syndrome (Yarom et al., 1986, 1987). Interestingly, it has been recently reported that SOD-1 mutations such as those found in FALS selectively inactivate the glial glutamate transporter GLT1 (Trotti et al., 1999). Because GLT1 is the predominant mechanism for the clearing of glutamate from the synaptic cleft and protecting neurons from glutamate toxicity, it is suspected that its inactivation would result in neuronal degeneration via excitotoxic mechanisms (Rothstein et al., 1996). Although there are no known reports of Down syndrome individuals exhibiting FALS-like pathology, this study indicates that oxidative stress, through SOD-1 overexpression, leads directly to molecular perturbations of specific cellular targets and subsequent neuronal death.

AMYLOID PRECURSOR PROTEIN

Most Down syndrome patients exhibit dementia and Alzheimer's-like pathology by middle age. Consequently, researchers have investigated a number of parallels between these two disorders. In both, intracellular and extracellular deposits of proteins in tangles, neuropil threads, and neuritic plaques are present and correlate with neuronal dysfunction leading to dementia (Hardy, 1997). Both Aβ deposition and oxidative stress are implicated in the pathogenesis of AD. It has been shown that Aβ, which is derived from the amyloid precursor protein (APP), is overproduced in the brain of these individuals and forms insoluble fibrillar aggregates (senile or neuritic plaques) that promote neuronal degeneration (Hardy, 1997). Because the APP gene resides on chromosome 21, Down syndrome patients would have an increased production of this protein, in addition to increased oxidative stress resulting from overexpression of SOD-1.

A possible neurotoxic effect of Aβ deposition may involve the generation of ROS, which disrupts cellular calcium homeostasis, and render neurons more susceptible to excitotoxicity and apoptosis. In vitro studies have shown that Aβ peptides can be toxic to a wide variety of neuronal cell types through the disruption of Ca^{2+} homeostasis (Mattson et al., 1992, 1993). Disturbances in Ca^{2+} homeostasis, through the accumulation of Aβ, could lead to increased intracellular Ca^{2+} (Koizumi et al., 1998; Silei et al., 1999). This would cause depolarization of the mitochondrial membrane, thereby disturbing the respiratory chain and production of ATP (Cassarino and Bennet, 1999; Mark et al., 1995). In turn, ATP deficiency could interfere with the function of plasma membrane Ca^{2+} ATPase resulting in a further elevation of cytosolic Ca^{2+} and subsequent cell death (Mark et al., 1995). It is interesting to note that cortical and hippocampal neurons from trisomy 16 mice, TS16 (a mouse model for Down syndrome), also display elevated levels of basal Ca^{2+} (Schuchmann et al., 1998) and increased glutamate-induced vulnerability to cell death in TS16 neuronal cultures. Although it is unclear from this study if susceptibility to glutamate-induced excitotoxicity is linked to Ca^{2+} homeostasis or increased mitochondrial ROS production, other studies have clearly demonstrated a relationship between high concentrations of Aβ and the production of ROS (Behl, 1994; Hensley et al., 1994). Furthermore, other studies have also demonstrated a correlation between increased oxidative stress and the appearance of amyloid plaques and neurofibrillar tangles in AD (Yan et al., 1996; Mattson, 1997), Lewy bodies in PD, and Lewy body-associated dementia (Lyras et al., 1998).

Interestingly, some experimental evidence suggests that oxidative stress may actually exacerbate Aβ aggregation (Multhaup et al., 1997), whereas fibrillar Aβ may increase intraneuronal generation of free radicals (Yan et al., 1995). This, in turn, could provide a trigger for an inflammatory reaction, causing further oxidative damage. Several studies have implicated hydrogen peroxide as being the major source of oxidative stress generated by Aβ peptides in in vitro systems (Behl et al., 1994; Goodman and Mattson, 1994; Yan et al., 1995). Lending support to this notion are in vivo studies that show that SOD-1 levels in the brains of Alzheimer's patients (Bruce et al., 1997) and aged APP mutant mice are elevated, especially around Aβ deposits (Papolla et al, 1998).

Finally, oxidative stress impairs mitochondrial function, resulting in energy depletion. Impaired cellular energy metabolism, in turn, increases the amyloidogenic processing of APP potentially leading to increased Aβ production and subsequent oxidative stress (Gabuzda et al., 1994). This could likely lead to a vicious cycle resulting in a further imbalance in ROS homeostasis and subsequent neuronal damage. The studies described earlier strengthen support for the involvement of these two chromosome 21-located genes in the neuropathology observed in Down syndrome, as well as provide some important insights as to the molecular pathways that may be operating in this disorder.

INITIATION OF AN APOPTOTIC PATHWAY VIA CHROMOSOME 21-LINKED GENES

The studies presented here support the notion that ROS and oxidative stress are implicated in neuronal abnormalities. Importantly, the role of SOD-1 appears to be central to this process with significant relevance to the brain, which appears to be particularly susceptible to oxidative stress. We have also described how APP may contribute to oxidative stress. However, it has only been through the understanding of the biological roles of other chromosome 21 genes that has now allowed insight into which molecular pathways may be operating and involved in the neurological abnormalities associated with Down syndrome. A striking example of this is the possible role of the transcription factor Ets-2 in the neurodegenerative pathology. Ets-2 is located on the distal part of chromosome 21 and is overexpressed in various tissues of individuals with Down syndrome. (Baffico et al., 1989; Tymms and Kola, 1994). We have previously shown that mice overexpressing Ets-2 develop craniofacial and/or bone abnormalities similar to those found in Down syndrome and trisomy 16 mice (Sumarsono et al., 1996). An important recent finding is that the overexpression of Ets-2 was found to be associated with an increase in p53 and Bax expression and that this coincided with a reduction in Bcl-2 (Wolvetang et al., personal communications). Their data place Ets-2

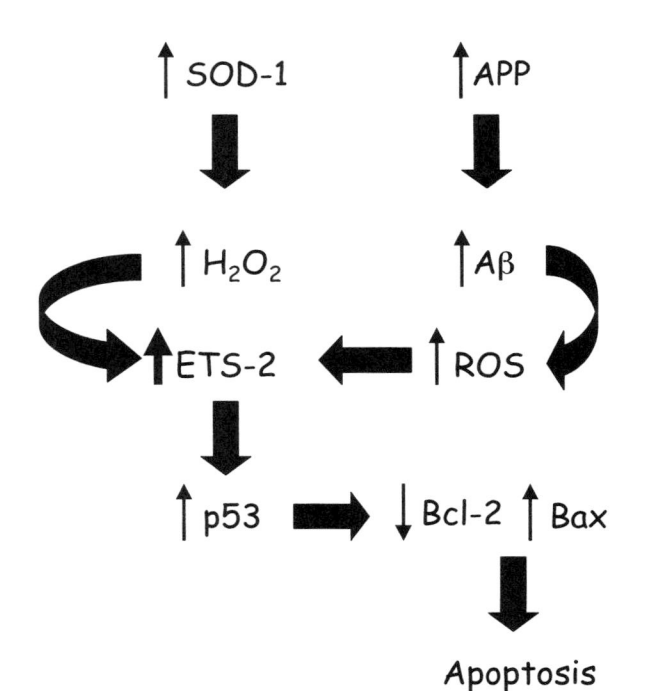

Fig. 2. Representation of how perturbations in three chromosome 21 genes can signal and augment the activation of an apoptotic pathway driven by p53 and Bax. Central to this pathway is ETS-2, which itself is elevated in Down syndrome tissue, but can also be activated through ROS generated as a result elevated levels of SOD-1 and APP.

upstream of p53 and Bcl-2 in an apoptosis-inducing pathway. Notably, ROS play an important upstream role in this pathway because Ets-2 is induced in 3T3 and primary neuronal cells following the addition of hydrogen peroxide to those cultures. It is suspected, therefore, that the extra copy of SOD-1 on chromosome 21 enhances the apoptotic sensitivity of Down syndrome neuronal cells by further upregulating Ets-2. Additional evidence for this is that thymocytes and primary cells from overexpressing SOD-1 transgenic mice were found to overexpress Ets-2 (Wolvetang and De Haan, personal communication).

CONCLUSIONS

The imbalance of ROS homeostasis resulting from elevated levels of SOD-1, overexpression of APP, and the increase in Ets-2 with the concomitant activation of an apoptotic pathway lends support for a prominent role for oxidative stress in the neuropathogenesis of Down syndrome. The compounding effects of overexpressing these genes with respect to specific pathophysiological responses can only be appreciated when placed in context of a molecular pathway, as depicted in Fig. 2. Oxidative stress represents one pathological response contributing to the neurological perturbations seen in Down syndrome. The identification and characterization of the potentially many molecular pathways involved in the neuropathology of Down syndrome will be an exciting challenge for researchers working in this area.

REFERENCES

Anneren, G. and Epstein, C. J. (1987) Lipid peroxidation and superoxide dismutase-1 and glutathione peroxidase activities in trisomy 16 fetal mice and human trisomy 21 fibroblasts. *Pediatr. Res.* **21,** 88–92.

Avraham, K. B., Sugarman, H., Rotshenker, S., and Groner, Y. (1991) Down's syndrome: morphological remodeling and increased complexity in the neuromuscular junction of transgenic CuZn-superoxide dismutase mice. *J. Neurocytol.* **20,** 208–215.

Baffico, M., Perroni, L., Rasore-Quartino, A., and Scartezzini, P. (1989) Expression of the human ETS-2 oncogene in normal fetal tissues and in the brain of a fetus with trisomy 21. *Hum. Genet.* **83,** 295–296.

Behl, C., Davis, J. B., Lesley, R., and Schubert, D. (1994) Hydrogen peroxide mediates amyloid beta protein toxicity. *Cell* **77,** 817–827.

Bogdanov, M. B., Ramos, L. E., Xu, Z., and Beal, M. F. (1998) Elevated "hydroxyl radical" generation in vivo in an animal model of amyotrophic lateral sclerosis. *J. Neurochem.* **71,** 1321–1324.

Brodsky, G., Otterson, G. A., Parry, B. B., Hart, I., Patterson, D., and Kaye, F. J. (1995) Localization of STCH to human chromosome 21q11. 1. *Genomics* **30,** 627–628.

Brooksbank, B. W. L. and Balazs, R. (1984) Superoxide dismutase, glutathione peroxidase and lipid peroxidation in Down's syndrome fetal brain. *Dev. Brain Res.* **16,** 37–44.

Bruce, A. J., Bose, S., Fu, W., Butt, C. M., Mirault, M.-E., Taniguchi, T. N., and Mattson, M. P. (1997) Amyloid β-peptide alters the profile of antioxidant enzymes in hippocampal cultures in a manner similar to that observed in Alzheimer's disease. *Pathogenesis* **1,** 15–30.

Bruijn, L. I. Beal, M. F., Becher, M. W., Schulz, J. B., Wong, P. C., Price, D. L., and Cleveland, D. W. (1997) Elevated free nitrotyrosine levels, but not protein-bound nitrotyrosine or hydroxyl radicals, throughout amyotrophic lateral sclerosis (ALS)-like disease implicate tyrosine nitration as an aberrant in vivo property of one familial ALS-linked superoxide dismutase 1 mutant. *Proc. Natl. Acad. Sci. USA* **94,** 7606–7611.

Busciglio, J. and Yankner, B. A. (1995) Apoptosis and increased generation of reactive oxygen species in Down's syndrome neurons *in vitro. Nature* **378,** 776–779.

Cardasis, C. A. (1983) Ultrastructural evidence of continued reorganization at the aging (11–26 months) rat soleus neuromuscular junction. *Anat. Rec.* **207,** 399–415.

Cassarino, D. S. and Bennett, J. P., Jr. (1999) An evaluation of the role of mitochondria in neurodegenerative diseases: mitochondrial mutations and oxidative pathology, protective nuclear responses, and cell death in neurodegeneration. *Brain Res. Brain Res. Rev.* **29,** 1–25.

Caviedes, P., Ault, B., and Rapoport, S. I. (1990) The role of altered sodium currents in action potential abnormalities of cultured dorsal root ganglion neurons from trisomy 21 (Down syndrome) human fetuses. *Brain Res.* **510,** 229–236.

Caviedes, P., Koistinaho, J., Ault, B., and Rapoport, S. I. (1991) Effects of nerve growth factor on electrical membrane properties of cultured dorsal root ganglia neurons from normal and trisomy 21 human fetuses. *Brain Res.* **556,** 285–291.

Colon, E. J. (1972) The structure of the cerebral cortex in Down's syndrome: a quantitative analysis. *Neuropadiatrie* **3,** 362–376.

Coyle, J. T., Oster-Granite, M. L., and Gearhart, J. D. (1986) The neurobiologic consequences of Down syndrome. *Brain Res. Bull.* **16,** 773–787.

Cristiano, F., de Haan, J. B., Iannello, R. C., and Kola, I. (1995) Changes in the levels of enzymes which modulate the antioxidant balance occur during aging and correlate with cellular damage. *Mech. Ageing Dev.* **80,** 93–105.

Crome, L. and Stern, J. (1972) Down syndrome, in *Pathology of Mental Retardation* (Crome, L. and Stern, J., eds.), Churchill Livingstone, Edinburgh, pp. 200–224.

de Haan, J. B., Cristiano, F., Iannello, R. C., and Kola, I. (1995) Cu/Zn-superoxide dismutase and glutathione peroxidase during aging. *Biochem. Mol. Biol. Int.* **35,** 1281–1297.

de Haan, J. B., Cristiano, F., Iannello, R., Bladier, C., Kelner, M. J., and Kola, I. (1996) Elevation in the ratio of Cu/Zn-superoxide dismutase to glutathione peroxidase activity induces features of cellular senescence and this effect is mediated by hydrogen peroxide. *Hum. Mol. Genet.* **5,** 283–292.

de Haan, J. B., Newman, J. D., and Kola, I. (1992) Cu/Zn superoxide dismutase mRNA and enzyme activity, and susceptibility to lipid peroxidation, increases with aging in murine brains. *Brain Res. Mol. Brain Res.* **13,** 179–187.

de Haan, J. B., Wolvetang, E. J., Cristiano, F., Iannello, R., Bladier, C., Kelner, M. J., and Kola, I. (1997) Reactive oxygen species and their contribution to pathology in Down syndrome. *Adv. Pharmacol.* **38,** 379–402.

Epstein, C. J. (1986) *The Consequences of Chromosome Imbalance: Principles, Mechanisms and Models.* Cambridge University Press, New York.

Fahim, M. A. and Robbins, N. (1982) Ultrastructural studies of young and old mouse neuromuscular junctions. *J. Neurocytol.* **11,** 641–656.

Feaster, W. W., Kwok, L. W., and Epstein, C. J. (1977) Dosage effects for superoxide dismutase-1 on nucleated cells aneuploid for chromosome 21. *Am. J. Hum. Genet.* **29,** 563–570.

Fridovich, I. (1986) Superoxide dismutases. *Adv. Enzymol. Rel. Areas Mol. Biol.* **58,** 61–97.

Fuentes, J. J., Pritchard, M. A., Planas, A. M., Bosch, A., Ferrer, I., and Estivill, X. (1995) A new human gene from the Down syndrome critical region encodes a proline-rich protein highly expressed in fetal brain and heart. *Hum. Mol. Genet.* **4,** 1935–1944.

Gabuzda, D., Busciglio, J., Chen, L. B., Matsudaira, P., and Yankner, B. A. (1994) Inhibition of energy metabolism alters the processing of amyloid precursor protein and induces a potentially amyloidogenic derivative. *J. Biol. Chem.* **269,** 13,623–13,628.

Gahtan, E., Auerbach, J. M., Groner, Y., and Segal, M. (1998) Reversible impairment of long–term potentiation in transgenic Cu/Zn-SOD mice. *Eur. J. Neurosci.* **10,** 538–544.

Galaburda, A. M. and Kemper, T. L. (1979) Cytoarchitectonic abnormalities in developmental dyslexia: a case study. *Ann. Neurol.* **6,** 94–100.

Goodman, Y. and Mattson, M. P. (1994) Secreted forms of β-amyloid precursor protein protect hippocampal neurons against amyloid β-peptide-induced oxidative injury. *Exp. Neurol.* **128,** 1–12.

Gurney, M. E., Pu, H., Chiu, A. Y., Dal Canto, M. C., Polchow, C. Y., Alexander, D. D., et al. (1994) Motor neuron degeneration in mice that express a human Cu/Zn superoxide dismutase mutation. *Science* **264,** 1772–17725.

Hardy, J. (1997) Amyloid, the presenilins and Alzheimer's disease. *TINS* **20,** 154–159.

Hensley, K., Carney, J. M., Mattson, M. P., Aksenova, M., Harris, M., Wu, J. F., et al. (1994) A model for Beta-amyloid aggregation and neurotoxicity based on free radical generation by the Peptide-relevance to alzheimer disease. *Proc. Natl. Acad. Sci. USA* **91,** 3270–3274.

Kehrer, J. P. (1993) Free radicals, mediators of tissue injury and disease. *Crit. Rev. Toxicol.* **23,** 21–48.

Kelner, M. J., Estes, L., Rutherford, M., Uglik, S. F., and Peitzke, J. A. (1997) Heterologous expression of carbonyl reductase: demonstration of prostaglandin 9-ketoreductase activity and paraquat resistance. *Life Sci.* **61,** 2317–2322.

Kesslak, J. P., Nagata, S. F., Lott, I., and Nalcioglu, O. (1994) Magnetic resonance imaging analysis of age-related changes in the brains of individuals with Down's syndrome. *Neurology* **44,** 1039–1045.

Khan, A. U. and Wilson, T. (1995) Reactive oxygen species as cellular messengers. *Chem. Biol.* **2,** 437–445.

Koizumi, S., Ishiguro, M., Ohsawa, I., Morimoto, T., Takamura, C., Inoue, K., and Kohsaka, S. (1998) The effect of a secreted form of beta-amyloid-precursor protein on intracellular Ca2+ increase in rat cultured hippocampal neurons. *Br. J. Pharmacol.* **123,** 1483–1489.

Kola, I. (1997) Simple minded mice from "in vivo" libraries. *Nat. Genet.* **16,** 8–9.

Kola, I. and Hertzog, P. J. (1997) Animal models in the study of the biological function of genes on human chromosome 21 and their role in the pathophysiology of Down syndrome. *Hum. Mol. Genet.* **6,** 1713–1727.

Korenberg, J. R. (1995) Mental modeling. *Nat. Genet.* **11,** 109–111.

Kraus, J. P., Oliveriusova, J., Sokolova, J., Kraus, E., Vlcek, C., de Franchis, R., et al. (1998) The human cystathionine β-synthase (cbs) gene: complete sequence, alternative splicing, and polymorphisms. *Genomics* **52,** 312–324.

Lyras, L., Perry, R. H., Perry, E. K., Ince, P. G., Jenner, A., Jenner, P., and Halliwell, B. (1998) Oxidative damage to proteins, lipids, and DNA in cortical brain regions from patients with dementia with Lewy bodies. *J. Neurochem.* **71,** 302–312.

Marin-Padilla, M. (1972) Structural abnormalities of the cerebral cortex in human chromosomal aberrations: a Golgi study. *Brain Res.* **44,** 625–629.

Marin-Padilla, M. (1976) Pyramidal cell abnormalities in the motor cortex of a child with Down's aberrations: a Golgi study. *J. Comp. Neurol.* **167,** 63–81.

Mark, R. J., Hensley, K., Butterfield, D. A., and Mattson, M. P. (1995) Amyloid β-peptide impairs ion-motive ATPase activities: evidence for a role in loss of neuronal Ca2+ homeostasis and cell death. *J. Neurosci.* **15,** 6239–6249.

Mattson, M. P. (1997) Cellular actions of β-amyloid precursor protein, and its soluble and fibrillogenic peptide derivatives. *Physiol. Rev.* **77,** 1081–1132.

Mattson, M. P., Cheng, B., Culwell, A. R., Esch, F. S., Lieberburg, I., and Rydel, R. E. (1993) Evidence for excitoprotective and intraneuronal calcium-regulating roles for secreted forms of the beta-amyloid precursor protein. *Neuron* **10,** 243–254.

Mattson, M. P., Cheng, B., Davis, D., Bryant, K., Lieberburg, I., and Rydel, R. E. (1992) beta-Amyloid peptides destabilize calcium homeostasis and render human cortical neurons vulnerable to excitotoxicity. *J. Neurosci.* **12,** 376–389.

Multhaup, G., Ruppert, T., Schlicksupp, A., Hesse, L., Beher, D., Masters, C. L., and Beyreuther, K. (1997) Reactive oxygen species and Alzheimer's disease. *Biochem. Pharmacol.* **54,** 533–539.

Pappolla, M. A., Chyan, Y. J., Omar, R. A., Hsiao, K., Perry, G., Smith, M. A., and Bozner, P. (1998) Evidence of oxidative stress and in vivo neurotoxicity of beta-amyloid in a transgenic mouse model of Alzheimer's disease: a chronic oxidative paradigm for testing antioxidant therapies in vivo. *Am. J. Pathol.* **152,** 871–877.

Patterson, D. H. (1987) The cause of Down syndrome. *Sci. Am.* **257,** 42–49.

Pueschel, S. (1990) Clinical aspects of Down's syndrome from infancy to adulthood. *Am. J. Med. Genet.* **7,** 52–56.

Ross, M. H., Galaburda, A. M., and Kemper, T. L . (1984) Down's syndrome: is there a decreased population of neurons? *Neurology* **34,** 909–916.

Rothstein, J. D., Dykes-Hoberg, M., Pardo, C. A., Bristol, L. A., Jin, L., Kuncl, R. W., et al. (1996) Knockout of glutamate transporters reveals a major role for astroglial transport in excitotoxicity and clearance of glutamate. *Neuron* **16,** 675–686.

Schuchmann, S., Muller, W., and Heinemann, U. (1998) Altered Ca2+ signaling and mitochondrial deficiencies in hippocampal neurons of trisomy 16 mice: a model of Down's syndrome. *J. Neurosci.* **18,** 7216–7231.

Scott, B. S., Petit, T. L., Becker, L. E., and Edwards, B. A. (1981) Abnormal electric membrane properties of Down's syndrome DRG neurons in cell culture. *Brain Res.* **254,** 257–270.

Silei, V., Fabrizi, C., Venturini, G., Salmona, M., Bugiani, O., Tagliavini, F., and Lauro, G. M. (1999) Activation of microglial cells by PrP and beta-amyloid fragments raises intracellular calcium through L-type voltage sensitive calcium channels. *Brain Res.* **818,** 168–170.

Sumarsono, S. H., Wilson, T. J., Tymms, M. J., Venter, D. J., Corrick, C. M., Kola, R., et al. (1996) Down's syndrome-like skeletal abnormalities in Ets2 transgenic mice. *Nature* **379,** 534–537.

Trotti, D., Rolfs, A., Danbolt, N. C., Brown, R. H., and Hediger, M. A. (1999) SOD1 mutants linked to amyotrophic lateral sclerosis inactivate a glial glutamate transporter. *Nat. Neurosci.* **2,** 427–433.

Tymms, M. J. and Kola I. (1994) Regulation of gene expression by transcription factors Ets-1 and Ets-2. *Mol. Reprod. Dev.* **39,** 208–214.

Wermeth, B. (1982) in *Enzymology of Carbonyl Metabolism* (Liss, A. R., ed.), New York, pp. 261–274.

Winyard, P. G. and Blake, D. R. (1997) Antioxidants, redox-regulated transcription factors, and inflammation. *Adv. Pharmacol.* **38,** 403–421.

Yan, S. D., Chen, X., Fu, J., Chen, M., Zhu, H., Roher, A., et al. (1996) RAGE and amyloid-beta peptide neurotoxicity in Alzheimer's disease. *Nature* **382,** 685–691.

Yan, S. D., Yan, S. F., Chen, X., Fu, J., Chen, M., Kuppusamy, P., et al. (1995) Non-enzymatically glycated tau in Alzheimer's disease induces neuronal oxidant stress resulting in cytokine gene expression and release of amyloid beta-peptide. *Nat. Med.* **1,** 693–699.

Yarom, R., Sagher, U., Havivi, Y., Peled, I. J., and Wexler, M. R. (1986) Myofibers in tongues of Down's syndrome. *J. Neurol. Sci.* **73,** 279–287.

Yarom, R., Sapoznikov, D., Havivi, Y., Avraham, K. B., Schickler, M., and Groner, Y. (1988) Premature aging changes in neuromuscular junctions of transgenic mice with an extra human CuZnSOD gene: a model for tongue pathology in Down's syndrome. *J. Neurol. Sci.* **88,** 41–53.

Yarom, R., Sherman, Y., Sagher, U., Peled, I. J., Wexler, M. R., and Gorodetsky, R (1987) Elevated concentrations of elements and abnormalities of neuromuscular junctions in tongue muscles of Down's syndrome. *J. Neurol. Sci.* **79,** 315–326.

Yu, B. P. (1994) Cellular defenses against damage from reactive oxygen species. *Physiol. Rev.* **74,** 139–162.

8

The Pathogenesis of Amyotrophic Lateral Sclerosis

Stanley H. Appel and R. Glenn Smith

CLINICAL FEATURES

Amyotrophic lateral sclerosis (ALS) is a progressive disease of the voluntary motor system with clinical signs of both lower and upper motor-neuron compromise (Haverkamp et al., 1995). Weakness and muscle atrophy usually begin asymmetrically and distally in one limb, spread within the neuraxis to involve contiguous groups of motor neurons, and then involve more rostral motor neurons. For example, if weakness begins in the right leg it will spread to the left leg and subsequently to the right arm. If weakness begins in the right arm, it may next involve the left arm prior to involvement of speech. In 25% of patients, dysarthria and bulbar involvement are the presenting symptoms and signs. Bulbar onset is more common in older individuals and usually starts with speech dysfunction prior to swallowing compromise. The muscles of eye movement and the urinary sphincters are usually spared until very late in the disease. Respiratory involvement is usually a later feature in limb-onset patients, but occasionally it can be an early manifestation, especially in older patients with bulbar symptoms and signs.

The incidence of ALS is 1/100,000 and the prevalence is 3–5/100,000. Ninety percent of ALS cases are sporadic (SALS), whereas 10% are familial (FALS). A defect in the gene encoding copper-zinc superoxide dismutase-1 (Cu/Zn-SOD-1) is present in 20% of FALS cases (Rosen et al., 1993). The mean age of onset is 57 yr and the median survival is approx 3 yr, with younger patients surviving longer than older individuals. Males develop SALS more frequently than females (ratio 1.7:1). However, in older individuals, the frequency of female involvement is increased and the ratio approaches 1:1 over the age of 65, which is similar to the ratio in FALS.

PATHOGENESIS

Our understanding of the mechanisms of motor neuron injury and cell loss in SALS is quite limited. Studies from our own laboratory suggest a prominent role for increased intracellular calcium in motor-neuron degeneration (Engelhardt et al., 1997; Siklos et al., 1996, 1998). Other studies emphasize the importance of excitotoxic (Rothstein et al., 1990, 1992; Lin et al., 1998) and free-radical (Bowling et al., 1993; Yim et al, 1996) mechanisms in mediating cell death in ALS. However, none of these explanations are mutually exclusive, and altered calcium homeostasis, free radicals, and glutamate

From: *Pathogenesis of Neurodegenerative Disorders* Edited by: M. P. Mattson © Humana Press Inc., Totowa, NJ

excitotoxicity may all participate in the cell-injury cascade leading to motor-neuron death. In fact, alterations in one parameter can lead to alterations in other parameters, and each can enhance and propagate the injury cascade. Increased intracellular calcium can enhance free radical production (Dykens, 1994) and glutamate release (Coyle and Putterfarcken, 1993; Tymianski et al., 1993; Dugan et al., 1995; Carriedo et al., 1998), which in turn can further increase intracellular calcium (Sheen et al., 1992). Increased free radicals can alter the glial uptake of glutamate, increasing the glutamate available to interact with neuronal AMPA/kainate receptors, which in turn can increase intracellular calcium and/or free radicals. Free radicals can enhance lipid peroxidation leading to increased intracellular calcium (Kakkar et al., 1992).

These changes of increased intracellular calcium, increased production of free radicals, and enhanced glutamate excitotoxicity could critically impair motor-neuron structures such as mitochondria or neurofilaments, and compromise energy production and axoplasmic flow. Once initiated, the changes could become self-propagating and induce an irreversible cascade resulting in cell death. Key questions in SALS are which of these changes initiates cell injury, which amplifies cell injury, and which is a secondary late marker of an injured motor neuron. In FALS, at least in those cases due to mutations in the SOD-1 gene, free-radical damage is possibly an early event (Siddique, et al., 1991; Rosen et al., 1993). However there is presently no evidence that genetic defects in free radical, calcium, or glutamate homeostasis can explain the majority of cases of SALS; and the initiating factors in sporadic disease remain unknown. Nevertheless, any of the defined alterations (i.e., in calcium, free radicals, glutamate excitotoxicity, or in mitochondria or neurofilaments) could represent early events in the pathogenesis of motor-neuron injury.

GLUTAMATE EXCITOTOXICITY

A potential role for glutamate excitotoxicity in ALS motor-neuron injury has been hypothesized since glutamate was first demonstrated to induce neuronal injury in both in vitro and in vivo experimental models (Choi, 1988). Since that time, numerous studies have supported the importance of glutamate in motor-neuron injury, beginning with the demonstration of elevated levels of glutamate in cerebrospinal fluid (CSF) and brain in some cases of SALS (Rothstein et al., 1990). Subsequently, a decrease in glutamate uptake was demonstrated in synaptosomes prepared from SALS spinal cord and motor cortex. (Rothstein et al., 1992). In organotypic spinal cord cultures, inhibition of glutamate uptake produced slow degeneration of motor neurons (Rothstein et al., 1993), which could be prevented by AMPA/kainate receptor antagonists but not by *N*-methyl-D-aspartate (NMDA) receptor antagonists.

Impaired glutamate transport in ALS spinal cord has been explained by the selective loss of the astroglial glutamate transporter EAAT2 (Rothstein et al., 1995; Bristol and Rothstein, 1996), possibly resulting from defects in EAAT2 mRNA processing (Lin et al., 1998). However, at least one recent report questions the specificity of EAAT2 mRNA splicing for ALS because of similar results in normal tissues (Nagai et al., 1998). One of the products of lipid peroxidation shown to be elevated in SALS tissues, namely, 4-hydroxynonenal (4HNE), has been demonstrated to modify the human EAAT2 glial transporter (Pedersen et al., 1998), and to inhibit glutamate transport in astrocytes in vitro (Blanc et al., 1998). In rat spinal cord organotypic cultures, antisense knockout of glutamate transporters demonstrated a predominant role for astroglial glutamate transport in excitotoxicity and clearance of extracellular glutamate (Rothstein et al, 1996). Glutamate toxicity is thought to arise from activation of motor-neuron AMPA/kainate receptors (Carriedo et al., 1996) leading to depolarization, entry of calcium into the cell, and activation of enzymes that enhance formation of free radicals leading to cell death. Calcium influx through the AMPA/kainate receptors is modulated by the presence of the GluR2 subunit. In ALS spinal cord, there is a decrease of the GluR2 subunit mRNA (Shaw et al., 1999; Takuma et al., 1999), and an increase in unedited GluR2 mRNA (i.e., persistent CAG [glutamine] instead of CGG [arginine]), which could result in increased calcium permeability. At present, it is unclear whether these changes in GluR2 are the consequence of motor-neuron injury or a cause of motor-neuron injury. Furthermore, despite considerable evidence implicating glutamate excitotoxicity, antiglutamatergic therapeutic strategies have produced extremely modest benefits both in human SALS patients (Lacomblez et al., 1996) and transgenic mouse models of human SOD-1 mutations (Gurney et al., 1996).

FREE RADICALS AND SOD-1 MUTATIONS

Over 60 different mutations in Cu/Zn-SOD-1 have now been reported in FALS, and many transgenic models expressing mutant human SOD-1 have been generated (Cleveland, 1999). These models have proven extremely useful because many of them develop a clinical phenotype and neuropathological changes that resemble changes seen in human FALS. SOD-1 is a metalloenzyme that catalyzes the conversion of superoxide anion to hydrogen peroxide. H_2O_2 itself is toxic and in the presence of Fe or Cu can lead to the formation of the highly reactive hydroxyl radical. The superoxide anion may also combine with nitric oxide (NO) to generate peroxynitrite, which can modify proteins as well further increase reactive hydroxyl radicals. These free-radical species can react with DNA, lipids, and proteins, and can disrupt mitochondria.

However, it is still unclear how mutant SOD-1 can lead to motor-neuron injury and cell death. The lack of correlation of dismutase activity with motor-neuron disease in either patients or transgenic mice expressing mutant SOD-1, and the lack of an ALS-like phenotype in SOD-1 knockout mice suggest a toxic gain-of-function as the mechanism by which mutant SOD-1 can induce motor-neuron injury (Borchelt et al., 1994; Reaume et al., 1996). Thus, the mutant enzyme may be toxic through a mechanism not dependent on dismutase catalytic activity, but still possibly involving oxidative damage. In fact, in transgenic mouse models with mutant SOD-1, markers for oxygen radical production are increased prior to motor-neuron degeneration (Liu et al., 1998; Bogdanov et al., 1998); and levels of free 3-nitrotyrosine (Bruijn et al., 1997; Ferrante et al., 1997b) and protein carbonyl groups (Andrus et al., 1998) are increased. Further, H_2O_2 and the hydroxyl anion free radicals are increased while superoxide anions are decreased in microdialysis studies of transgenic mutant SOD-1 mouse spinal cord (Liu et al., 1999a). However, the changes in this latter study are relatively modest, and may not clearly explain subsequent motor-neuron injury.

In support of potentially toxic properties of mutant SOD-1, the half-life of the mutant protein is diminished (Borchelt et al., 1994), and the peroxidase activity of the mutant SOD-1 protein is enhanced (Wiedau-Pazos et al., 1996; Yim et al., 1996). Mutation at the catalytic site can reduce the production of the hydroxyl anion and in vitro cytotoxicity, and has been postulated to support enhanced peroxidative function (Liu et al., 1999b). Additional proposed properties of the mutant SOD-1 include the ability to promote nitration (Beckman et al., 1993); an alteration in metal binding properties (Carri et al., 1994), possibly leading to aberrant copper or zinc interactions; and protein aggregation (including mutant SOD-1 aggregates) in motor neurons as well as astrocytes Brown, 1998). Recent emphasis on aberrant copper chemistry of the mutant SOD-1 was sparked by the demonstration of a specific copper chaperone for SOD-1 in yeast (Culotta et al., 1997) and subsequently in mammalian species (Rothstein et al., 1999), but further studies are needed to determine the relevance of these findings to either pathogenesis or therapy. The mutant SOD-1 also has a decreased affinity for zinc, and the loss of zinc can induce apoptosis in cultured motor neurons that required copper bound to SOD and the endogenous production of NO (Estevez et al, 1999). It was recently reported that spinal-cord motor neurons from CuZn-SOD-1 mutant mice exhibit enhanced basal oxyradical production, lipid peroxidation, and perturbed calcium homeostasis (Kruman et al., 1999). Moreover, cultured motor neurons from the Cu/Zn-SOD-1 mutant mice were more vulnerable to glutamate toxicity than were motor neurons from wild-type mice. The endangering effect of the Cu/Zn-SOD-1 mutation was ameliorated by treatment of the cells with antioxidants and NO-suppressing agents, indicating roles for oxidative stress and nitric oxide in the adverse effect of the mutant Cu/Zn-SOD-1. Whether any or all of these mechanisms contribute significantly to motor neuron injury in the SOD-1 transgenic models is still unresolved.

HUMAN STUDIES

These studies in transgenic mouse models prompted an examination of oxidative stress as a key intermediate in motor-neuron injury in human FALS and SALS, given the clinical similarity between FALS and SALS. Interestingly, initial studies of SALS and FALS brains documented significant increases in protein carbonyl groups in SALS, but not FALS specimens (Bowling et al., 1993). These results were confirmed in further studies of brain homogenates with the demonstration of increased protein carbonyl groups and oxidized nucleic acids (8-hydroxy-2'-deoxyguanosine [OH8DG]) in SALS, but not FALS (Ferrante et al., 1997a). Increased protein carbonyl groups have also been demonstrated in spinal cords of SALS patients compared to controls or patients with FALS (Shaw et al., 1995). This increase in markers of oxidative stress throughout brain tissue in SALS, but not FALS, was also noted in a marker for lipid peroxidation, 4-hydroxynonenal (4-HNE) (Smith et al., 1998). 4-HNE was significantly increased in the CSF of SALS patients but not FALS patients. In SALS spinal-cord tissue, proteins modified by 4-HNE were present and could contribute to motor-neuron injury (Pedersen et al, 1998; Keller and Mattson, 1998). Interestingly, one of the proteins most heavily modified by 4-HNE in ALS spinal cord tissue is apolipoprotein E. Particularly interesting is evidence that the E4 isoform of apolipoprotein E is much less effective in binding 4-HNE than are the E2 or E3 isoforms (Pedersen et al., 2000a). The latter study further provided evidence that apoE2 and apoE3 can protect neurons against oxidative insults relevant to the pathogenesis of ALS, whereas E4 is ineffective. These findings may provide a mechanistic explanation for the increased neurodegeneration associated with apoE4 genotype.

With immunohistochemistry, increased staining for hemeoxygenase 1, malondial-dehyde-modified protein, and OH8DG has been demonstrated in both FALS and SALS specimens (Ferrante et al., 1997a). Thus, oxidative stress is increased in both FALS and SALS, but the pattern and distribution of reactive oxygen species is significantly different. 3-Nitrotyrosine has been reported to be increased in both SALS and FALS spinal cord, but this result is not specific because injuries such as ischemia can also elevate 3-nitrotyrosine levels (Beal et al., 1997).

Further evidence for oxidative stress in motor-neuron injury has been based on the demonstration that fibroblasts from ALS patients with mutant SOD-1 appear more sensitive to oxidative stress caused by H_2O_2 (Aguirre et al., 1998). However, lymphoblastoid cells from FALS patients with a SOD-1 gene mutation do not have increased susceptibility to radiation-induced free radicals (Mithal et al., 1999). Glutathione peroxidase activity, which can reduce H_2O_2, has been reported to be reduced 40% in one study of SALS premotor cortex tissue (Przedborski et al., 1996), to be increased 26% in another study of spinal-cord SALS tissue (Ince et al., 1994), and to be unchanged in a third study of spinal-cord SALS tissue (Fujita et al., 1996). Thus, there is no evidence in either FALS or SALS of increased susceptibility to free-radical stress.

The most intriguing aspect of the human mutant SOD-1 studies is that increased reactive oxygen species compared to SALS was anticipated, but not found. What was found was a significant increase in modified proteins, nucleic acids, and lipids in SALS compared to FALS. Only with immunohistochemistry, which is far from quantitative, could evidence of increased oxidative stress in FALS be demonstrated. Thus, despite the clinical similarities, in SALS and FALS the distribution and therefore the mechanism of production of increased reactive oxygen species must be distinctly different. Free-radical stress is more widespread in SALS and more circumscribed in FALS, with the latter results being in accord with the data from transgenic mouse models of human mutant SOD-1.

INCREASED INTRACELLULAR CALCIUM

The excessive influx of Ca possibly triggered by stimulation of glutamate receptors has been postulated to be the common denominator underlying cell death in a number of neurodegenerative diseases, including ALS. Abnormally high and sustained increases in the intracellular levels of Ca following glutamate receptor activation may activate a series of Ca-stimulated catabolic enzymes including protein kinase C (PKC), calpains, phospholipase A2, phospholipase C, nitric oxide synthase (NOS), and endonucleases (Choi, 1988; Lipton et al., 1993). Under physiological conditions, intracellular Ca is maintained at precise levels by Ca ATPases, Na–Ca exchangers, and mitochondria (Orrenius et al., 1989) as well as by calcium-binding proteins (Heizmann and Braun, 1992). In motor neurons, the relevant glutamate receptor appears to be the AMPA/kainate receptor, which, when activated, may enhance Ca entry (Carriedo et al., 1996).

Calcium entry may be also enhanced by mechanisms other than activation of AMPA/kainate receptors. Immunoglobulins from 8 of 10 ALS patients induced transient increases in intracellular Ca in a motor-neuron cell line as monitored with Ca imaging (Colom et al., 1997) and increased Ca currents as monitored with electrophysiological techniques (Mosier et al., 1995). Furthermore, following intraperitoneal injections of mice with antimotoneuronal IgG or ALS IgG, calcium was specifically increased in synaptic vesicles and mitochondria of motor-axon terminals as visualized with the oxalate pyroantimonate technique (Engelhardt et al., 1995, 1997). Studies of experimental models of ALS further support a role for perturbed calcium homeostasis in the degeneration of motor neurons. For example, oxidative insults relevant to ALS impair glutamate and glucose transport and disrupt calcium homeostasis, and expression of mutant Cu/Zn-SOD-1 in transgenic mice similarly disrupts calcium homeostasis and increases vulnerability of neurons to excitotoxic and oxidative insults (Pedersen et al., 1998; Kruman et al., 1999).

The oxalate–pyroantimonate technique has also been used to characterize morphometric alterations in intracellular calcium in motor-nerve terminals from patients with SALS and with different neurological disorders (Siklos et al., 1996). Muscle biopsy specimens from 7 patients with ALS, 10 nondenervating control subjects, and 5 patients with denervating neuropathies were analyzed. The motor-nerve terminals from ALS specimens contained significantly increased calcium, increased mitochondrial volume, and increased numbers of synaptic vesicles compared to any of the disease controls, without exhibiting excess Schwann-cell envelopment specific to denervating terminals. These results parallel the effect of ALS IgG passively transferred to mice, and demonstrate that neuronal calcium is, in fact, increased in motor neurons from SALS patients in vivo. Since glutamate receptors and specifically AMPA/kainate receptors are not present at the mammalian neuromuscular junction, it is clear that mechanisms other than glutamate receptor activation may also participate in increasing Ca entry and intracellular levels in motor neurons. One of these mechanisms may involve interaction of ALS IgG with the FcγR receptor with subsequent increases in intracellular Ca (Habib, personal communication).

MITOCHONDRIAL DYSFUNCTION

Mitochondrial alterations have been described in SALS as well as in models of FALS. Increased cytoplasmic free radicals as well as increased cytoplasmic calcium can disrupt the function of mitochondria, particularly in neurons (Bolanos et al., 1995). Swollen mitochondria together with increased intramitochondrial calcium have been demonstrated in SALS motor-axon terminals (Siklos et al., 1996). Furthermore, there is an increase in mitochondria in proximal axons of motor neurons and in muscle of SALS patients (Sasaki et al., 1990; Wiedemann et al., 1998). The mitochondrial changes have not been confined to the central nervous system (CNS) since large mitochondria with inclusions have been described in liver biopsies from SALS patients, suggesting that tissue outside the CNS may be compromised in ALS (Nakano et al., 1987). In transgenic mouse models of mutant SOD-1, mitochondrial swelling, and vacuolization have been described as early pathological lesions (Dal Canto and Gurney, 1994; Wong et al., 1995). Expression of the mutant enzyme in neuroblastoma cells results in a loss of mitochondrial membrane potential and elevated cytosolic calcium concentrations (Carri et al., 1997). In addition, motor neurons from transgenic mice expressing the G93A Cu/Zn-SOD-1 mutation exhibit impaired mitochondrial function characterized by decreased intramitochondrial calcium levels and transmembrane potential under basal conditions and exaggerated mitochondrial dysfunction following excitotoxic insults (Kruman et al., 1999). Of interest is the demonstration that cybrids (cells in which mitochondria from SALS platelets are fused with neuroblastoma cells previously depleted of mitochondrial DNA) possess abnormal electron transport, altered mitochondrial ultrastructure, altered calcium homeostasis, and increased free-radical scavenger activities (Swerdlow et al., 1998). In a recent study, SALS motor neurons demonstrate a significant decrease in mitochondrial DNA-mediated cytochrome *c* oxidase activity (Borthwich et al., 1999). These data suggest the presence of mitochondrial DNA alterations in vivo, possibly the presence of 8-hydroxy-2'-deoxyguanosine (OH8DG), which could have resulted either from increased cytoplasmic calcium and/or increased oxidative damage, and which could become self-propagating and contribute to progressive motor-neuron injury.

NEUROFILAMENT ALTERATIONS

Neurofilaments (NF) are 10 nm intermediate filaments specific to neurons that accumulate in proximal axons and cell bodies in ALS motor neurons (Delisle and Carpenter, 1984; Hirano et al., 1984). In transgenic mice, point mutations of the NF-light rod (Lee et al., 1994), overexpression of human NF-heavy subunit (Julien et al., 1995), or overexpression of mouse NF-light chain develop motor-neuron pathology (Xu et al., 1993). Furthermore, transgenic human mutant SOD-1 mice develop neurofilament accumulations (Tu et al., 1997). Of considerable importance is the demonstration that decreasing expression of neurofilaments in mice expressing a SOD-1 mutation significantly delayed onset and slowed progression of disease (Williamson et al., 1998). Furthermore, increasing expression of the NF-heavy subunit traps neurofilaments in the cell body, decreases their presence in axons, and markedly extends life span (Couillard-Despres et al., 1998). These experiments have been interpreted to suggest that axonal neurofilament content may contribute to motor-neuron vulnerability (Cleveland, 1999). However, neurofilaments can bind intracellular calcium, and an alternative explanation for the protective effect may not be the decrease of neurofilaments in axons but the removal of the excess toxic calcium by binding to NF in the cytoplasm.

Abnormalities in NF have also been described in SALS patients. Several patients were described with NFH variants (Figlewicz et al., 1994) and recently four novel NFH tail deletions were reported in three patients with sporadic ALS a family with autosomal dominant ALS, as well as two young controls (Al-Chalabi et al., 1999). It is not clear whether any of these NF-H variants segregate with disease but, at the very least, variants in NF may be a risk factor in SALS.

NF alterations could clearly impair axoplasmic transport and interfere with the traffic of macromolecules destined for the axon terminal. Such axoplasmic transport deficits have now been described as an early event in mutant SOD-1 transgenic mice (Warita et al., 1999; Williamson and Cleveland, 1999) and have long been recognized in SALS (Breuer et al., 1987).

APOPTOTIC CELL DEATH

Many studies of in vitro models suggest that degeneration of motor neurons in ALS may involve apoptotic or programmed cell death mechanisms, but initial studies of human ALS tissue have been less convincing. Early reports suggested the presence of nick-end labeling in spinal motor neurons in ALS (Yoshiyama et al, 1994), as well as altered expression of bcl-2 and Bax mRNA (Mu et al., 1996). Upregulation of the cell death-promoting Bax protein and increased DNA degradation were recently confirmed in ALS spinal-cord motor neurons (Ekegren et al., 1999). However the report by Migheli et al. (1997) questioned the involvement of the apoptotic pathway in ALS tissue. The JNK/APK-c-Jun pathway was overexpressed in ALS spinal cord primarily in astrocytes, and was accompanied by NF-κB activation, possibly as a protective response to oxidant stress. Motor neurons, on the other hand, had an unusually low expression of the pathway. A more recent analysis by Martin (1999) demonstrated that the proapoptotic proteins Bax and Bak were elevated in the mitochondrial-enriched membrane compartment, but were reduced or unchanged in the cytosol of selectively vulnerable regions in ALS tissue compared to controls. In addition, levels of prostate apoptosis response 4 (PAR-4) are increased in spinal cords of ALS patients and Cu/Zn-SOD-1 mutant mice, strongly supporting a role for apoptosis of motor neurons in the pathogenesis of ALS (Pedersen et al., 2000b). The antiapoptotic protein Bcl-2, on the other hand, was decreased in the mitochondrial-enriched membrane compartment, but increased in the cytosol. DNA fragmentation was present as was the presence of the DNA fragmentation factor-45/40 activation and increased caspase-3 activity. This redistribution of proapoptotic and antiapoptotic proteins is consistent with the involvement of apoptosis in motor-neuron degeneration. However, further work is necessary to define both the initiating factors and the proapoptotic proteins specifically required for cell death, and to determine whether targeting specific components of the relevant apoptotic pathway is a meaningful therapeutic strategy. At present, the evidence of an activated classical apoptotic pathway is far from convincing.

POTENTIAL ROLE OF INITIATING FACTORS IN SALS

Despite the explosion of studies in ALS prompted by the discovery of mutant SOD-1 in 20% of the FALS cases, minimal insight has been gained into the factors that could initiate SALS. There is no clear understanding of what initiates motor-neuron degeneration in models of FALS. Yet the hope remains that the mechanisms of motor-neuron degeneration in FALS will help explain motor-neuron degeneration in SALS. The consistent difference in the distribution of free-radical species in SALS and FALS would suggest that the mechanisms might be different, and the initiating factors might be different. As a starting hypothesis, we suggest that FALS could represent injury from the "inside out," while active cellular injury could proceed from the "outside in." In this case the "inside" would refer to the motor neuron. Mutant proteins such as SOD-1 in FALS patients could lead to perturbed motor-neuron function by altering the internal milieu, while cell damage at the cell body or nerve terminal initiated from the outside may be an effector of perturbed function in SALS. However, initiating events for an "outside in" process need not be solely directed from outside the motor neuron. Certainly, genetic factors might help to predispose a patient population to develop motor-neuron disease, even though active disease initiation might rely upon a subsequent "second hit," such as from trauma, toxin, or infection. Although data suggesting that such secondary processes may help to initiate motor-neuron disease have been extremely limited and controversial, recent evidence from two groups describing a specific association between the presence of enteroviral RNA in motor neurons of 74% (Berger et al., 2000) and 89% (Woodall et al., 1994) of patients with SALS by reverse transcriptase polymerase chain reaction (RT-PCR) and direct *in situ* PCR is intriguing. Although it is not yet clear whether such infection represents an initiating event or a secondary infection in stressed cells, a pathological assessment of autopsied spinal cords from patients with FALS could help answer this critical point.

A major response in mammals to injury, toxin exposure, or infection is activation of the immune system, which, in a susceptible host, may precipitate autoimmune disease. For example, streptococcal infection may lead, in genetically appropriate hosts, to secondary, immune-mediated rheumatic fever. Likewise, enterovirus induced immune-mediated disease has been implicated in human idiopathic dilated cardiomyopathy and diabetes mellitus (Juhela et al., 1999; Knip and Akerblom, 1999; Pauchinger et al., 1999), while staphylococcal toxins have been linked to the pathogenesis of a number of autoimmune diseases (Krakauer, 1999). An altered immune state could potentially disturb motor-neuron function through the action of immunoglobulins, cytokines, lymphocytes, and/or activated microglia.

Our own suggestion for the role of an immune/inflammatory process was based on the presence of other immune disorders, paraproteinemias, as well as lymphomas in many patients with SALS. The presence of inflammatory cells, or cytokines, in SALS spinal cord (Engelhardt and Appel, 1990; Engelhardt et al., 1993) could provide a possible explanation for the spread of clinical manifestations from a focal onset, in one extremity, to the involvement of a contralateral extremity, and then in a rostral direction to involve other extremities and bulbar musculature. This clinical progression has no simple explanation, but certainly suggests the involvement of potentially toxic factors (such as cytokines released from inflammatory cells or products of lipid peroxidation, such as 4-hydroxynonenal) diffusing across the spinal cord to involve contralateral motor neurons followed by rostral diffusion of factors to involve more rostral motor neurons. Whether the inflammatory reaction represents an attempt to repair altered motor neurons or contributes to motor-neuron injury is presently unclear. Nevertheless, it is of interest that peripheral trauma, which is cited as a risk factor in ALS (Kurtzke and Kurland, 1983), can injure motor axons and, in turn, initiate central microglial (Streit et al., 1989) and lymphocytic (Raivich et al., 1998) responses that could contribute to motor-neuron injury through contact or diffusible cytokines.

CNS INFLAMMATION IN ALS

Although early studies of ALS tissue did not describe the presence of inflammatory cells in ALS, reports beginning in 1988 suggested that inflammatory cell infiltration might be more common than previously suspected (Lampson et al., 1988; Troost et al., 1988, 1989; McGeer et al., 1988). Further studies (by Lampson et al., 1990) demonstrated activated microglia and small numbers of T cells in degenerating white matter of ALS cords. Our own studies (Engelhardt et al., 1993) documented the presence of lymphocytes in the spinal cord in 18 of 27 consecutive ALS autopsies. Lymphocytes were predominantly CD4-positive in the vicinity of degenerating cortical spinal tracts. More significantly, CD4 and CD8 cells were found in ventral horns. Activated microglia were also prominent in the ventral horn of spinal cords from ALS patients. Kawamata et al. (1992) demonstrated the presence of significant numbers of CD8-reactive T cells and, to a lesser extent, CD4 T cells marginating along capillaries in the parenchyma of spinal cord and brains of 13 ALS patients. Cells including lymphocytes and activated microglia expressing MHC Class I and Class II, leukocyte common anti-gen, and FcγRI and β-2 integrins were present in ALS tissue. Recent studies on periph-eral blood lymphocytes of ALS patients have also shown that two rare subsets expressing the FcγRIII (CD16) CD8 with or without CD57 are significantly increased in the blood of ALS patients with upper motor-neuron and bulbar signs (Schubert and Schwan, 1995).

The diversity of T cells in ALS spinal cord has also been examined employing the RT-PCR with variable region sequence-specific oligonucleotide primers to amplify Vβ T-cell receptor transcripts (Panzara et al., 1999). A greater expression of Vβ$_2$ tran-scripts was detected in ALS specimens, independent of the HLA genotype of the indi-vidual. As additional confirmation, cells were assayed from CSF of 22 consecutive ALS patients for the presence of Vβ$_2$ transcripts. This specific T-lymphocyte receptor was demonstrated in 17 of 22 ALS patients' CSF, whereas only 4 of 19 control patients had similar expression. It is of interest that Vβ$_2$ is a T-cell receptor known to respond to superantigens. Whether superantigen stimulation of lymphocytes with Vβ$_2$ receptors is a primary factor in motor-neuron injury or the consequence of motor-neuron injury due to some other cause is not known, but the presence of T-lymphocyte restriction does suggest the involvement of immune mechanisms.

MICROGLIA: MEDIATOR OF IMMUNE/INFLAMMATORY REACTIONS IN THE CNS

Microglia comprise about 10% of the brain parenchyma and are distributed through-out the CNS, where they play important roles as mediators of immune inflammatory responses (Fedoroff, 1995; Streit, 1993). CNS microglia can proliferate and become activated after inflammatory, mechanical, or neurotoxic insults to the peripheral nerve (reviewed by Zielasek and Hartung, 1996), suggesting that different kinds of tissue damage can trigger anatomically limited microglial responses within the anterior horn of relevant spinal-cord levels. Following peripheral axonal injury, changes in microglia around motor-neuron cell bodies appear first, followed by the appearance of components of complement, infiltration of IgG, astroglial proliferation, and T-cell recruitment (reviewed by Kreutzberg, 1996). These reactions then gradually subside. The specific signals from motor neurons that activate microglia following peripheral injury are far from clear, although γ-interferon (IFN-γ) has been implicated. Activated microglia are fully competent antigen-presenting cells (Matsumoto et al., 1992) and may play a role in the recognition, uptake, processing, and presentation of foreign antigens in the CNS. These microglia have been described as having both positive and negative effects associated with cell injury, i.e., they may facilitate repair or enhance injury (Kreutzberg, 1996; Merrill and Benveniste, 1996). The negative effects involve the production of cytotoxic mediators such as free radicals (Banati et al., 1994; Gehrmann et al., 1995; Hu et al., 1995, 1996), tumor necrosis factor-α (TNF-α), quinolinic acid, as well as other factors. The signals that trigger these negative responses have not been well-delineated in any of the neurodegenerative diseases including ALS, although in Alzheimer's disease (AD), the β-amyloid in senile plaques has been demonstrated to activate microglia and trigger neuronal injury (Banati et al., 1993). Activated microglia have been well-delineated to participate in the pathogen-esis of demyelination in multiple sclerosis (MS) (Sriram and Rodriguez, 1997); and the peripheral counterpart, the macrophage, has been documented to contribute to the pathogenesis of demyelination in inflammatory neuropathies (Hall et al., 1992; Hartung et al., 1995).

The fact that an immune/inflammatory reaction is present in ALS spinal cord and that it can occur in the setting of distal (peripheral) trauma provides possible relevance for immune/inflammatory reactions in ALS. Peripheral trauma is a well-known risk factor in ALS (Schubert and Schwan, 1995) and could initiate a central response. However, because trauma is a common occurrence and ALS is relatively uncommon (Annegers et al., 1991; Haverkamp et al., 1995), trauma alone might not be sufficient to produce a sustained response without implicating prior genetic or environmental sensitizing factors.

ASSOCIATED AUTOIMMUNE DISEASES, PARAPROTEINEMIAS, AND LYMPHOMAS

Circumstantial evidence for immune mechanisms has included a higher incidence of immune disorders in patients with ALS, the presence of paraproteinemias and lymphomas, the presence of lymphocytes and activated macrophages in ALS spinal cord, and the presence of immunoglobulin G (IgG) within ALS motor neurons. In our series of more than 1,200 patients, 21% had thyroid disease, as determined either by history or by laboratory investigations (Haverkamp et al., 1995). Shy et al. (1986) reported that 5.6% of 202 patients with motor-neuron disease had paraproteinemia, compared to only 1% of control subjects. Subsequently, paraproteins were demonstrated in 9% of patients using immunofixation electrophoresis compared to 3% of patients using cellulose acetate electrophoresis (Younger et al., 1990). Employing more sensitive techniques, Duarte et al. (1991) demonstrated that 60% of ALS patients had serum monoclonal immunoglobulins, whereas only 13% of control subjects had such abnormalities. In 9 patients, lymphoma was associated with motor-neuron disease (Younger et al., 1991), and a similar association was present in 25 additional cases from the literature. Although the presence of lymphoma is most commonly associated with lower motor-neuron disease, 6 of the first 25 published cases of lymphoma and motor-neuron disease had both upper and lower motor-neuron signs. In none of the cases could direct infiltration of the meninges, nerve roots, or nerves with lymphomatous cells be demonstrated. The presence of paraproteinemias raises the possibility that the paraproteins could directly cause motor-neuron injury (Hays et al., 1990). However, no passive-transfer experiments document that such paraproteins directly lead to motor neuron injury. A more plausible explanation is that paraproteins as well as lymphoma reflect alterations in the immune system, which give rise to motor-neuron injury by diverse humoral and/or cell-mediated mechanisms.

IGG IN ALS MOTOR NEURONS

The demonstration of IgG in ALS motor neurons provides additional circumstantial evidence for immune mechanisms. In 13 of 15 consecutive ALS autopsies, IgG was present in spinal motor neurons in a patchy and coarse granular cytoplasmic localization (Engelhardt and Appel, 1990). In the motor cortices of 11 ALS patients, IgG was demonstrated in pyramidal motor neurons in 6 patients. Minimal IgG reactivity was found in spinal motor neurons and pyramidal motor neurons in 10 disease-control patients. Although limited uptake of IgG can be demonstrated in motor neurons following intraperitoneal injection (Fabian, 1988), the intraperitoneal injection of ALS IgG results in much greater uptake of IgG in motoneurons than following injection of disease-control IgG. In addition, following ALS IgG injection, IgG is found in proximity to microtubules, endoplasmic reticulum, and Golgi apparatus, whereas disease-control IgG has a lysosomal localization (Engelhardt et al., 1990). However, both ALS IgG and disease-control IgG can be detected in motoneuron lysosomes. In human spinal cords, IgG can be detected in association with microtubules in ALS specimens, but not in disease-control specimens. Despite this localization, the significance of IgG reactivity within motor neurons is unclear, and there is no specific correlation of IgG reactivity in ALS specimens with the rate of progression or stage of the disease, nor any indication that the IgG within the motor neuron contributes to cell degeneration. As noted previously, IgG accumulates in spinal-cord ventral horn following peripheral trauma to motor axons and the association of IgG with motor neurons in ALS could be a reflection of motor-axon injury. However, IgG could possibly contribute to motor-neuron injury.

Clearly, any or all of these the immune/inflammatory changes (microglial activation, the presence of IgG, T-cell infiltration) could be functioning to repair rather than injure motor neurons. Thus, cellular infiltrates within ALS spinal-cord ventral horns could be secondary to motor neuron destruction rather than its cause. Furthermore, the increased incidence of paraproteinemias and lymphomas could be due to a common factor that could also independently compromise motor neurons.

ANIMAL MODELS OF IMMUNE-MEDIATED MOTOR NEURON DISEASE: EAMND AND EAGMD

Two distinct models of immune-mediated motor-neuron destruction provide evidence that immune/inflammatory reactions do have the capacity to mediate motor-neuron injury in ALS. Experimental autoimmune motor-neuron disease (EAMND) is a lower motor-neuron syndrome induced in guinea pigs by the inoculation of purified bovine motor neurons in Freund's adjuvant (Engelhardt et al, 1989). In EAMND, the affected animals demonstrated gradual onset of hind-limb weakness associated with electromyographic and morphologic evidence of denervation and spinal motor-neuron injury and cell death. Experimental autoimmune gray-matter disease (EAGMD) is a more acute disorder involving lower and upper motor neurons induced in guinea pigs by the inoculation of spinal-cord ventral horn homogenates (Engelhardt et al., 1990). EAGMD is clinically characterized by the relatively rapid onset of extremity weakness as well as bulbar signs. The latter presented in approx 25–30% of cases. There is evidence of denervation by electromyogram and morphologic criteria. Scattered foci of perivascular inflammation are present within the CNS, as well as injury and loss of spinal-cord motor neurons in large pyramidal cells in the motor cortex. In both disorders, early in the disease there is an increased resting frequency of spontaneous miniature end-plate potentials (MEPP), while the muscle membrane resting potential and MEPP amplitude and time course are normal (Garcia et al., 1990). Such data suggest enhanced release of acetylcholine from motor-nerve terminals, possibly due to increased intracellular calcium. The immune-mediated animal models of motor-neuron destruction, especially EAGMD, resemble human ALS, with respect to the loss of upper and lower motor neurons, the presence of inflammatory foci, including activated microglia within the spinal cord, the presence of IgG within upper and lower motor neurons, and the presence of physiologic changes at the neuromuscular junction (Maselli et al., 1993).

The animal models clearly suggest the potential relevance of immune/inflammatory mechanisms in the pathogenesis of ALS, but in no sense suggest that the specific antigens involved in immune-mediated models of motor-neuron injury are necessarily involved in human ALS.

SELECTIVE VULNERABILITY OF MOTOR NEURONS

Our understanding of the factors dictating selective vulnerability of motor neurons is incomplete. It is unlikely to be explained by the interaction of ALS IgG with unique calcium channels in motor neurons, because such channels are present in the plasma membrane of most neurons, and antibodies to VGCC should therefore affect most neurons. Purkinje cells, for example, have an abundance of P-type calcium channels and ALS IgG can be documented to enhance currents through P-type calcium channels (Llinas et al., 1993). However, Purkinje cells are not compromised in ALS. Furthermore, in our passive transfer experiments, increased calcium could not be detected within Purkinje cells, suggesting that regulation of calcium homeostasis may be more active or more finely adjusted in Purkinje cells than in motor neurons. Thus, a susceptible type of calcium channel, per se, cannot explain the pattern of selective vulnerability, and other factors may also influence the ability of neurons to cope with an increased calcium load.

In addition, despite extensive investigations of SOD-1 mutations in familial ALS and excitotoxicity in sporadic ALS, neither mechanism provides a definitive explanation for motor neuron selective vulnerability in ALS. The SOD-1 mutations are present in all motor neurons, yet oculomotor neurons are spared in SOD-1 mutant transgenic mice (Siklos et al., 1998). Specific glutamate receptors have also been proposed as an explanation for selective vulnerability, but there is no evidence for different AMPA/kainate receptor subunits in motor neurons innervating the resistant eye motor neurons compared to susceptible spinal motor neurons. A paper by Morrison et al. (1998), demonstrated no changes in the distribution of the GluR2 subunits of the AMPA receptor in motor neurons that are affected in ALS compared to sensory neurons that are resistant. However there is a selective decrease in the levels and the editing of GluR2 in SALS spinal ventral gray matter, which could result in increased Ca^{2+} entry (Takuma et al., 1999). These effects could be secondary (*see* section on glutamate excitotoxicity), and there is still presently no cogent evidence to support specific glutamate receptor or transporter localization as major determinants of selective vulnerability.

CALCIUM-BINDING PROTEINS

Another potential approach to selective vulnerability involves the calcium-binding proteins calbindin-D_{28K} and parvalbumin, which can influence calcium homeostasis (Celio et al., 1988, 1990). Calcium-binding proteins are elevated in Purkinje cells, and physiologically absent in adult motor neurons. In fact, the lack of calbindin-D_{28K} and/or parvalbumin immunohistochemical reactivity in motor neurons parallels the known selective vulnerability of motor neurons in ALS. Motor neurons affected early in the disease (spinal motor neurons and cranial nerve XII) lack immunoreactivity for these calcium-binding proteins. Motor neurons controlling eye muscles (cranial nerves III, IV, VI), as well as Onuf's nucleus motor neurons controlling bladder muscles, are relatively spared and have high levels of calcium-binding proteins (Alexianu et al., 1994). In our motor neuron cell line, VSC4.1, only differentiated cells are injured by ALS IgG, and such differentiated cells lack calbindin-D_{28K} and parvalbumin. Undifferentiated VSC4.1 cells possess ample calbindin-D_{28K} and parvalbumin and are relatively resistant to cytotoxic effects of ALS IgG. Transfection of VSC4.1 cells with calbindin cDNA with a phosphoglycerate kinase promoter yielded cells that, when differentiated, still maintained high levels of calbindin. These differentiated transfected cells were resistant to ALS-IgG mediated cytotoxicity (Ho et al., 1996). Furthermore, enhanced expression of calbindin-D_{28K} in motoneurons cultured from mutant SOD-1 mice conferred resistance to CA^{2+}-dependent injury (Roy et al., 1998). In the SOD-1 mutant mouse, spinal motor neurons (which lack calbindin and parvalbumin) undergo degenerative changes with vacuoles filled with calcium, whereas oculomotor neurons, which possess ample parvalbumin, undergo no degenerative changes and have normal calcium homeostasis (Siklos et al., 1996).

The function of calbindin-D_{28K} and parvalbumin is still unclear, but these calcium-binding proteins appear to enhance calcium homeostasis by altering calcium entry through voltage-gated calcium channels (VGCCs), as well as by promoting calcium extrusion and compartmentalization (Lledo et al., 1992). Thus, regardless of whether the damage to motor neurons is mediated by oxidant stress (Bowling et al., 1993; Yim, et al., 1996), by immune mechanisms and/or excitotoxic mechanisms (Rothstein et al., 1990; Appel, 1993), the regulatory effects of calbindin-D_{28K} and parvalbumin may influence the selective resistance and vulnerability in motor neurons to cell death in ALS.

As evidence for the importance of calcium (and the lack of calcium-binding proteins, such as parvalbumin) in triggering spinal-motor neuron injury, we have developed a parvalbumin transgenic mouse. This mouse expresses parvalbumin in large motor neurons under control of a calmodulin II promoter. Passive transfer of sera from ALS patients, which increased calcium and enhanced spontaneous transmitter release from motoneuron terminals of control mice, produced no alterations in calcium or spontaneous release from motoneuron terminals of parvalbumin transgenic mice. Moreover, progeny of parvalbumin transgenic mice bred to mutant SOD-1 (G93A) transgenic mice, a model of familial ALS, had delayed disease onset and longer survival compared to mutant SOD-1 control mice. These results suggest a protective effect of parvalbumin overexpression in motor-neuron injury.

TRIGGERING OF INCREASED CALCIUM IN MOTOR NEURONS

Although motor neurons may be selectively vulnerable to an increased calcium load in ALS because of the physiologic absence of the calcium-binding proteins, calbindin-D_{28K} and parvalbumin, it is still not clear what triggers the increase in motor-neuron calcium. Our own data suggest a prominent role for immune mechanisms in initiating the increase in motor-neuron calcium and subsequent free-radical formation. Alterations in glutamate transport perhaps initiated by the immune mechanisms and mediated by 4-hydroxynonenal could aggravate the initial motor-neuron injury. However, even with repeated intraperitoneal injections of ALS IgG, significant motor-neuron cell death does not occur despite changes in motor-neuron calcium and early changes of motor-neuron injury. An additional factor, which is presently unknown, appears to be required, possibly mediated by the inflammatory cells known to be present in ALS spinal cord or mediated by a breakdown of the blood–brain barrier, which permits higher concentrations of IgG to enter the CNS.

An instructive model may well be experimental allergic encephalomyelitis produced in marmosets (Genain et al., 1995). In this model, inflammation and demyelination occur following immunization with myelin oligodendrocyte glycoprotein (MOG). If myelin basic protein (MBP) is employed, a central inflammatory process occurs but minimal demyelination is noted. If animals are immunized with MBP followed by the intravenous administration of antibodies to MOG, then marked demyelination occurs. The explanation for these events is that the antibodies, per se, cannot penetrate the blood–brain barrier in sufficient doses to contribute to demyelination unless the barrier has been broken by a prior inflammatory response. A similar process may be going on in immune-mediated motor-neuron injury. The immunoglobulins that lead to increased intracellular calcium and enhanced acetylcholine release by action at the motor-nerve terminal are clearly not sufficient to trigger motor-neuron cell death. However, the increasing changes in the motor neurons with chronic immune attack may lead to central inflammatory processes as noted following axotomy of cranial nerves (Kristensson et al., 1994). Subsequently, downregulation of the glial glutamate transporter in addition to the presence of an inflammatory response of T cells as well as microglia may enhance the oxidant stress and free radicals; and increase access of immune/inflammatory components to the CNS. Certainly, this hypothesis merits testing, and could lead to new therapeutic approaches.

ACKNOWLEDGMENT

Supported by grants from the Muscular Dystrophy Association (MDA).

REFERENCES

Aguirre, T., Van Den Bosch, L., Goetschalckx, K., Tilkin, P., Mathijs, G., Cassiman, J. J., and Robberecht, W. (1998) Increased sensitivity of fibroblasts from amyotrophic lateral sclerosis patients to oxidative stress. *Ann. Neurol.* **43,** 452–457.

Al-Chalabi, A., Andersen, P. M., Nilsson, P., Chioza, B., Andersson, J. L., Russ, C., et al. (1999) Deletions of the heavy neurofilament subunit tail in amyotrophic lateral sclerosis. *Hum. Mol. Genet.* **8,** 157–164.

Alexianu, M. E., Ho, B-K., Mohamed, A. H., La Bella, V., Smith, R. G., and Appel, S. H. (1994) The role of calcium-binding proteins in selective motor neuron vulnerability in amyotrophic lateral sclerosis. *Ann. Neurol.* **36,** 846–858.

Andrus, P. K., Fleck, T. J., Gurney, M. E., and Hall, E. D. (1998) Protein oxidative damage in a transgenic mouse model of familial amyotrophic lateral sclerosis. *J. Neurochem.* **71,** 2041–2048.

Annegers, J. F., Appel, S., Lee, J. R., and Perkins, P. (1991) Incidence and prevalence of amyotrophic lateral sclerosis in Harris Country, Texas, 1985–1988. *Arch. Neurol.* **48,** 589–593.

Appel, S. H. (1993) Excitotoxic neuronal cell death in amyotrophic lateral sclerosis. *Trends Neurosci.* **16,** 3–4.

Banati, R. B., Gehrmann, J., Czech, C., Monning, U., Jones, L. L., Konig, G., et al. (1993) Early and rapid de novo synthesis of Alzheimer beta A4-amyloid precursor protein (APP) in activated microglia. *Glia* **9,** 199–210.

Banati, R. B., Schubert, P., Rothe, G., Gehrmann, J., Rudolphi, K., Valet, G., and Kreutzberg, G. W. (1994) Modulation of intracellular formation of reactive oxygen intermediates in peritoneal macrophages and microglia/brain macrophages by propentofylline. *J. Cereb. Blood Flow Metab.* **14,** 145–149.

Beal, M. F., Ferrante, R. J., Browne, S. E., Matthews, R. T., Kowall, N. W., and Brown, R. H., Jr. (1997) Increased 3-nitrotyrosine in both sporadic and familial amyotrophic lateral sclerosis. *Ann. Neurol.* **42,** 644–654.

Beckman, J. S., Carson, M., Smith, C. D., and Koppenol, W. H. (1993) ALS, SOD and peroxynitrite. *Nature* **364,** 584.

Berger, M. M., Kopp, N., Vital, C., Redl, B., Aymard, M., and Lina, B. (2000) Detection and cellular localization of enterovirus RNA sequences in spinal cord of patients with ALS. *Neurology* **54,** 20–25.

Blanc, E. M., Keller, J. N., Fernandez, S., and Mattson, M. P. (1998) 4-Hydroxynonenal, a lipid peroxidation product, impairs glutamate transport in cortical astrocytes. *Glia* **22,** 149–160.

Bogdanov, M. B., Ramos, L. E., Xu, Z., and Beal, M. F. (1998) Elevated "hydroxyl radical" generation in vivo in an animal model of amyotrophic lateral sclerosis. *J. Neurochem.* **71,** 1321–1324.

Bolanos, J. P., Heales, S. J., Land, J. M., and Clark, J. B. (1995) Effect of peroxynitrite on the mitochondrial respiratory chain: differential susceptibility of neurons and astrocytes in primary culture. *J. Neurochem.* **64,** 1965–1972.

Borchelt, D. R., Lee, M. K., Slunt, H. S., Guarnieri, M., Xu, Z. S., Wong, P. C., et al. (1994) Superoxide dismutase 1 with mutations linked to familial amyotrophic lateral sclerosis possesses significant activity. *Proc. Natl. Acad. Sci. USA* **91,** 8292–8296.

Borthwick, G. M., Johnson, M. A., Ince, P. G., Shaw, P. J., and Turnbull, D. M. (1999) Mitochondrial enzyme activity in amyotrophic lateral sclerosis: implications for the role of mitochondria in neuronal cell death. *Ann. Neurol.* **46,** 787–790.

Bowling, A. C., Schulz, J. B., Brown, R. H. Jr., and Beal, M. F. (1993) Superoxide dismutase activity, oxidative damage, and mitochondrial energy metabolism in familial and sporadic amyotrophic lateral sclerosis. *J. Neurochem.* **61,** 2322–2325.

Breuer, A. C., Lynn, M. P., Atkinson, M. B., Chou, S. M., Wilbourn, A. J., Marks, K. E., et al. (1987) Fast axonal transport in amyotrophic lateral sclerosis: an intra-axonal organelle traffic analysis. *Neurology* **37,** 738–48.

Bristol, L. A. and Rothstein, J. D. (1996) Glutamate transporter gene expression in amyotrophic lateral sclerosis motor cortex. *Ann. Neurol.* **39,** 676–679.

Brown, R. H., Jr. (1998) SOD1 aggregates in ALS: cause, correlate, or consequence? *Nat. Med.* **4,** 1362–1364.

Bruijn, L. I., Beal, M. F., Becher, M. W., Schulz, J. B., Wong, P. C., Price, D. L., and Cleveland, D. W. (1997) Elevated free nitrotyrosine levels, but not protein-bound nitrotyrosine or hydroxyl radicals, throughout amyotrophic lateral sclerosis (ALS)-like disease implicate tyrosine nitration as an aberrant in vivo property of one familial ALS-linked superoxide dismutase 1 mutant. *Proc. Natl. Acad. Sci. USA* **94,** 7606–7611.

Carri, M. T., Battistoni, A., Polizio, F., Desideri, A., and Rotilio, G. (1994) Impaired copper binding by the H46R mutant of human Cu,Zn superoxide dismutase, involved in amyotrophic lateral sclerosis. *FEBS Lett.* **356,** 314–316.

Carri, M. T., Ferri, A., Battistoni, A., Famhy, L., Gabbianelli, R., Poccia, F., and Rotilio, G. (1997) Expression of a Cu,Zn superoxide dismutase typical of familial amyotrophic lateral sclerosis induces mitochondrial alteration and increase of cytosolic Ca2+ concentration in transfected neuroblastoma SH-SY5Y cells. *FEBS Lett.* **414,** 365–368.

Carriedo, S. G., Yin, H. Z., and Weiss, J. H. (1996) Motor neurons are selectively vulnerable to AMPA/kainate receptor-mediated injury in vitro. *J. Neurosci.* **16,** 4069–4079.

Carriedo, S. G., Yin, H. Z., Sensi, S., and Weiss, J. H. (1998) Rapid Ca2+ entry through Ca2+-permeable AMPA/Kainate channels triggers marked intracellular Ca2+ rises and consequent oxygen radical production. *J. Neurosci.* **18,** 7727–7738.

Celio, M. R., Baier, W., Scharer, L., de Viragh, P. A., and Gerday, C. (1988) Monoclonal antibodies directed against the calcium-binding protein parvalbumin. *Cell Calcium* **9,** 81–86.

Celio, M. R., Baier, W., Scharer, L., Gregersen, H. J., de Viragh, P. A., and Norman, A. W. (1990) Monoclonal antibodies directed against the calcium-binding protein calbindin-D$_{28K}$. *Cell Calcium* **11,** 599–602.

Choi, D. W. (1988) Glutamate neurotoxicity and diseases of the nervous system. *Neuron* **8,** 623–634.

Cleveland, D. W. (1999) From Charcot to SOD1: mechanisms of selective motor neuron death in ALS. *Neuron* **24,** 515–520.

Colom, L. V., Alexianu, M. E., Mosier, D. R., Smith, R. G., and Appel, S. H. (1997) Amyotrophic lateral sclerosis immunoglobulins increase intracellular calcium in a motoneuron cell line. *Exp. Neurol.* **146,** 354–360.

Couillard-Despres, S., Ahu, Q., Wong, P. C., Price, D. L., Cleveland, D. W., and Julien, J. P. (1998) Protective effect of neurofilament heavy gene overexpression in motor neuron disease induced by mutant superoxide dismutase. *Proc. Natl. Acad. Sci. USA* **95,** 9626–9630.

Coyle, J. T. and Putterfarcken, P. (1993) Oxidative stress, glutamate and neurodegenerative disorders. *Science* **262,** 689–696.

Culotta, V. C., Klomp, L. W., Strain, J., Casareno, R. L., Krems, B., and Gitlin, J. D. (1997) The copper chaperone for superoxide dismutase. *J. Biol. Chem.* **272,** 23,469–23,472.

Dal Canto, M. C. and Gurney, M. E. (1994) Development of central nervous system pathology in a murine transgenic model of human amyotrophic lateral sclerosis. *Am. J. Pathol.* **145,** 1271–1279.

Delisle, M. B. and Carpenter, S. (1984) Neurofibrillary axonal swellings and amyotrophic lateral sclerosis. *J. Neurol. Sci.* **63,** 241–250.

Duarte, F., Binet, S., Lacomblex, L., Bouche, P., Breudhome, J. L., and Meininger, V. (1991) Quantitative analysis of monoclonal immunoglobulins in serum of patients with amyotrophic lateral sclerosis. *J. Neurol. Sci.* **104,** 88–91.

Dugan, L. L., Sensi, S., Canzoniero, L. M. T., Handran, S. D., Rothman, S. M., Lin, T. S., et al. (1995) Mitochondrial production of reactive oxygen species in cortical neurons following exposure to N-methyl-D-aspartate. *J. Neurosci.* **15,** 6377–6388.

Dykens, J. A. (1994) Isolated cerebral and cerebellar mitochondria produce free radicals when exposed to elevated Ca2+ and Na+: implications for neurodegeneration. *J. Neurochem.* **63,** 584–591.

Ekegren, T., Grundstrom, E., Lindholm, D., and Aquilonius, S. M. (1999) Upregulation of Bax protein and increased DNA degradation in ALS spinal cord motor neurons. *Acta. Neurol. Scand.* **100,** 317–321.

Engelhardt, J. I. and Appel, S. H. (1990) IgG reactivity in the spinal cord and motor cortex in amyotrophic lateral sclerosis. *Arch. Neurol.* **47,** 1210–1216.

Engelhardt, J. I., Appel, S. H., and Killian, J. M. (1989) Experimental autoimmune motor neuron disease. *Ann. Neurol.* **26,** 368–376.

Engelhardt, J. I., Appel, S. H., and Killian, J. M. (1990) Motor neuron destruction in guinea pigs immunized with bovine spinal cord ventral horn homogenate: experimental autoimmune gray matter disease. *J. Neuroimmunol.* **27,** 21–31.

Engelhardt, J. I., Siklos, L., and Appel, S. H. (1997) Altered calcium homeostasis and ultrastructure in motoneurons of mice caused by passively transferred anti-motoneuronal IgG. *J. Neuropath. Exp. Neurol.* **56,** 21–39.

Engelhardt, J. I., Siklos, L., Komuves, L., Smith, R. G., and Appel, S. H. (1995) Antibodies to calcium channels from ALS patients passively transferred to mice selectively increase intracellular calcium and induce ultrastructural changes in motoneurons. *Synapse* **20,** 185–199.

Engelhardt, J. I., Tajti, J., and Appel, S. H. (1993) Lymphocytic infiltrates in the spinal cord in amyotrophic lateral sclerosis. *Arch. Neurol.* **50,** 30–36.

Estevez, A. G., Crow, J. P., Sampson, J. B., Reiter, C., Zhuang, Y., Richardson, G. J., et al. (1999) Induction of nitric oxide-dependent apoptosis in motor neurons by zinc-deficient superoxide dismutase. *Science* **286,** 2498–2500.

Fabian, R. (1988) Uptake of plasma IgG by CNS motor neurons: comparison of anti-neuronal and normal IgG. *Neurology* **38,** 1775–1780.

Fedoroff, S. (1995) Development of microglia, in *Neuroglia* (Kettenmann, H. and Ransom, B. R., eds.), Oxford University Press, New York, pp. 162–181.

Ferrante, R. J., Browne, S. E., Shinobu, L. A., Bowling, A. C., Baik, M. J., MacGarvey, U., et al. (1997a) Evidence of increased oxidative damage in both sporadic and familial amyotrophic lateral sclerosis. *J. Neurochem.* **69,** 2064–2074.

Ferrante, R. J., Shinobu, L. A., Schulz, J. B., Matthews, R. T., Thomas, C. E., Kowall, N. W., et al. (1997b) Increased 3-nitrotyrosine and oxidative damage in mice with a human copper/zinc superoxide dismutase mutation. *Ann. Neurol.* **42,** 326–334.

Figlewicz, D. A., Krizus, A., Martinoli, M. G., Meininger, V., Dib, M., Rouleau, G. A., and Julien, J. P. (1994) Variants of the heavy neurofilament subunit are associated with the development of amyotrophic lateral sclerosis. *Hum. Mol. Genet.* **3,** 1757–1761.

Fujita, K., Yamauchi, M., Shibayama, K., Ando, M., Honda, M., and Nagata, Y. (1996) Decreased cytochrome c oxidase activity but unchanged superoxide dismutase and glutathione peroxidase activities in the spinal cords of patients with amyotrophic lateral sclerosis. *J. Neurosci. Res.* **45,** 276–281.

Garcia, J., Engelhardt, J. I., Appel, S. H., and Stefani, E. (1990) Increased mepp frequency as an early sign of experimental immune-mediated motor neuron disease. *Ann. Neurol.* **28,** 329–334.

Gehrmann, J., Banati, R., Wiessner, C., Hossmann, K. A., and Kreutzberg, G. W. (1995) Reactive microglia in cerebral ischemia: an early mediator of tissue damage? *Neuropathol. Appl. Neurobiol.* **21,** 277–289.

Genain, C. P., Nguyen, M. H., Letvin, N. L., Pearl, R., Davis, R. L., Adelman, M., et al. (1995) Antibody facilitation of multiple sclerosis-like lesions in a nonhuman primate. *J. Clin. Invest.* **96,** 2966–2974.

Guo, Z., Kindy, M. S., Kruman, I., and Mattson, M. P. (2000) ALS-linked Cu/Zn-SOD mutation impairs cerebral synaptic glucose and glutamate transport and exacerbates ischemic brain injury. *J. Cereb. Blood Flow Metab.* **20,** 463–468.

Gurney, M. E., Cutting, F. B., Zhai, P., Doble, A., and Taylor, C. P., Andrus, P. K., and Hall, E. D. (1996) Benefit of vitamin E, riluzole, and gabapentin in a transgenic model of familial amyotrophic lateral sclerosis. *Ann. Neurol.* **39,** 147–157.

Hall, S. M., Hughes, R. A., Atkinson, P. F., McColl, I., and Gale, A. (1992) Motor nerve biopsy in severe Guillain-Barre syndrome. *Ann. Neurol.* **31,** 441–444.

Hartung, H. P., Pollard, J. D., Harvey, G. K., and Toyka, K. V. (1995) Immunopathogenesis and treatment of the Guillain-Barre syndrome. *Muscle Nerve* **18,** 137–153.

Haverkamp, L. J., Appel, V., and Appel, S. H. (1995) Natural history of amyotrophic lateral sclerosis in a database population: validation of a scoring system and a model for survival prediction. *Brain* **118,** 707–719.

Hays, A. P., Roxas, A., Sadiq, S. A., Vallejos, H., D'Agati, V., Thomas, F. P., et al. (1990) A monoclonal IgA in a patient with amyotrophic lateral sclerosis reacts with neurofilaments and surface antigen on neuroblastoma cells. *J. Neuropathol. Exp. Neurol.* **49,** 383–398.

Heizmann, C. W. and Braun, K. (1992) Changes in Ca(2+)-binding proteins in human neurodegenerative disorders. *Trends Neurosci.* **15,** 259–264.

Hirano, A., Donnenfeld, H., Sasaki, S., and Nakano, I. (1984) Fine structural observations of neurofilamentous changes in amyotrophic lateral sclerosis. *J. Neuropathol Exp. Neurol.* **43**, 461–470.

Ho, B-K., Alexianu, M. E., Colom, L. V., and Mohamed, A. H., Serrano, F., and Appel, S. H. (1996) Expression of calbindin-D_{28K} in motoneuron-hybrid cells following retroviral infection with calbindin-D_{28K} cDNA prevents amyotrophic lateral sclerosis IgG-mediated cytotoxicity. *Proc. Natl. Acad. Sci. USA* **93**, 6796–6801.

Hu, S., Chao, C. C., Khanna, K. V., Gekkeer, G., Peterson, P. K., and Molitor, T. W. (1996) Cytokine and free radical production by porcine microglia. *Clin. Immunol. Immunopathol.* **78**, 93–96.

Hu, S., Sheng, W. S., Peterson, P. K., and Chao, C. (1995) Cytokine modulation of murine microglial cell superoxide production. *Glia* **13**, 45–50.

Ince, P. G., Shaw, P. J., Candy, J. M., Mantle, D., Tandon, L., Ehmann, W. D., and Markesbery, W. R. (1994) Iron, selenium and glutathione peroxidase activity are elevated in sporadic motor neuron disease. *Neurosci. Lett.* **182**, 87–90.

Juhela, S., Hyoty, H., Hinkkanen, A., Elliott, J., Roivainen, M., Kulmala, P., et al. (1999) T-cell responses to enterovirus antigens and to beta cell autoantigens in unaffected children positive for IDDM-associated autoantibodies. *J. Autoimmun.* **12**, 269–278.

Julien, J. P., Cote, F., and Collard, J. F. (1995) Mice overexpressing the human neurofilament heavy gene as a model of ALS. *Neurobiol. Aging* **16**, 487–490.

Kakkar, P., Mehrota, S., and Viaswanathan, P. N. (1992) Interrelation of active oxygen species, membrane damage and altered calcium functions. *Mol. Cell Biochem.* **111**, 11–15.

Kawamata, T., Akiyama, H., Yamada, T., and McGeer, P. L. (1992) Immunologic reactions in amyotrophic lateral sclerosis, brain and spinal cord. *Am. J. Pathol.* **140**, 691–707.

Keller, J. N. and Mattson, M. P. (1998) Roles of lipid peroxidation in modulation of cellular signaling pathways, cell dysfunction, and death in the nervous system. *Rev. Neurosci.* **9**, 105–116.

Knip, M. and Akerblom, H. K. (1999) Environmental factors in the pathogenesis of type 1 diabetes mellitus. *Exp. Clin. Endocrinol. Diabetes* **107(Suppl. 3),** 93–100.

Krakauer, T. (1999) Immune response to staphylococcal superantigens. *Immunol. Res.* **20**, 163–173.

Kreutzberg, G. W. (1996) Microglia: a sensor for pathological events in the CNS. *Trends Neurosci.* **19**, 312–318.

Kristensson, K., Aldskogius, M., Peng, Z. C., Olsson, T., Aldskogius, H., and Bentivoglio, M. (1994) Co-induction of neuronal interferon-gamma and nitric oxide synthase in rat motor neurons after axotomy: A role in nerve repair or death? *J. Neurocytol.* **23**, 453–459.

Kruman, II, Pedersen, W. A., Springer, J. E., and Mattson, M. P. (1999) ALS-linked Cu/Zn-SOD mutation increases vulnerability of motor neurons to excitotoxicity by a mechanism involving increased oxidative stress and perturbed calcium homeostasis. *Exp. Neurol.* **160**, 28–39.

Kurtzke, J. F. and Kurland, L. T. (1983) The epidemiology of neurologic disease, in *Clinical Neurology, vol. 4* (Joint, R. J., ed.), J. B. Lippincott, Philadelphia, pp. 1–43.

Lacomblez, L., Bensimon, G., Leigh, P. N., Guillet, P., and Meininger, V. (1996) Dose-ranging study of riluzole in amyotrophic lateral sclerosis. Amyotrophic Lateral Sclerosis/Riluzole Study Group II. *Lancet* **347**, 1425–1431.

Lampson, L. A., Kushner, P. D., and Sobel, R. A. (1988) Strong expression of class II major histocompatibility complex (MHC) antigens in the absence of detectable T-cell infiltration in amyotrophic lateral sclerosis. *J. Neuropathol. Exp. Neurol.* **47**, 353.

Lampson, L. A., Kushner, P. D., and Sobel, R. A. (1990) Major histocompatibility complex antigen expression in the affected tissues in amyotrophic lateral sclerosis. *Ann. Neurol.* **28**, 365–372.

Lee, M. K., Marszalek, J. R., and Cleveland, D. W. (1994) A mutant neurofilament subunit causes massive, selective motor neuron death: implications for the pathogenesis of human motor neuron disease. *Neuron* **13,** 975–988.

Lin, C. L., Bristol, L. A., Jin, L., Dykes-Hoberg, M., Crawford, T., Clawson, L., and Rothstein, J. D. (1998) Aberrant RNA processing in a neurodegenerative disease: the cause for absent EAAT2, a glutamate transporter, in amyotrophic lateral sclerosis. *Neuron* **20,** 589–602.

Lipton, S. A., Choi, Y. B., Pan, Z. H., Lei, S. Z., Chen, H. S., Sucher, N. J., et al. (1993) A redox-based mechanism for the neuroprotective and neurodestructive effects of nitric oxide and related nitroso-compounds. *Nature* **364,** 626–632.

Liu, D., Wen, J., Liu, J., and Li, L. (1999a) The roles of free radicals in amyotrophic lateral sclerosis: reactive oxygen species and elevated oxidation of protein, DNA, and membrane phospholipids. *FASEB J.* **13,** 2318–2328.

Liu, R., Althaus, J. S., Elelrbrock, B. R., Becker, D. A., and Gurney, M. E. (1998) Enhanced oxygen radical production in a transgenic mouse model of familial amyotrophic lateral sclerosis. *Ann. Neurol.* **44,** 763–770.

Liu, R., Narla, R. K., Kurinov, I., Li, B., and Uckun, F. M. (1999b) Increased hydroxl radical production and apoptosis in PC12 neuron cells expressing the gain-of-function mutant G93A SOD1 gene. *Radiat. Res.* **151,** 133–141.

Lledo, P-M., Somasundaram, B., Morton, A. J., Emson, P. C., and Mason, W. T. (1992) Stable transfection of calbindin-D_{28K} into GH_3 cell line alters calcium currents and intracellular calcium in homeostasis. *Neuron* **9,** 943–954.

Llinas, R., Sugimori, M., Cherksey, B. D., Smith, R. G., Delbono, O., Stefani, E., and Appel, S. H. (1993) IgG from amyotrophic lateral sclerosis patients increases current through P-type calcium channels in mammalian cerebellar Purkinje cells and in isolated channel protein in lipid bilayers. *Proc. Natl. Acad. Sci. USA* **90,** 11,743–11,747.

Martin, L. J. (1999) Neuronal death in amyotrophic lateral sclerosis is apoptosis: possible contribution of a programmed cell death mechanism. *J. Neuropathol. Exp. Neurol.* **58,** 459–471.

Maselli, R., Wollman, R., Leung, C., Distad, B., Palombi, S., Richman, D. P., et al. (1993) Neuromuscular transmission in amyotrophic lateral sclerosis. *Muscle Nerve* **16,** 1193–1203.

Matsumoto, Y., Ohmori, K., and Fujiwara, M. (1992) Immune regulation by brain cells in the central nervous system: microglia but not astrocytes present myelin basic protein to encephalitogenic T cells under in vivo-mimicking conditions. *Immunology* **76,** 209–216.

McGeer, P. L., Itagaki, S., and McGeer, E. G. (1988) Expression of histocompatibility glyco-protein HLA-DR in neurologic disease. *Acta Neuropathol. (Berl.)* **76,** 550–557.

Merrill, J. E. and Benveniste, E. N. (1996) Cytokines in inflammatory brain lesions: helpful and harmful. *TINS* **19,** 331–338.

Migheli, A., Piva, R., Atzori, C., Troost, D, and Schiffer, D. (1997) C-Jun, JNK/SAPK kinases and transcription factor NF-kappa B are selectively activated in astrocytes, but not motor neurons, in amyotrophic lateral sclerosis. *J. Neuropathol. Exp. Neurol.* **56,** 1314–1322.

Mithal, N. P., Radunovic, A., Figlewicz, D. A., McMillan, T. J., and Leigh, P. N. (1999) Cells from individuals with SOD-1 associated familial amyotrophic lateral sclerosis do not have an increased susceptibility to radiation-induced free radical production or DNA damage. *J. Neurol. Sci.* **164,** 89–92.

Morrison, B. M., Janssen, W. G. M., Gordon, J. W., and Morrison, J. H. (1998) Light and electron microscopic distribution of the AMPA receptor subunit, Glu R2, in the spinal cord of control and G68R mutant superoxide dismutase transgenic mice. *J. Comp. Neurol.* **395,** 523–534.

Mosier, D. R., Baldelli, P., Delbono, O., Smith, R. G., Alexianu, M. E., Appel, S. H., and Stefani, E. (1995) Amyotrophic lateral sclerosis immunoglobulins increase Ca2+currents in a motoneuron cell line. *Ann. Neurol.* **37,** 102–109.

Mu, X., He, J., Anderson, D. W., Trojanowski, J. Q., and Springer, J. E. (1996) Altered expression of bcl-2 and bax mRNA in amyotrophic lateral sclerosis spinal cord motor neurons. *Ann. Neurol.* **40,** 379–386.

Nagai, M., Abe, K., Okamoto, K., and Itoyama, Y. (1998) Identification of alternative splicing forms of GLT-1 mRNA in the spinal cord of amyotrophic lateral sclerosis patients. *Neurosci. Lett.* **244,** 165–168.

Nakano, Y., Hirayama, K., and Terao, K. (1987) Hepatic ultrastructural changes and liver dysfunction in amyotrophic lateral sclerosis. *Arch. Neurol.* **44,** 103–106.

Orrenius, S., McConkey, D. J., Bellomo, G., and Nicotera, P. (1989) Role of Ca2+ in toxic cell killing. *Trends Pharmacol. Sci.* **10,** 281–285.

Panzara, M. A., Gussoni, E., Begovich, A. B., Murray, R. S., Zang, Y. Q., Appel, S. H., et al. (1999) T-cell receptor BV gene rearrangements in the spinal cords and cerebrospinal fluids of patients with amyotrophic lateral sclerosis. *Neurobiol. Dis.* **6,** 392–405.

Pauchinger, M., Bowles, N. E., Fuentes-Garcia, F. J., Pham, V., Kuhl, U., Schwimmbeck, P. L., et al. (1999) Detection of adenoviral genome in the myocardium of adult patients with idiopathic left ventricular dysfunction. *Circulation* **99,** 1348–1354.

Pedersen, W. A., Chan, S. L., and Mattson, M. P. (2000a) A mechanism for the neuroprotective effect of apolipoprotein E: isoform-specific modification by the lipid peroxidation product 4-hydroxynonenal. *J. Neurochem.* **74,** 1426–1433.

Pedersen, W. A., Fu, W., Keller, J. N., Markesbery, W. R., Appel, S., Smith, R. G., et al. (1998) Protein modification by the lipid peroxidation product 4-hydroxynonenal in the spinal cords of amyotrophic lateral sclerosis patients. *Ann. Neurol.* **44,** 819–824.

Pedersen, W. A., Luo, H., Fu, W., Guo, Q., Sells, S. F., Rangnekar, V., and Mattson, M. P. (2000b) Evidence that Par-4 participates in motor neuron death in amyotrophic lateral sclerosis. *FASEB J.* **14,** 913–924.

Przedborski, S., Donaldson, D., Jakowec, M., Kish, S. J., Guttman. M., Rosoklija, G., and Hays, A. P. (1996) Brain superoxide dismutase, catalase, and glutathione peroxidase activities in amyotrophic lateral sclerosis. *Ann. Neurol.* **39,** 158–165.

Raivich, G., Jones, L. L., Kloss, C. U. A., Werner, A., Neumann, H., and Kreutzberg, G. W. (1998) Immune surveillance in the injured nervous system: T-lymphocytes invade the axotomized mouse facial motor nucleus and aggregate around sites of neuronal degeneration. *J. Neurosci.* **18,** 5804–5816.

Reaume, A. G., Elliott, J. L., Hoffman, E. K., Kowall, N. W., Ferrante, R. J., Siwek, D. F., et al. (1996) Motor neurons in Cu/Zn superoxide dismutase: different mice develop normally but exhibit enhanced cell death after axonal injury. *Nat. Genet.* **13,** 43–47.

Rosen, D. R., Siddique, T., Patterson, D., Figlewicz, D. A., Sapp, P., Hentati, A., et al. (1993) Mutations in Cu/Zn superoxide dismutase genes are associated with familial amyotrophic lateral sclerosis. *Nature* **362,** 59–62.

Rothstein, J. D., Dykes-Hoberg, M., Corson, L. B., Becker, M., Cleveland, D. W., Price, D. L., et al. (1999) The copper chaperone CCS is abundant in neurons and astrocytes in human and rodent brain. *J. Neurochem.* **72,** 422–429.

Rothstein, J. D., Dykes-Hoberg, M., Pardo, C. A., Bristol, L. A., Jin, L., Kuncl, R. W., et al. (1996) Knockout of glutamate transporters reveals a major role for astroglial transport in excitotoxicity and clearance of glutamate. *Neuron* **16,** 675–686.

Rothstein, J. D., Jin, L., Dykes-Hoberg, M., and Kuncl, R. W. (1993) Chronic inhibition of glutamate uptake produces a model of slow neurotoxicity. *Proc. Natl. Acad. Sci. USA* **90,** 6591–6595.

Rothstein, J. D., Martin, L. J., and Kuncl, R. W. (1992) Decreased glutamate transport by the brain and spinal cord in amyotrophic lateral sclerosis. *N. Engl. J. Med.* **326,** 1464–8.

Rothstein, J. D., Tsai, G., Kuncl, R. W., Clawson, L., Cornblath, D. R., Drachman, D. B., et al. (1990) Abnormal excitatory amino acid metabolism in amyotrophic lateral sclerosis. *Ann. Neurol.* **28,** 18–25.

Rothstein, J. D., Van Kammen, M., Levy, A. I., Martin, L. J., and Kuncl, R. W. (1995) Selective loss of glial glutamate transporter (GLT-1) in amyotrophic lateral sclerosis. *Ann. Neurol.* **38,** 73–84.

Roy, J., Minotti, S., Dong, L., Figlewicz, D. A., and Durham, H. D. (1998) Glutamate potentiates the toxicity of mutant Cu/Zn-superoxide dismutase in motor neurons by postsynaptic calcium-dependent mechanisms. *J. Neurosci.* **18,** 9673–9684.

Sasaki, S., Maruyama, S., Yamane, K., Sakuma, H., and Takeishi, M. (1990) Ultrastructure of swollen proximal axons of anterior horn neurons in motor neuron disease. *J. Neurol Sci.* **97,** 233–240.

Schubert, W. and Schwan, H. (1995) Detection by 4-parameter microscopic imaging and increase of rare mononuclear blood leukocyte types expressing the FcγR III receptor (CD16) for immunoglobulin G in human sporadic amyotrophic lateral sclerosis. *Neurosci. Lett.* **198,** 29–32.

Shaw, P. J., Ince, P. G., Falkous, G., and Mantle, D. (1995) Oxidative damage to protein in sporadic motor neuron disease spinal cord. *Ann. Neurol.* **38,** 691–695

Shaw, P. J., Williams, T. L., Slade, J. Y., Eggett, C. J., and Ince, P. G. (1999) Low expression of GluR2 AMPA receptor subunit protein by human motor neurons. *Neuroreport* **10,** 261–265.

Sheen, V. L., Dreyer, E. B., and Macklis, J. D. (1992) Calcium-mediated neuronal degeneration following singlet oxygen production. *Neuroreport* **3,** 705–708.

Shy, M. E., Rowland, L. P., Smith, T., Trojaborg, W., Latov, N., Sherman, W., et al. (1986) Motor neuron disease and plasma cell dyscrasia. *Neurology* **36,** 1429–1436.

Siddique, T., Figlewicz, D., Pericak-Vance, M. A., Haines, J. L., Rouleau, G., Jeffers, A. J., Sapp, P., Hung, W. Y., Bebout, J., McKenna-Yasek, D., et al. (1991) Linkage of a gene causing familial amyotrophic lateral sclerosis to chromosome 21, and evidence of genetic locus heterogeneity. *N. Engl. J. Med.* **324,** 1381–1384.

Siklos, L., Engelhardt, J. I., Alexianu, M. E., Gurney, M. E., Siddique, T., and Appel, S. H. (1998) Intracellular calcium parallels motoneuron degeneration in SOD-1 mutant mice. *J. Neuropathol. Exp. Neurol.* **57,** 571–587.

Siklos, L., Engelhardt, J. I., Harati, Y., Smith, R. G., Joo, F., and Appel, S. H. (1996) Ultrastructural evidence for altered calcium in motor nerve terminals in amyotrophic lateral sclerosis. *Ann. Neurol.* **39,** 203–216.

Smith, R. G., Henry, K. Y., Mattson, M. P., and Appel, S. H. (1998) Presence of 4-hydroxynonenal in the cerebrospinal fluid of patients with sporadic amyotrophic lateral sclerosis. *Ann. Neurol.* **44,** 696–699.

Sriram, S. and Rodriguez, M. (1997) Indictment of the microglia as the villain in multiple sclerosis. *Neurology* **48,** 464–470.

Streit, W. J. (1993) Microglia-neuronal interactions. *J. Chem. Neuroanat.* **6,** 261–266.

Streit, W. J., Graeber, M. B., and Kreutzberg, G. W. (1989) Expression of Ia antigen on perivascular and microglial cells after sublethal and lethal motor neuron injury. *Exp. Neurol.* **105,** 115–126.

Swerdlow, R. H., Parks, J. K., Cassarino, D. S., Trimmer, P. A., Miller, S. W., Maguire, D. J., et al. (1998) Mitochondria in sporadic amyotrophic lateral sclerosis. *Exp. Neurol.* **153,** 135–142.

Takuma, H., Kwak, S., Yoshizawa, T., and Kanazawa, I. (1999) Reduction of GluR2 RNA editing, a molecular change that increases calcium influx through AMPA receptors, selective in the spinal ventral gray of patients with amyotrophic lateral sclerosis. *Ann. Neurol.* **46,** 806–815.

Troost, D., Van den Oord, J. J., and de Jong, J. M. B. V. (1988) Analysis of the inflammatory infiltrate in amyotrophic lateral sclerosis. *J. Neuropathol. Appl. Neurobiol.* **14,** 255–256.

Troost, D., Van den Oord, J. J., de Jong, J. M. B. V., and Swaab, D. F. (1989) Lymphocytic infiltration in the spinal cord of patients with amyotrophic lateral sclerosis. *Clin. Neuropathol.* **8,** 289–294.

Tu, P. H., Gurney, M. E., Julien, J. P., Lee, V. M., and Trojanowski, J. Q. (1997) Oxidative stress, mutant SOD1, and neurofilament pathology in transgenic mouse models of human motor neuron disease. *Lab. Invest.* **76,** 441–456.

Tymianski, M., Charlton, M. P., Carlen, P. L., and Tator, C. H. (1993) Source specificity of early calcium neurotoxicity in cultured embryonic spinal neurons. *J. Neurosci.* **13,** 2085–2104.

Warita, H., Itoyama, Y., and Abe, K. (1999) Selective impairment of fast anterograde axonal transport in the peripheral nerves of asymptomatic transgenic mice with a G93A mutant SOD1 gene. *Brain Res.* **819,** 120–131.

Wiedau-Pazos, M., Goto, J. J., Rabizadeh, S., Gralla, E. B., Roe, J. A., Lee, M. K., et al. (1996) Altered reactivity of superoxide dismutase in familial amyotrophic lateral sclerosis. *Science* **271,** 515–518.

Wiedemann, F. R., Winkler, K., Kuznetsov, A. V., Bartels, C., Vielhaber, S., Feistner, H., and Kunz, W. S. (1998) Impairment of mitochondrial function in skeletal muscle of patients with amyotrophic lateral sclerosis. *J. Neurol. Sci.* **156,** 65–72.

Williamson, T. L. and Cleveland, D. W. (1999) Slowing of axonal transport is a very early event in the toxicity of ALS-linked SOD1 mutants to motor neurons. *Nat. Neurosci.* **2,** 50–60.

Williamson, T. L., Bruijn, L. I., Zhu, Q., Anderson, K. L., Anderson, S. D., Julien, J. P., and Cleveland, D. W. (1998) Absence of neurofilaments reduces the selective vulnerability of motor neurons and slows disease caused by a familial amyotrophic lateral sclerosis-linked superoxide dismutase 1 mutant. *Proc. Natl. Acad. Sci. USA* **95,** 9631–9636.

Wong, P. C., Pardo, C. A., Borchelt, D. R., Lee, M. K., Copeland, N. G., Jenkins, N. A., et al. (1995) An adverse property of a familial ALS-linked SOD1 mutation causes motor neuron disease characterized by vacuolar degeneration of mitochondria. *Neuron* **14,** 1105–1116.

Woodall, C. J., Riding, M. H., Graham, D. I., and Clements, G. B. (1994) Sequences specific for enterovirus detected in spinal cord from patients with motor neurone disease. *BMJ* **308,** 1541–1543.

Xu, Z., Cork, L. C., Griffin, J. W., and Cleveland, D. W. (1993) Increased expression of neurofilament subunit NF-L produces morphological alterations that resemble the pathology of human motor neuron disease. *Cell* **73,** 23–33.

Yim, M. B., Kang, J-H., Kwak, H-S., Chock, P. B., and Stadtman, E. R. (1996) A gain-of-function of an amyotrophic lateral sclerosis-associated Cu, Zn superoxide dismutase mutant: an enhancement of free radical formation due to a decreased in Km for hydrogen peroxide. *Proc. Natl. Acad. Sci. USA* **93,** 5709–5714.

Yoshiyama, Y., Yamada, T., Asanuma, K., and Asahi, T. (1994) Apoptosis related antigen, Le(Y) and nick-end labeling are positive in spinal motor neurons in amyotrophic lateral sclerosis. *Acta. Neuropathol. (Berl.)* **88,** 207–211.

Younger, D. S., Rowland, L. P., Latov, N., Hays, A. P., Lange, D. J., Sherman, W., et al. (1991) Lymphoma, motor neuron diseases, and amyotrophic lateral sclerosis. *Ann. Neurol.* **29,** 78–86.

Younger, D. S., Rowland, L. P., Latov, N., Sherman, W., Pesce, M., Lange, D. J., et al. (1990) Motor neuron disease and amyotrophic lateral sclerosis: relation of high CSF protein content to paraproteinemia in clinical syndromes. *Neurology* **40,** 595–599.

Zielasek, J. and Hartung, H-P. (1996) Molecular mechanisms of microglial activation. *Adv. Neuroimmunol.* **6,** 191–222.

9

Experimental Genetics as a Tool for Understanding Pathogenesis of ALS

Philip C. Wong, Donald L. Price, and Jamuna Subramaniam

INTRODUCTION

The identification of genes linked to a variety of neurodegenerative diseases allows the examination of the pathogenic mechanisms underlying these illnesses. The use of transgenic and gene-targeted mice approaches has allowed investigators to perturb the normal pattern of gene expression in mammals and to determine the phenotypic consequences of expressing wild-type/mutant genes linked to neurodegenerative diseases. Over the past decade, transgenic mice strategies have allowed researchers to reproduce features of human neurodegenerative disease in mice, and among the various models, the mutant superoxide dismutase-1 (SOD-1) mice—a model of familial amyotrophic lateral sclerosis (FALS)—has become a prototype for investigations of mechanisms of selective neuronal degeneration. In this chapter, we will discuss briefly the clinical, genetic, and neuropathological features of ALS, and illustrate how transgenic approaches are valuable tools for testing the role of copper in the pathogenesis of SOD-1-linked ALS. The delivery of copper to SOD-1 in yeast is mediated through a soluble protein called CCS (copper chaperone for SOD-1) and by generating CCS-deficient mice, we document that CCS is necessary to selectively deliver copper to SOD-1 to activate this mammalian metalloenzyme. These CCS knockout mice are valuable, in crossbreeding studies using mutant SOD-1 mice, to test directly whether copper in mutant SOD-1 is critical to cause motor-neuron degeneration in SOD-1-linked FALS. Finally, we will review the lessons we have learned from the SOD-1 transgenic mice models regarding the pathogenic mechanisms of FALS.

AMYOTROPHIC LATERAL SCLEROSIS

Clinical Features

ALS and FALS, with a worldwide prevalence of approx 5/100,000, usually occur in mid-to-late life and manifest as muscle weakness accompanied by hyperreflexia and spasticity. Electrophysiological examinations show fibrillations, fasciculations, and giant polyphasic potentials; muscle biopsies demonstrate denervation and muscle atrophy (Kuncl et al., 1992; Munsat, 1992; Harding, 1993; Williams and Windebank, 1993; Bennett and Scheller, 1994; Rowland, 1994). The disease is associated with progressive paralysis, and patients usually die of intercurrent illnesses.

From: *Pathogenesis of Neurodegenerative Disorders* Edited by: M. P. Mattson © Humana Press Inc., Totowa, NJ

Neuropathology

Muscle atrophy and weakness reflect selective degeneration of large motor neurons of the brainstem and spinal cord; spasticity, hyperreflexia, and extensor plantar signs are attributable to lesions of upper motor neurons. Lower motor neurons show several abnormalities including the presence of phosphorylated neurofilament (NF) and ubiquitin immunoreactivities in perikaryal inclusions; Lewy body-like intracytoplasmic inclusions; NF swellings of perikarya and proximal axons; fragmentation of Golgi; and attenuation of dendrites (Banker, 1986; Gonatas et al., 1992; Hirano and Kato, 1992). In some cases of FALS, intracytoplasmic inclusions contain SOD-1, ubiquitin, and NF immunoreactivities (Shibata et al., 1993, 1996; Chou et al., 1996; Rouleau et al., 1996), and NF accumulate in cell bodies and neuronal processes (Mizusawa et al., 1989; Takahashi et al., 1992; Rouleau et al., 1996). Motor nerves show axonal atrophy, distal axonal degeneration, reduction in number of large-caliber axons, and Wallerian degeneration (Bradley et al., 1983). In ALS and FALS, there are reduced numbers of motor neurons in brainstem nuclei and spinal cord, as well as loss of large pyramidal neurons in motor cortex; these lesions are accompanied by degeneration of axons in peripheral motor nerves and corticospinal tracts, respectively, and denervation of target fields of these inputs (Metcalf and Hirano, 1971; Kato et al., 1993; Lowe, 1994).

Genetics of FALS

SOD-1 *Gene*

Approximately 10% of ALS cases are familial and, in almost all cases, inheritance exhibits an autosomal dominant pattern (Siddique et al., 1989, 1991). Approximately 15–20% of patients with autosomal dominant FALS have missense point mutations in the gene that encodes cytosolic Cu/Zn-SOD-1 (Deng et al., 1993; Rosen et al., 1993; Enayat et al., 1995; Al-Chalabi et al., 1998), the enzyme that catalyzes the conversion of O_2^-· to O_2 and H_2O_2 (Fridovich, 1986). To date, more than 60 different mutations have been identified throughout the *SOD-1* gene (Brown, 1997; Al-Chalabi et al., 1998). There is clear evidence of allelic heterogeneity (Cudkowicz et al., 1997, 1998; Rowland, 1998) with phenotypes associated with different mutations sometimes showing differences; for example, cases of A4V SOD-1 FALS have limited involvements of the corticospinal tract (Cudkowicz et al., 1998), whereas other *SOD-1* mutations are associated with a more classic ALS syndrome.

Other Genetic Loci Relevant to ALS

A variety of chromosomal loci have been linked to autosomal dominant, autosomal recessive, and X-linked dominant forms of ALS (Hentati et al., 1994; Chance et al., 1998; Hong et al., 1998). Notably, investigators have described a large family with autosomal dominant juvenile ALS (JALS) linked to a loci at the 9q34 region (Chance et al., 1998). Manifested by the presence of slowly progressive distal atrophy, weakness, and corticospinal signs, these cases show a marked loss of spinal motor neurons and less pronounced degeneration of the corticospinal tracts (Rabin et al., 1999). The peripheral nerves and dorsal and ventral roots show prominent axonal swellings, axonal degeneration, with marked loss of both large and small myelinated fibers. These findings extend the spectrum of FALS and JALS to include a slowly progressive, autosomal dominant, nonfatal but debilitating disease.

To date, deletion/insertion mutations in the KSP repeat motif of the NF-H tail domain were identified in 10/1047 patients with sporadic ALS and in 1/295 FALS patients (Figlewicz et al., 1994; Tomkins et al., 1998; Al-Chalabi et al., 1999). There is no direct evidence that mutations of NF genes are a primary cause of ALS (Rooke et al., 1996; Vechio et al., 1996). In a subset of patients with sporadic ALS (SALS), levels of EAAT2 protein were shown to be reduced as compared to controls and to individuals who died of other neurodegenerative diseases. In 17 of 28 cases of ALS, aberrant EAAT2 mRNA were found; significantly, lower levels of the EAAT2 protein were observed in these same tissue samples (Lin et al., 1998). Functional studies indicate that the mutant RNA species led to a dominant downregulation of EAAT2 protein synthesis (Lin et al., 1998).

COPPER-MEDIATED MUTANT SOD-1 NEUROTOXICITY

As discussed above, mutations in SOD-1 cause up to 20% of cases of FALS (Rosen et al., 1993; Deng et al., 1993), a fatal adult-onset motor-neuron disease characterized by the selective loss of spinal and cortical motor neurons. A variety of in vivo and in vitro studies have demonstrated that the mutant enzyme causes selective neuronal degeneration through a gain of toxic property rather than a loss of SOD activity, consistent with FALS displaying an autosomal dominant pattern of inheritance. While some FALS-linked SOD-1 mutants show reduced enzymatic activities, others retain nearly wild-type levels (Borchelt et al., 1994) and mutant SOD-1 subunits do not appear to alter the metabolism/activity of wild-type SOD-1 in a dominant-negative fashion (Borchelt et al., 1995). Transgenic mice expressing different FALS-linked SOD-1 mutants exhibit progressive motor-neuron disease resembling that occurring in cases of ALS (Gurney et al., 1994; Dal Canto and Gurney, 1994; Morrison et al., 1998; Ripps et al., 1995; Morrison et al., 1996; Wong et al., 1995; Borchelt et al., 1998; Bruijn et al., 1997b), and mice deficient in SOD-1 do not develop a motor-neuron disease-like phenotype (Reaume et al., 1996). Taken together, these results are consistent with the view that mutant SOD-1 causes disease through a gain of neurotoxic property.

Although the role of SOD-1 in scavenging superoxide free radicals is well-established (McCord and Fridovich, 1969), other functions such as peroxidase activity (Hodgson and Fridovich, 1975; Ischiropoulos et al., 1992; Wiedau-Pazos et al., 1996) or buffering against Cu toxicity (Culotta et al., 1995) have been documented. However, the molecular mechanisms whereby mutant SOD-1 causes selective motor-neuron death remain uncertain (Wong et al., 1998; Cleveland, 1999). One hypothesis is that the Cu bound to mutant SOD-1 plays a key role in generating the toxic property in SOD-1-linked FALS, i.e., mutations induce conformational changes in SOD-1 to facilitate the interactions of the catalytic Cu, with small molecules such as peroxynitrite (Beckman et al., 1993) or hydrogen peroxide (Wiedau-Pazos et al., 1996) to generate toxic free radicals that damage a variety of cell constituents important for the maintenance and survival of motor neurons. Consistent with the peroxynitrite hypothesis are results showing increased levels of free nitrotyrosine in G37R mutant SOD-1 transgenic mice (Bruijn et al., 1997a), and in both sporadic and FALS (Beal et al., 1997; Ferrante et al., 1997), as well as the documentation of nitric oxide (NO)-dependent apoptosis induced by zinc-deficient SOD-1 in cultured motor neurons (Estevez et al., 1999). Similarly, the demonstration that the glutamate transporter GLT1 is inactivated by oxida-

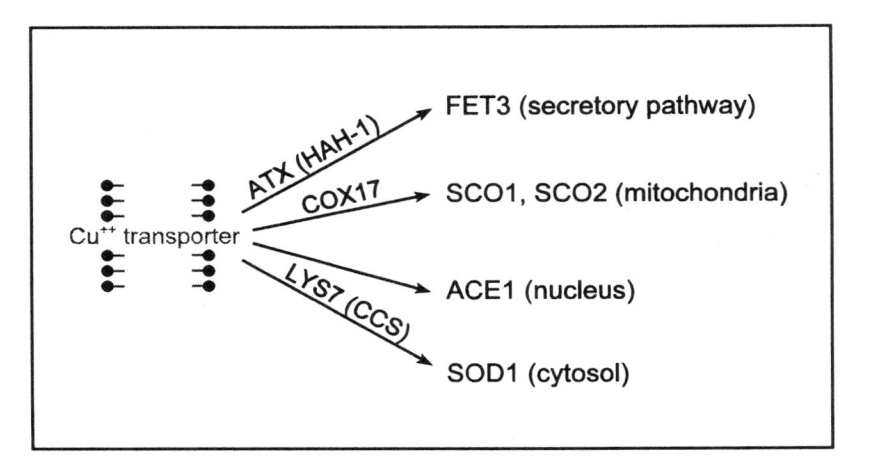

Fig. 1. Copper chaperones in yeast and mammalian cells. Copper is taken up in the cell by specific copper chaperones and delivered to their specific targets in various compartments of the cell.

tive reactions initiated by hydrogen peroxide and catalyzed by FALS-linked mutant SOD-1 (Trotti et al., 1999). However, no in vivo proof for these aberrant Cu chemistries, proposed to be central to the pathogenesis of FALS, has yet been established. Indeed, recent studies showed that onset and progression of disease in the G85R mutant mice is independent of the level of wild-type SOD-1 activities; results interpreted as inconsistent with the peroxynitrite hypothesis because the absence of wild-type SOD-1 would expected to exacerbate toxicity, whereas increasing SOD-1 would slow disease (Bruijn et al., 1998). In addition, disease progression in G93A mutant mice lacking neuronal nitric oxide synthase (NOS) gene showed no significant difference when compared to that of G93A mice, results that do not support a critical role for neuronal-derived NO in the pathogenesis of FALS (Facchinetti et al., 1999). Furthermore, recent studies questioned the significance of the peroxidase hypothesis and its relationship to FALS (Bruijn et al., 1997a; Marklund et al., 1997; Singh et al., 1998; Sankarapandi and Zweier, 1999). However, data obtained from studies of motor neurons in mixed spinal-cord cultures from G93A mice (Kruman et al., 1999) and studies of vulnerability of cortical neurons to ischemic injury in vivo (Guo et al., 2000) suggest that FALS mutations endanger neurons by a mechanism involving increased oxidative stress. Although the Cu-mediated toxicity models remain controversial, the discovery of CCS (Culotta et al., 1997) now provides the opportunity to test directly whether Cu within mutant SOD-1 mediates motor neuron degeneration in SOD-1-linked FALS.

COPPER TRAFFICKING AND CCS

The delivery of Cu to specific proteins is mediated through distinct intracellular pathways of copper trafficking (Culotta et al., 1997; Pufahl et al., 1997; Valentine and Gralla, 1997; Culotta et al., 1999). A family of soluble proteins termed Cu chaperones (Fig. 1) are required to deliver Cu to specific intracellular metalloproteins (Lin and Cizewski Culotta, 1995; Glerum et al., 1996a, b; Klomp et al., 1997; Pufahl et al., 1997). Recently, the yeast Cu chaperone, termed lys7 or its mammalian homolog CCS

(Fig. 1) was shown to be necessary to deliver copper to SOD-1 in yeast (Culotta et al., 1997). CCS is able to rescue the lys7 null mutant and is shown to interact physically with SOD-1 (Casareno et al., 1998; Rae et al., 1999). Biochemical and structural analysis of yeast CCS (Lamb et al., 1999; Schmidt et al., 1999) indicated that the insertion of Cu into SOD-1 requires the interactions among three distinct domains of the copper chaperone: an Atx1p-like amino-terminal domain responsible for Cu uptake; an SOD-1-like central domain functions in SOD-1 recognition; and a carboxyl-terminal domain unique to copper chaperones mediates Cu incorporation into SOD-1. Moreover, yeast CCS is sufficient to incorporate Cu into SOD-1 through a direct transfer of Cu from the metallochaperone to its target enzyme under restricted concentrations of intracellular free Cu, indicating that a pool of intracellular free Cu is not used to activate SOD-1 (Corson et al., 1998; Rae et al., 1999).

TESTING THE COPPER HYPOTHESIS

Recent studies showed that one common property of both wild-type and FALS-linked mutant SOD-1 is that Cu incorporation is CCS-dependent. Thus, with regard to mutant SOD-1, aberrant Cu chemistry may mediate degeneration of motor neurons in FALS (Corson et al., 1998). Consistent with this view are findings that CCS physically interacts with SOD-1 (Casareno et al., 1998; Rae et al., 1999) and that both proteins are co-localized in many cell types, including motor neurons (Rothstein et al., 1999). To determine whether CCS is necessary to incorporate Cu into SOD-1, and to test the role of Cu in the pathogenesis of FALS using a genetic approach to alter the amount of Cu bound to mutant SOD-1, we generated mice with targeted disruption of *CCS* loci (*CCS$^{-/-}$* mice), which are viable and showed reduced SOD-1 activity (Wong et al., 2000) and these data are summarized below.

Generation of CCS-*Deficient Mice*

We inactivated the mouse *CCS* gene using a standard homologous recombination strategy in embryonic stem (ES) cells (Wong et al., 2000). In constructing the *CCS* targeting vector, we replaced a 2.5 kb fragment containing the first two exons, part of the promoter and the first two introns of the *CCS* gene with a neomycin-resistance gene (Fig. 2A). The *CCS* targeting vector was transfected into CJ7 ES cells by electroporation and G418 resistant colonies were screened by genomic Southern blot analysis with a 5' probe (Fig. 2A). The efficiency of targeting was 16 in 209 clones. *CCS*-targeted ES cells were used to generate the *CCS$^{-/-}$* mice. We confirmed that the targeting event led to inactivation of the *CCS* gene by Western blot analysis of tissue extracts with a highly specific CCS antibody (Rothstein et al., 1999). For brain and spinal cord of *CCS$^{+/-}$* mice, we observed that CCS accumulates to approx 50% the level of control littermates, whereas the same tissues from *CCS$^{-/-}$* mice showed no detectable level of CCS (Fig. 2B). These results demonstrated the inactivation of *CCS*.

CCS Is Essential for SOD-1 Enzymatic Activity
and Copper Incorporation into SOD-1

To determine whether CCS is necessary for SOD-1 activity, we examined the enzymatic activity of SOD-1 in Triton X-100-soluble cell lysates from various tissues of *CCS$^{-/-}$* mice using activity gel assay. Compared to lysates from *CCS$^{+/-}$* mice or control

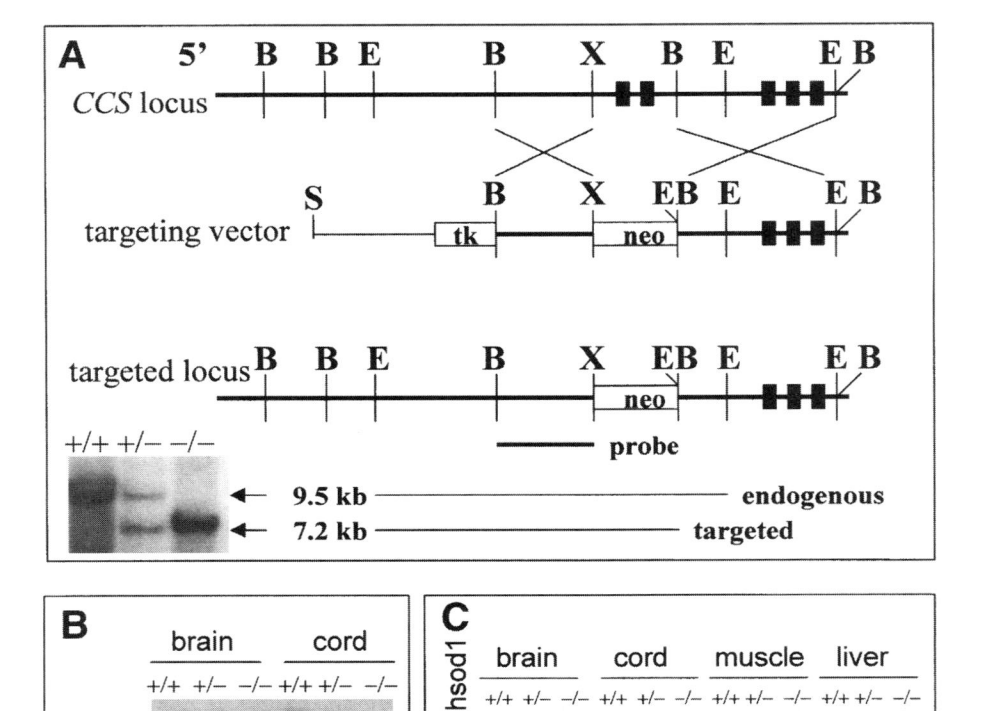

Fig. 2. *CCS* knockout mice possess diminished SOD-1 enzymatic activity. (**A**) Abbreviated restriction maps of the wild-type *CCS* allele, the targeting vector, and the targeted *CCS* locus. Exons of the *CCS* gene are denoted by black boxes. The targeting vector shows the replacement of exons 1 and 2 and flanking genomic sequences including portions of the promoter by the neomycin gene (neo) and the HSV thymidine kinase gene (tk). Arrows below indicate expected sizes from a Southern blot for *Eco*RI-digested fragments detected with a 5'-probe from targeted and endogenous *CCS* genes. B, *Bam*HI; E, *Eco*RI; H, *Hind*III; X, *Xho*I. (**B**) Protein extracts (20 µg) from brains and spinal cords of wild-type (+/+), heterozygous (+/–), and homozygous (–/–) *CCS* knockout mice were immunoblotted using antisera specific for CCS, SOD-1 and α-tubulin. Bound antibodies were detected using an enhanced chemiluminescent detection method. (**C**) SOD-1 activity assay gel of protein extracts (40 µg) from various tissues indicated of wild-type (+/+), heterozygous (+/–), and homozygous (–/–) *CCS* knockout mice. Purified human erythrocyte SOD-1 is shown in the left lane.

littermates, lysates from *CCS*⁻/⁻ mice showed a large reduction of SOD-1 activity using *in situ* native gel assay (Fig. 2C); we estimated that all *CCS*⁻/⁻ tissues examined retain 10–20% of normal SOD-1 activity with the exception of liver which is approx 30%. However, protein blotting analysis using a highly specific SOD-1 antibody (Pardo et al., 1995) showed that the levels of SOD-1 polypeptide from lysates of *CCS*⁻/⁻ mice are similar to those of *CCS*⁺/⁻ mice or control littermates (Fig. 2B). These results demonstrated that the superoxide scavenging activity in CCS-deficient mice is significantly

Table 1
$CCS^{-/-}$ Mice Resemble $SOD\text{-}1^{-/-}$ Mice

Phenotypes	$SOD\text{-}1^{-/-}$	$CCS^{-/-}$
1. Viable	+	+
2. SOD-1 activity	None	Low
3. More sensitive to paraquat	+	+
4. Reduced fertility in females	+	+
5. More sensitive to axonal injury	+	N.D.

N.D. = not determined.

diminished. We further showed that the reduced SOD-1 activity observed in CCS-deficient mice was owing to an abnormality in copper incorporation into SOD-1 polypeptide using in vivo ^{64}Cu metabolic labeling analysis (Wong et al., 2000). These results further demonstrated that CCS is essential to incorporate Cu into mammalian SOD-1 to activate the enzyme.

Increased Sensitivity to Paraquat and Reduced Fertility in CCS$^{-/-}$ Mice

Although no FALS-like motor neuron disease was observed in SOD-1-deficient mice, these mice show a distal neuropathy during aging (Shefner et al., 1999) and increased sensitivity to axonal injury (Reaume et al., 1996) and to paraquat toxicity (Huang et al., 1997; Ho et al., 1998). In addition, female $CCS^{-/-}$ mice exhibit reduced fertility (Ho et al., 1998; Matzuk et al., 1998). Because CCS-deficient mice possess diminished SOD-1 activity and SOD-1-deficient mice are hypersensitive to paraquat, we tested whether $CCS^{-/-}$ mice are also more susceptible to paraquat as compared to control mice. As expected, $CCS^{-/-}$ mice showed dramatic increased sensitivity to paraquat (Wong et al., 2000). However, compared to CCS-deficient mice, $SOD\text{-}1^{-/-}$ mice require a lower dose of paraquat to elicit its toxic effects (Ho et al., 1998). This is consistent with the observation that there are residual levels of SOD-1 activity in the $CCS^{-/-}$ mice.

Because female SOD-1-deficient mice showed subnormal reproductive capacity, we examined the fertility of CCS-deficient mice. Mating of 4 heterozygous CCS females resulted in 8 litters with average of 6.5 pups/litter over a period of 3 mo, whereas 4 homozygous CCS females gave 4 litters with an average of 3.5 pups/litter. These observations indicated that female CCS null mice exhibit subfertility, a phenotype that is characteristic of the SOD-1-deficient mice. We traced this phenotype to a defect in development of follicles (Wong et al., 2000), a phenotype characteristic of mice lacking SOD-1 (Matzuk et al., 1998). Taken together, the phenotypes observed in CCS null mice are remarkably similar to those of the SOD-1 null mice (*see* Table 1), and these results provide strong proof for the view that CCS is essential to activate efficiently mammalian SOD-1 in vivo.

Generation of Mutant SOD-1 Mice Deficient in CCS

As discussed above, although the view that Cu bound to mutant SOD-1 mediates toxicity in SOD-1-linked FALS remains controversial, our demonstration that CCS null mice are viable and possess marked reductions in SOD-1 activity now offers an opportunity to test directly whether Cu in mutant SOD-1 plays a key role in the pathogenesis of mutant SOD-1-induced FALS. Crossbreeding strategies using these CCS null mice and several lines of FALS mice (G37R, G93A, and G85R; reviewed below) are now currently underway and outcomes of these studies should be instructive in deciphering the role of Cu in SOD-1-linked FALS. Over the past several years, these mutant SOD-1 transgenic mice have been valuable in testing other hypotheses regarding the molecular mechanisms of selective motor neuron degeneration induced by mutant SOD-1 and the results of these investigations are reviewed below.

MUTANT SOD-1 TRANSGENIC MICE

G93A SOD-1 Mice

The G1H line of mice that expresses G93A human SOD-1 at approx 5× endogenous level develop hindlimb weakness at 3 mo. By 5 mo, these mice are moribund. The number of choline acetyltransferase-positive motor neurons are reduced (Gurney et al., 1994; Chiu et al., 1995), and the Golgi is fragmented (Mourelatos et al., 1996). Vacuoles, seen in cell bodies/dendrites (Dal Canto and Gurney, 1994), are now thought to represent degenerating mitochondria (Kong and Xu, 1997). Some axons are swollen, showing accumulations of neurofilaments (Tu et al., 1996), and other axons are reduced in caliber; fast and slow transport appear to be impaired (Zhang et al., 1997). In late stage, there is a loss of motor neurons (Dal Canto and Gurney, 1994). Another line of mice expressing low levels of SOD-1, ubiquitin immunoreactivity, and Lewy body-like inclusions are present in cell bodies of motor neurons; these mice also show axonal swellings, loss of motor neurons, and local astrogliosis (Dal Canto and Gurney, 1997).

The G93A SOD-1 transgenic mice have been used to test a variety of therapeutic approaches. Vitamin E and selenium modestly delay both the onset/progression of disease without affecting survival; in contrast, riluzole and gabapentin do not influence the onset/progression but do increase survival slightly (Gurney et al., 1996). Oral administration of d-penicillamine (Hottinger et al., 1997) or creatine (Klivenyi et al., 1999) significantly delays the onset of disease (Hottinger et al., 1997). Overexpression of Bcl-2 in these transgenic mice extends the survival of these mice, but the presence of the gene does not change the progression of the disease (Kostic et al., 1997). Maintenance of G93A mice on a dietary restriction regimen, a manipulation known to extend lifespan, did not delay age of disease onset and did not reduce disease progression (Pedersen and Mattson, 1999). Recently, the level of Bax, molecules that promote apoptosis, was shown to be increased in G93A mice (Vukosavic et al., 1999). However, this increase in Bax was augmented by an increase in Bcl-2/Bax heterodimers in G93A mice overexpressing Bcl-2 (Vukosavic et al., 1999). Interestingly, there is evidence indicating that an apoptotic mechanism may be involved in motor-neuron loss in ALS (Martin, 1999). In a small group of G93A SOD-1 mice overexpressing a dominant-negative inhibitor of interleukin-1β converting enzyme, a cell-death gene, there is a modest slowing of progression of disease (Friedlander et al., 1997). To date, it appears that agents that act in an antioxidant/antiapoptotic manner will delay disease onset, whereas antiexcitotoxic agents do not affect onset of disease, but do significantly prolong survival.

G37R SOD-1 Mice

These mice accumulate the G37R SOD-1 to 3–12× levels of endogenous SOD-1 in the spinal cord (Wong et al., 1995); the mutant SOD-1 retains full specific activity. Levels of the mutant transgene product determine the age of onset. At 5–7 wk of age (approx 2–3 mo before the appearance of clinical signs), SOD-1 accumulates in irregular swollen intraparenchymal portions of motor axon, the axonal cytoskeleton is abnormal, and vacuoles are present in the axon (Borchelt et al., 1998). Radiolabeling studies demonstrate that both endogenous SOD-1 and G37R SOD-1 are transported anterogradly in axons (Borchelt et al., 1998). Deficits in slow or fast axonal transport prior to disease onset have been documented in G37R and G85R mice (Williamson and Cleveland, 1999) or in G93A mice (Warita et al., 1999), respectively. Thus, toxic mutant SOD-1 is transported anterograde, accumulates early in the disease, and is associated with early structural pathology in axons. Small vacuoles are also present in dendrites (Wong et al., 1995), and some vacuoles appear to originate in the space between the outer and inner mitochondrial membranes with prominent distention of the outer mitochondrial membrane, displacement of the inner membrane, and disruption of the cristae. Both axonal and dendritic abnormalities occur months before the onset of clinical signs. By 20 wk of age, motor axons show vacuoles, cytoskeletal alterations, and Wallerian degeneration. The cell bodies of some neurons showed ubiquitin-immunoreactive inclusions and phosphorylated NF-H immunoreactivity. Recently, motor-axon inclusions immunoreactive with peripherin, an intermediate filament protein that when overexpressed in transgenic mice causes motor neuron death (Beaulieu et al., 1999), have been documented in presymptomatic G37R mice (Beaulieu et al., 1999). The number of motor neurons was reduced. Astrogliosis is present in proximity to abnormal neurons. These findings suggest that the accumulation of mutant SOD-1 is associated with subtle and early damage to organelles within axons/dendrites of motor neurons. If these early axonal and dendritic lesions occur in ALS, then they would be virtually impossible to detect, because they are early and subtle; in humans, they are masked by end-stage processes, immediate antemortem events, and postmortem delays.

NFs have been implicated in the pathogenesis of ALS (*see* Julien, 1999; Cleveland, 1999 for reviews). To test the role of NFs in motor-neuron disease caused by SOD-1 mutations, transgenic mice expressing G37R SOD-1 were crossbred to: (1) transgenic mice that accumulate NF-H-β-galactosidase fusion protein (NF-H-lacZ), which crosslinks neurofilaments in neuronal perikarya and limits their export to axons (Eyer and Peterson, 1994); and (2) transgenic mice expressing human NF-H subunits (Côté et al., 1993). In G37R SOD-1 mice expressing NF-H-lacZ, NFs are withheld from the axonal compartment, but there is no influence on the progression of disease (Eyer et al., 1998), implying that neither initiation nor progression of disease requires an axonal NF cytoskeleton and that alterations in NF biology observed in some forms of motor-neuron disease may be secondary responses. By contrast, the expression of wild-type human NF-H transgenes in the SOD-1 mutant mice significantly increased the mean life-span of the G37R SOD-1 mice (Couillard-Després et al., 1998). In contrast to the striking axonal degeneration observed in 1-yr-old transgenic mice expressing G37R SOD-1, the compound G37R NF-H transgenic mice showed remarkable sparing of motor neurons (Couillard-Després et al., 1998). The reasons for this protection of G37R

SOD-1 mice by increased levels of human NF-H, particularly in cell bodies but not in axons (where NF content is reduced) are not known. It could be related to the ability of NF-H to bind calcium and, therefore, influence calcium-mediated damage (Couillard-Després et al., 1998). Consistent with a role for calcium in disease is the observation that overexpression of the calcium-binding protein, calbindin D (28K) confers protection against mutant SOD-1-mediated death of PC12 cells (Ghadge et al., 1997). Moreover, cultured motor neurons from mutant SOD-1 mice exhibit a disturbance in regulation of calcium homeostasis such that calcium responses to glutamate are greatly increased (Kruman et al., 1999).

G85R SOD-1 Mice

In multiple lines with different copy numbers (2–15 copies), the accumulation of mutant G85R proteins ranged from approx 20% of endogenous mouse SOD-1 (in the lowest expressing line) to equal to endogenous levels (in the highest expressing line). The age of onset of weakness varied from 8–10 mo for the highest expressing line and 12–14 mo for the lowest expressing line (Bruijn et al., 1997b). Two weeks after initial weakness in one hindlimb, these mice were completely paralyzed. In the preclinical period, astrocytes contained SOD-1 and ubiquitin immunoreactive Lewy body-like inclusions (Bruijn et al., 1997b); at later stages, motor neurons also exhibited SOD-1 and ubiquitin-positive inclusions. At 6.5 mo (coincident with the earliest pathology), there was no significant loss of axons in the motor roots, but, 1–2 wk after the onset of clinical signs, many axons had degenerated (Bruijn et al., 1997b). Similar pathology has been reported in two individuals with FALS who had a two-basepair deletion in codon 126 of SOD-1, leading to a frameshift and truncation of the final 27 amino acids of SOD-1 (Kato et al., 1996). To test the role of neurofilaments in disease, NF-L null mice were mated to G85R SOD-1 transgenic mice. G85R mutant mice without NFs show delay in onset of disease and extension of survival (Williamson et al., 1998). Thus, although assembled neurofilaments are not required for mutant SOD-1-induced disease, absence of NFs does appear to slow the G85R SOD-1-mediated disease. To further examine disease mechanisms, G85R SOD-1 transgenic mice were mated to mice expressing wild-type SOD-1 mice or to SOD-1 null mice; elevation or elimination of wild-type SOD-1 did not influence the SOD-1 mutant-associated disease (Bruijn et al., 1998). These studies indicated that SOD mimetic may not be useful and is inconsistent with the view that mutant SOD-1-mediated toxicity derives from superoxide-associated oxidative stress (Bruijn et al., 1998)

G86R SOD-1 Mice

Mice expressing this mutant transgene, the murine equivalent of the G85R SOD-1 mutation, developed progressive paralysis at 3 mo (Ripps et al., 1995; Morrison et al., 1998); some neurons exhibit cytological evidence of degeneration (Ripps et al., 1995; Morrison et al., 1996). At end-stage disease, there is accumulation of phosphorylated NF inclusions in surviving motor neurons, reduction in numbers of motor neurons and interneurons, and reactive gliosis (Morrison et al., 1998). Vacuolar pathology was not observed (Morrison et al., 1996). Calbindin-containing nerve cells appeared to be relatively spared (Morrison et al., 1996, 1998).

CONCLUSION

A number of transgenic mice studies showed clear evidence that is consistent with the view that motor-neuron disease is caused by a toxic property of the mutant SOD-1 enzymes (Gurney et al., 1994; Ripps et al., 1995; Wong et al., 1995; Bruijn et al., 1997b). Furthermore, this view is supported by the fact that mice homozygous for the deletion of the SOD-1 gene do not show ALS-like motor-neuron disease (Reaume et al., 1996). One key question regarding the pathogenesis of SOD-1-linked ALS is: what is the nature of this toxic property imparted by the mutant SOD-1? Several hypotheses for the aberrant property of the mutant enzyme involve the deleterious copper chemistry generated by the mutant SOD-1. In these hypotheses, Cu bound to mutant SOD-1 is proposed to play a critical role, but it has been difficult to test the Cu theory. However, the discovery of CCS provides the opportunity to alter selectively the amount of Cu bound to mutant SOD-1 in vivo and thus allowing us to test (using a genetic approach to selectively remove Cu from mutant SOD-1) whether the aberrant SOD-1 Cu chemistry is important for mediating motor neuron degeneration in mutant SOD-1 mice. We have now created mice that are deficient in CCS and they are viable and show no overt phenotype. Importantly, the SOD-1 activity is markedly reduced in the CCS null mice. With CCS null mice, we are now in a position to test the critical question as to whether Cu bound to mutant SOD-1 is necessary to mediate motor-neuron degeneration induced by FALS-linked mutant SOD-1. To test this hypothesis, we propose to crossbreed a series of mutant SOD-1 mice to CCS null mice. If Cu participates in the pathogenesis of FALS mice, then CCS null mice that express mutant SOD-1 should show amelioration of motor-neuron disease. These studies will allow the examination of the role aberrant copper chemistry plays in SOD-1-linked FALS. Results of these efforts, which will have the potential to identify novel therapeutic targets (i.e., CCS, SOD-1, and Cu trafficking pathways), may have important implications for design of drug treatments for FALS.

ACKNOWLEDGMENTS

The authors thank Valeria Culotta, Jonathan Gitlin, Jeffrey Rothstein, Darrel Waggoner, Lino Tessarollo, and Thomas Bartnikas for discussions and contributions to some of the work mentioned in this chapter. Parts of this work were supported by NIH grant NS37771 and the Amyotrophic Lateral Sclerosis Association.

REFERENCES

Al-Chalabi, A., Andersen, P. M., Chioza, B., Shaw, C., Sham, P. C., Robberecht, W., et al. (1998) Recessive amyotrophic lateral sclerosis families with the D90A SOD1 mutation share a common founder: evidence for a linked protective factor. *Hum. Mol. Genet.* **7,** 2045–2050.

Al-Chalabi, A., Andersen, P. M., Nilsson, P., Chioza, B., Anderson, J. L., Russ C., et al. (1999) Deletions of the heavy neurofilament subunit tail in amyotrophic lateral sclerosis. *Hum. Mol. Genet.* **8,** 157–164.

Banker, B. Q. (1986) The pathology of motor neuron disorders, in *Myology,* vol. II (Engel, A. G. and Banker, B. Q., eds.), McGraw Hill, New York, pp. 2031–2067.

Beal, M. F., Ferrante, R. J., Browne, S. E., Matthews, R. T., Kowall, N. W., and Brown, R. H., Jr. (1997) Increased 3-nitrotyrosine in both sporadic and familial amyotrophic lateral sclerosis. *Ann. Neurol.* **42,** 646–654.

Beaulieu, J.-M., Nguyen, M. D., and Julien, J.-P. (1999) Late onset death of motor neurons in mice overexpressing wild-type peripherin. *J. Cell Biol.* **147,** 531–544.

Beckman, J. S., Carson, M., Smith, C. D., and Koppenol, W. H. (1993) ALS, SOD and peroxynitrite. *Nature* **364,** 584.

Bennett, M. K. and Scheller, R. H. (1994) Molecular correlates of synaptic vesicle docking and fusion. *Curr. Opin. Neurobiol.* **4,** 324–329.

Borchelt, D. R., Guarnieri, M., Wong, P. C., Lee, M. K., Slunt, H. S., Xu, Z.-S., et al. (1995) Superoxide dismutase 1 subunits with mutations linked to familial amyotrophic lateral sclerosis do not affect wild-type subunit function. *J. Biol. Chem.* **270,** 3234–3238.

Borchelt, D. R., Lee, M. K., Slunt, H. H., Guarnieri, M., Xu, Z.-S., Wong, P. C., et al. (1994) Superoxide dismutase 1 with mutations linked to familial amyotrophic lateral sclerosis possesses significant activity. *Proc. Natl. Acad. Sci. USA* **91,** 8292–8296.

Borchelt, D. R., Wong, P. C., Becher, M. W., Pardo, C. A., Lee, M. K., Xu, Z.-S., et al. (1998) Axonal transport of mutant superoxide dismutase 1 and focal axonal abnormalities in the proximal axons of transgenic mice. *Neurobiol. Dis.* **5,** 27–35.

Bradley, W. G., Good, P., Rasool, C. G., and Adelman, L. S. (1983) Morphometric and biochemical studies of peripheral nerves in amyotrophic lateral sclerosis. *Ann. Neurol.* **14,** 267–277.

Brown, R. H., Jr. (1997) Amyotrophic lateral sclerosis. Insights from genetics. *Arch. Neurol.* **54,** 1246–1250.

Bruijn, L. I., Beal, M. F., Becher, M. W., Schulz, J. B., Wong, P. C., Price, D. L., and Cleveland, D. W. (1997a) Elevated free nitrotyrosine levels, but not protein-bound nitrotyrosine or hydroxyl radicals, throughout amyotrophic lateral sclerosis (ALS)-like disease implicate tyrosine nitration as an aberrant *in vivo* property of one familial ALS-linked superoxide dismutase 1 mutant. *Proc. Natl. Acad. Sci. USA* **94,** 7606–7611.

Bruijn, L. I., Becher, M. W., Lee, M. K., Anderson, K. L., Jenkins, N. A., Copeland, N. G., et al. (1997b) ALS-linked SOD1 mutant G85R mediates damage to astrocytes and promotes rapidly progressive disease with SOD1-containing inclusions. *Neuron* **18,** 327–338.

Bruijn, L. I., Houseweart, M. K., Kato, S., Anderson, A. G., Scott, R. W., and Cleveland, D. W. (1998) Aggregation and motor neuron toxicity of an ALS-linked SOD1 mutant independent from wild-type SOD1. *Science* **281,** 1851–1854.

Casareno, R. L. B., Waggoner, D., and Gitlin, J. D. (1998) The copper chaperone CCS directly interacts with copper/zinc superoxide dismutase. *J. Biol. Chem.* **273,** 23,625–23,628.

Chance, P. F., Rabin, B. A., Ryan, S. G., Ding, Y., Scavina, M., Crain, B., et al. (1998) Linkage of the gene for an autosomal dominant form of juvenile amyotrophic lateral sclerosis to chromosome 9q34. *Am. J. Hum. Genet.* **62,** 633–640.

Chiu, A. Y., Zhai, P., Dal Canto, M. C., Peters, T. M., Kwon, Y. W., Prattis, S. M., and Gurney, M. E. (1995) Age-dependent penetrance of disease in a transgenic mouse model of familial amyotrophic lateral sclerosis. *Mol. Cell. Neurosci.* **6,** 349–362.

Chou, S. M., Wang, H. S., and Komai, K. (1996) Colocalization of NOS and SOD1 in neurofilament accumulation within motor neurons of amyotrophic lateral sclerosis: an immunohistochemical study. *J. Chem. Neuroanat.* **10,** 249–258.

Cleveland, D. W. (1999) From Charcot to SOD1: mechanisms of selective motor neuron death in ALS. *Neuron* **24,** 515–520.

Corson, L. B., Culotta, V. C., and Cleveland, D. W. (1998) Chaperone-facilitated copper binding is a property common to several classes of familial amyotrophic lateral sclerosis-linked superoxide dismutase mutants. *Proc. Natl. Acad. Sci. USA* **95,** 6361–6366.

Couillard-Després, S., Zhu, Q., Wong, P. C., Price, D. L., Cleveland, D. W., and Julien, J.-P. (1998) Protective effect of neurofilament NF-H overexpression in motor neuron disease induced by mutant superoxide dismutase. *Proc. Natl. Acad. Sci. USA* **95,** 9626–9630.

Côté, F., Collard, J.-F., and Julien, J.-P. (1993) Progressive neuronopathy in transgenic mice expressing the human neurofilament heavy gene: a mouse model of amyotrophic lateral sclerosis. *Cell* **73,** 35–46.

Cudkowicz, M. E., McKenna-Yasek, D., Chen, C., Hedley-Whyte, E. T., and Brown R. H., Jr. (1998) Limited corticospinal tract involvement in amyotrophic lateral sclerosis subjects with the A4V mutation in the copper/zinc superoxide dismutase gene. *Ann. Neurol.* **43,** 703–710.

Cudkowicz, M. E., McKenna-Yasek, D., Sapp, P. E., Chin, W., Geller, B., Hayden, D. L., et al. (1997) Epidemiology of mutations in superoxide dismutase in amyotrophic lateral sclerosis. *Ann. Neurol.* **41,** 210–221.

Culotta, V. C., Joh, H. D., Lin, S. J., Slekar, K. H., and Strain, J. (1995) A physiological role for saccharomyces cerevisiae copper/zinc superoxide dismutase in copper buffering. *J. Biol. Chem.* **270,** 29,991–29,997.

Culotta, V. C., Klomp, L. W. J., Strain, J., Casareno, R. L. B., Krems, B., and Gitlin G. D. (1997) The copper chaperone for superoxide dismutase. *J. Biol. Chem.* **272,** 23,469–23,472.

Culotta, V. C., Lin, S. J., Schmidt, P., Klomp, L. W., Casareno, R. L., and Gitlin, J. (1999) Intracellular pathways of copper trafficking in yeast and humans. *Adv. Exp. Med. Biol.* **448,** 247–254.

Dal Canto, M. C. and Gurney, M. E. (1994) Development of central nervous system pathology in a murine transgenic model of human amyotrophic lateral sclerosis. *Am. J. Pathol.* **145,** 1271–1280.

Dal Canto, M. C. and Gurney, M. E. (1997) A low expressor line of transgenic mice carrying a mutant human Cu,Zn superoxide dismutase (*SOD1*) gene develops pathological changes that most closely resemble those in human amyotrophic lateral sclerosis. *Acta Neuropathol.* **93,** 537–550.

Deng, H.-X., Hentati, A., Tainer, J. A., Iqbal, Z., Cayabyab, A., Hung, W.-Y., et al. (1993) Amyotrophic lateral sclerosis and structural defects in Cu,Zn superoxide dismutase. *Science* **261,** 1047–1051.

Enayat, Z. E., Orrell, R. W., Claus, A., Ludolph, A., Bachus, R., Brockmüller, J., et al. (1995) Two novel mutations in the gene for copper zinc superoxide dismutase in UK families with amyotrophic lateral sclerosis. *Hum. Mol. Genet.* **4,** 1239–1240.

Estevez, A. G., Crow, J. P., Sampson, J. B., Reiter, C., Zhuang, Y., Richardson, G. J., et al. (1999) Induction of nitric oxide-dependent apoptosis in motor neurons by zinc-deficient superoxide dismutase. *Science* **286,** 2498–2500.

Eyer, J., Cleveland, D. W., Wong, P. C., and Peterson, A. C. (1998) Pathogenesis of two axonopathies does not require axonal neurofilaments. *Nature* **391,** 584–587.

Eyer, J. and Peterson, A. (1994) Neurofilament-deficient axons and perikaryal aggregates in viable transgenic mice expressing a neurofilament-β-galactosidase fusion protein. *Neuron* **12,** 389–405.

Facchinetti, F., Sasaki, M., Cutting, F. B., Zhai, P., MacDonald, J. E., Reif, D., et al. (1999) Lack of involvement of neuronal nitric oxide synthase in the pathogenesis of a transgenic mouse model of familial amyotrophic lateral sclerosis. *Neuroscience* **90,** 1483–1492.

Ferrante, R. J., Browne, S. E., Shinobu, L. A., Bowling, A. C., Baik, M. J., MacGarvey, U., et al. (1997) Evidence of increased oxidative damage in both sporadic and familial amyotrophic lateral sclerosis. *J. Neurochem.* **69,** 2064–2074.

Figlewicz, D. A., Krizus, A., Martinoli, M. G., Meininger, V., Dib, M., Rouleau, G. A., and Julien, J.-P. (1994) Variants of the heavy neurofilament subunit are associated with the development of amyotrophic lateral sclerosis. *Hum. Mol. Genet.* **3,** 1757–1761.

Fridovich, I. (1986) Superoxide dismutases. *Adv. Enzymol. Relat. Areas Mol. Biol.* **58,** 61–97.

Friedlander, R. M., Brown, R. H., Gagliardini, V., Wang, J., and Yuan, J. (1997) Inhibition of ICE slows ALS in mice. *Nature* **388,** 31.

Ghadge, G. D., Lee, J. P., Bindokas, V. P., Jordan, J., Ma, L., Miller, R. J., and Roos, R. P. (1997) Mutant superoxide dismutase-1-linked familial amyotrophic lateral sclerosis: molecular mechanisms of neuronal death and protection. *J. Neurosci.* **17,** 8756–8766.

Glerum, D. M., Shtanko, A., and Tzagoloff, A. (1996a) *SCO1* and *SCO2* act as high copy suppressors of a mitochondrial copper recruitment defect in *Saccharomyces cerevisiae. J. Biol. Chem.* **271,** 20,531–20,535.

Glerum, D. M., Shtanko, A., and Tzagoloff, A. (1996b) Characterization of *COX17*, a yeast gene involved in copper metabolism and assembly of cytochrome oxidase. *J. Biol. Chem.* **271,** 14,504–14,509.

Gonatas, N. K., Stieber, A., Mourelatos, Z., Chen, Y., Gonatas, J. O., Appel, S. H., et al. (1992) Fragmentation of the Golgi apparatus of motor neurons in amyotrophic lateral sclerosis. *Am. J. Pathol.* **140,** 731–737.

Guo, Z., Kindy, M. S., Kruman, I., and Mattson, M. P. (2000) ALS-linked Cu/Zn-SOD mutation impairs cerebral synaptic glucose and glutamate transport and exacerbates ischemic brain injury. *J. Cereb. Blood Flow Metab.* **20,** 463–468.

Gurney, M. E., Cuttings, F. B., Zhai, P., Doble, A., Taylor, C. P., Andrus, P. K., and Hall, E. D. (1996) Benefit of vitamin E, riluzole, and gabapentin in a transgenic model of familial amyotrophic lateral sclerosis. *Ann. Neurol.* **39,** 147–157.

Gurney, M. E., Pu, H., Chiu, A. Y., Dal Canto, M. C., Polchow, C. Y., Alexander, D. D., et al. (1994) Motor neuron degeneration in mice that express a human Cu/Zn superoxide dismutase mutation. *Science* **264,** 1772–1775.

Harding, A. E. (1993) Inherited neuronal atrophy and degeneration predominantly of lower motor neurons, in *Peripheral Neuropathy*, 3rd ed. (Dyck, P. J., et al., eds.), W. B. Saunders, Philadelphia, pp. 1051–1064.

Hentati, A., Bejaoui, K., Pericak-Vance, M. A., Hentati, F., Speer, M. C., Hung, W.-Y., et al. (1994) Linkage of recessive familial amyotrophic lateral sclerosis to chromosome 2q33-q35. *Nature Genet.* **7,** 425–428.

Hirano, A. and Kato S. (1992) Fine structural study of sporadic and familial amyotrophic lateral sclerosis, in *Handbook of Amyotrophic Lateral Sclerosis, Neurological Disease and Therapy*, vol. 12 (Smith, R. A., ed.), Marcel Dekker, New York, pp. 183–192.

Ho, Y. S., Gargano, M., Cao, J., Bronson, R. T., Heimler, I., and Hutz, R. J. (1998) Reduced fertility in female mice lacking copper-zinc superoxide dismutase. *J. Biol. Chem.* **273,** 7765–7769.

Hodgson, E. K. and Fridovich, I. (1975) The interaction of bovine erythrocyte superoxide dismutase with hydrogen peroxide: inactivation of the enzyme. *Biochemistry* **14,** 5294–5303.

Hong, S., Brooks, B. R., Hung, W. Y., Siddique, N. A., Rimmler, J., Fan, C., et al. (1998) X-linked dominant locus for late-onset familial amyotrophic lateral sclerosis. *Soc. Neurosci. Abstr.* **24,** 478.

Hottinger, A. F., Fine, E. G., Gurney, M. E., Zurn, A. D., and Aebischer, P. (1997) The copper chelator d-penicillamine delays onset of disease and extends survival in a transgenic mouse model of familial amyotrophic lateral sclerosis. *Eur. J. Neurosci.* **9,** 1548–1551.

Huang, T. T., Yasunami, M., Carlson, E. J., Gillespie, A. M., Reaume, A. G., and Hoffman, E. K. (1997) Superoxide-mediated cytotoxicity in superoxide dismutase-deficient fetal fibroblasts. *Arch. Biochem. Biophys.* **344,** 424–432.

Ischiropoulos, H., Zhu, L., Chen, J., Tsai, M., Martin, J. C., Smith, C. D., and Beckman, J. S. (1992) Peroxynitrite-mediated tyrosine nitration catalyzed by superoxide dismutase. *Arch. Biochem. Biophys.* **298,** 431–437.

Julien, J.-P. (1999) Neurofilament functions in health and disease. *Curr. Opin. Neurobiol.* **9,** 554–560.

Kato, S., Oda, M., and Hayashi, H. (1993) Neuropathology in amyotrophic lateral sclerosis patients on respirators: uniformity and diversity in 13 cases. *Neuropathology* **13,** 229–236.

Kato, S., Shimoda, M., Watanabe, Y., Nakashima, K., Takahashi, K., and Ohama, E. (1996) Familial amyotrophic lateral sclerosis with a two base pair deletion in superoxide dismutase 1 gene: multisystem degeneration with intracytoplasmic hyaline inclusions in astrocytes. *J. Neuropathol. Exp. Neurol.* **55,** 1089–1101.

Klivenyi, P., Ferrante, R. J., Matthews, R. T., Bogdanov, M. B., Klein, A. M., Andreassen, O. A., et al. (1999) Neuroprotective effects of creatine in a transgenic animal model of amyotrophic lateral sclerosis. *Nature Med.* **5,** 347–350.

Klomp, L. W. J., Lin, S.-J., Yuan, D. S., Klausner, R. D., Culotta, V. C., and Gitlin, J. D. (1997) Identification and functional expression of *HAH1*, a novel human gene involved in copper homeostasis. *J. Biol. Chem.* **272,** 9221–9226.

Kong, J. and Xu, Z. (1997) Massive mitochondrial degeneration in motor neurons triggers the onset of amyotrophic lateral sclerosis in mice expressing a mutat SOD1. *J. Neurosci.* **18,** 3241–3250.

Kostic, V., Jackson-Lewis, V., de Bilbao, F., Dubois-Dauphin, M., and Przedborski, S. (1997) Bcl-2: prolonging life in a transgenic mouse model of familial amyotrophic lateral sclerosis. *Science* **277,** 559–562.

Kruman, I., Pedersen, W. A., and Mattson, M. P. (1999) ALS-linked Cu/Zn-SOD mutation increases vulnerability of motor neurons to excitotoxicity by a mechanism involving increased oxidative stress and perturbed calcium homeostasis. *Exp. Neurol.* **160,** 28–39.

Kuncl, R. W., Crawford, T. O., Rothstein, J. D., and Drachman, D. B. (1992) Motor neuron diseases, in *Diseases of the Nervous System, Clinical Neurobiology*, 2nd ed., vol. II (Asbury, A. K., McKhann, G. M., and McDonald, W. I., eds.), W. B. Saunders, Philadelphia, pp. 1179–1208.

Lamb, A. L., Wernimont, A. K., Pufahl, R. A., Culotta, V. C., O'Halloran, T. V., and Rosenzweig, A. C. (1999) Crystal structure of the copper chaperone for superoxide dismutase. *Nature Struct. Biol.* **6,** 1.

Lin, C. L., Bristol, L. A., Jin, L., Dykes-Hoberg, M., Crawford, T., Clawson, L., and Rothstein, J. D. (1998) Aberrant RNA processing in a neurodegenerative disease: the cause for absent EAAT2, a glutamate transporter, in amyotrophic lateral sclerosis. *Neuron* **20,** 589–602.

Lin, S.-J. and Cizewski Culotta, V. (1995) The *ATX1* gene of *Saccharomyces cerevisiae* encodes a small metal homeostasis factor that protects cells against reactive oxygen toxicity. *Proc. Natl. Acad. Sci. USA* **92,** 3784–3788.

Lowe, J. (1994) New pathological findings in amyotrophic lateral sclerosis. *J. Neurol. Sci.* **124,** 38–51.

Marklund, S. L., Andersen, P. M., Forsgren, L., Nilsson, P., Ohlsson, P.-I., Wikander, G., and Öberg, A. (1997) Normal binding and reactivity of copper in mutant superoxide dismutase isolated from amyotrophic lateral sclerosis patients. *J. Neurochem.* **69,** 675–681.

Martin, L. J. (1999) Neuronal death in amyotrophic lateral sclerosis is apoptosis: possible contribution of a programmed cell death mechanism. *J. Neuropathol. Exp. Neurol.* **58,** 459–471.

Matzuk, M. M., Dionne, L., Guo, Q., Kumar, T. R., and Lebovitz, R. M. (1998) Ovarian function in superoxide dismutase 1 and 2 knockout mice. *Endocrinology* **139,** 4008–4011.

McCord, J. M. and Fridovich, I. (1969) Superoxide dismutase. An enzymic function for erythrocuprein (hemocuprein). *J. Biol. Chem.* **244,** 6049–6055.

Metcalf, W. C. and Hirano, A. (1971) Amyotrophic lateral sclerosis. Clinicopathological studies of a family. *Arch. Neurol.* **24,** 518–523.

Mizusawa H., Matsumoto S., Yen S.-H., Hirano A., Rojas-Corona R. R., and Donnefeld, H. (1989) Focal accumulation of phosphorylated neurofilaments within anterior horn cell in familial amyotrophic lateral sclerosis. *Acta Neuropathol.* **79,** 37–43.

Morrison, B. M., Gordon, J. W., Ripps, M. E., and Morrison, J. H. (1996) Quantitative immunocytochemical analysis of the spinal cord in G86R superoxide dismutase transgenic mice: neurochemical correlates of selective vulnerability. *J. Comp. Neurol.* **373,** 619–631.

Morrison, B. M., Janssen, W. G., Gordon, J. W., and Morrison, J. H. (1998) Time course of neuropathology in the spinal cord of G86R superoxide dismutase transgenic mice. *J. Comp. Neurol.* **391,** 64–77.

Mourelatos, Z., Gonatas, N. K., Stieber, A., Gurney, M. E., and Dal Canto, M. C. (1996) The Golgi apparatus of spinal cord motor neurons in transgenic mice expressing mutant Cu,Zn superoxide dismutase becomes fragmented in early, preclinical stages of the disease. *Proc. Natl. Acad. Sci. USA* **93,** 5472–5477.

Munsat, T. L. (1992) The natural history of amyotrophic lateral sclerosis, in *Handbook of Amyotrophic Lateral Sclerosis* (Smith, R. A., ed.), Marcel Dekker, New York, pp. 39–63.

Pardo, C. A., Xu, Z., Borchelt, D. R., Price, D. L., Sisodia, S. S., and Cleveland, D. W. (1995) Superoxide dismutase is an abundant component in cell bodies, dendrites, and axons of motor neurons and in a subset of other neurons. *Proc. Natl. Acad. Sci. USA* **92,** 954–958.

Pedersen, W. A. and Mattson, M. P. (1999) No benefit of dietary restriction on disease onset or progression in ALS Cu/Zn-SOD mutant mice. *Brain Res.* **833,** 28–39.

Pufahl, R. A., Singer, C. P., Peariso, K. L., Lin, S.-J., Schmidt, P. J., Fahrni, C. J., et al. (1997) Metal ion chaperone function of the soluble Cu(I) receptor Atx1. *Science* **278,** 853–856.

Rabin, B. A., Griffin, J. W., Crain, B. J., Scavina, M., Chance, P. F., and Cornblath, D. R. (1999) Autosomal dominant juvenile amyotrophic lateral sclerosis. *Brain* **122,** 1539–1550.

Rae, T. D., Schmidt, P. J., Pufahl, R. A., Culotta, V. C., and O'Holloran, T. V. (1999) Undetectable intracellular free copper: the requirement of a copper chaperone for superoxide dismutase. *Science* **284,** 805–808.

Reaume, A. G., Elliott J. L., Hoffman, E. K., Kowall, N. W., Ferrante, R. J., Siwek, D. F., et al. (1996) Motor neurons in Cu/Zn superoxide dismutase-deficient mice develop normally but exhibit enhanced cell death after axonal injury. *Nature Genet.* **13,** 43–47.

Ripps, M. E., Huntley, G. W., Hof, P. R., Morrison, J. H., and Gordon, J. W. (1995) Transgenic mice expressing an altered murine superoxide dismutase gene provide an animal model of amyotrophic lateral sclerosis. *Proc. Natl. Acad. Sci. USA* **92,** 689–693.

Rooke, K., Figlewicz, D. A., Han, F.-Y., and Rouleau, G. A. (1996) Analysis of the *KSP* repeat of the neurofilament heavy subunit in familial amyotrophic lateral sclerosis. *Neurology* **46,** 789–790.

Rosen, D. R., Siddique, T., Patterson, D., Figlewicz, D. A., Sapp, P., Hentati, A., et al. (1993) Mutations in Cu/Zn superoxide dismutase gene are associated with familial amyotrophic lateral sclerosis. *Nature* **362,** 59–62.

Rothstein, J. D., Dykes-Hoberg, M., Corson, L. B., Becher, M., Cleveland, D. W., Price, D. L., et al. (1999) The copper chaperone CCS is abundant in neurons and astrocytes in human and rodent brain. *J. Neurochem.* **72,** 422–429.

Rouleau, G. A., Clark, A. W., Rooke, K., Pramatarova, A., Krizus, A., Suchowersky, O., et al. (1996) SOD1 mutation is associated with accumulation of neurofilaments in amyotrophic lateral sclerosis. *Ann. Neurol.* **39,** 128–131.

Rowland, L. P. (1994) Natural history and clinical features of amyotrophic lateral sclerosis and related motor neuron diseases, in *Neurodegenerative Diseases* (Calne, D. B., ed.), W. B. Saunders, Philadelphia, pp. 507–521.

Rowland, L. P. (1998) What's in a name? Amyotrophic lateral sclerosis, motor neuron disease, and allelic heterogeneity. *Ann. Neurol.* **43,** 691–694.

Sankarapandi, S. and Zweier, J. L. (1999) Evidence against the generation of free hydroxyl radicals from the interaction of copper, zinc-superoxide dismutase and hydrogen peroxide. *J. Biol. Chem.* **274,** 34,576–34,583.

Schmidt, P. J., Rae, T. D., Pufahl, R. A., Hamma, T., Strain, J., O'Halloran, T. V., and Culotta, V. C. (1999) Multiple protein domains contribute to the action of the copper chaperone for superoxide dismutase. *J. Biol. Chem.* **274,** 23719–23725.

Shefner, J. M., Reaume, A. G., Flood, D. G., Scott, R. W., Kowall, N. W., Ferrante, R. J., et al. (1999) Mice lacking cytosolic copper/zinc superoxide dismutase display a distinctive motor axonopathy. *Neurology* **53,** 1239–1246.

Shibata, N., Hirano, A., Kobayashi, M., Siddique, T., Deng, H.-X., Hung, W.-Y., et al. (1996) Intense superoxide dismutase-1 immunoreactivity in intracytoplasmic hyaline inclusions of familial amyotrophic lateral sclerosis with posterior column involvement. *J. Neuropathol. Exp. Neurol.* **55**, 481–490.

Shibata, N., Hirano, M., and Kobayashi, K. (1993) Immunohistochemical demonstration of Cu/Zn superoxide dismutase in the spinal cord of patients with familial amyotrophic lateral sclerosis. *Acta Histochem. Cytochem.* **26**, 619–622.

Siddique, T., Figlewicz, D. A., Pericak-Vance, M. A., Haines, J. L., Rouleau, G., Jeffers, A. J., et al. (1991) Linkage of a gene causing familial amyotrophic lateral sclerosis to chromosome 21 and evidence of genetic-locus heterogeneity. *N. Engl. J. Med.* **324**, 1381–1384.

Siddique, T., Pericak-Vance, M. A., Brooks, B. R., Roos, R. P., Hung ,W.-Y., Antel J. P., et al. (1989) Linkage analysis in familial amyotrophic lateral sclerosis. *Neurology* **39**, 919–925.

Singh, R. J., Karoui, H., Gunther, M. R., Beckman, J. S., Mason, R. P., and Kalyanaraman, B. (1998) Reexamination of the mechanism of hydroxyl radical adducts formed from the reaction between familial amyotrophic lateral sclerosis-associated Cu,Zn superoxide dismutase mutants and H_2O_2. *Proc. Natl. Acad. Sci. USA* **95**, 6675–6680.

Takahashi, H., Oyanagi, K., Ohama, E., and Ikuta, F. (1992) Clarke's column in sporadic amyotrophic lateral sclerosis. *Acta Neuropathol.* **84**, 465–470.

Tomkins, J., Usher, P., Slade, J. Y., Ince, P. G., Curtis, A., Bushby, K., and Shaw, P. J. (1998) Novel insertion in the KSP region of the neurofilament heavy gene in amyotrophic lateral sclerosis (ALS). *Neuroreport* **9**, 3967–3970.

Trotti, D., Rolfs, A., Danbolt, N. C., Brown, R. H., and Hediger, M. A. (1999) SOD1 mutants linked to amyotrophic lateral sclerosis selectively inactivate a glial glutamate transporter. *Nature Neurosci.* **2**, 427–433.

Tu, P.-H., Raju, P., Robinson, K. A., Gurney, M. E., Trojanowski, J. Q., and Lee, V. M. Y. (1996) Transgenic mice carrying a human mutant superoxide dismutase transgene develop neuronal cytoskeletal pathology resembling human amyotrophic lateral sclerosis lesions. *Proc. Natl. Acad. Sci. USA* **93**, 3155–3160.

Valentine, J. S. and Gralla, E. B. (1997) Delivering copper inside yeast and human cells. *Science* **278**, 817–818.

Vechio, J. D., Bruijn, L. I., Xu, Z., Brown, R. H., Jr., and Cleveland, D. W. (1996) Sequence variants in human neurofilament proteins: absence of linkage to familial amyotrophic lateral sclerosis. *Ann. Neurol.* **40**, 603–610.

Vukosavic, S., Dubois-Dauphin, M., Romero, N., and Przedborski, S. (1999) Bax and Bcl-2 interaction in a transgenic mouse model of familial amyotrophic lateral sclerosis. *J. Neurochem.* **73**, 2460–2468.

Warita, H., Itoyama, Y., and Abe, K. (1999) Selective impairment of fast anterograde axonal transport in the peripheral nerves of asymptomatic transgenic mice with a G93A mutant SOD1 gene. *Brain Res.* **819**, 120–131.

Wiedau-Pazos, M., Goto, J. J., Rabizadeh, S., Gralla, E. B., Roe, J. A., Lee, M. K., Valentine, J. S., and Bredesen, D. E. (1996) Altered reactivity of superoxide dismutase in familial amyotrophic lateral sclerosis. *Science* **271**, 515–518.

Williams, D. B. and Windebank, A. J. (1993) Motor neuron disease, in *Peripheral Neuropathy*, 3rd ed. (Dyck, P. J., et al., eds.), W. B. Saunders, Philadelphia, pp. 1028–1050.

Williamson, T. L., Bruijn, L. I., Zhu, Q., Anderson, K. L., Anderson, S. D., Julien, J.-P., and Cleveland, D. W. (1998) Absence of neurofilaments reduces the selective vulnerability of motor neurons and slows disease caused by a familial amyotrophic lateral sclerosis-linked superoxide dismutase 1 mutant. *Proc. Natl. Acad. Sci. USA* **95**, 9631–9636.

Williamson, T. L. and Cleveland, D. W. (1999) Slowing of axonal transport is a very early event in the toxicity of ALS-linked SOD1 mutants to motor neurons. *Nature Neurosci.* **2**, 50–56.

Wong, P. C., Pardo, C. A., Borchelt, D. R., Lee, M. K., Copeland, N. G., Jenkins, N. A., et al. (1995) An adverse property of a familial ALS-linked SOD1 mutation causes motor neuron disease characterized by vacuolar degeneration of mitochondria. *Neuron* **14**, 1105–1116.

Wong, P. C., Rothstein, J. D., and Price, D. L. (1998) The genetic and molecular mechanisms of motor neuron disease. *Curr. Opin. Neurobiol.* **8,** 791–799.

Wong, P. C., Waggoner, D., Subramaniam, J., Tessarollo, L., Bartnikas, T. B., Culotta, V. C., et al. (2000) Copper chaperone for superoxide dismutase is essential to activate mammalian Cu/Zn superoxide dismutase. *Proc. Natl. Acad. Sci. USA* **97,** 2886–2891.

Zhang, B., Tu, P.-H., Abtahian, F., Trojanowski, J. Q., and Lee, V. M. Y. (1997) Neurofilaments and orthograde transport are reduced in ventral root axons of transgenic mice that express human SOD1 with a G93A mutation. *J. Cell Biol.* **139,** 1307–1315.

10
Pathogenesis of Ischemic Stroke

Mark P. Mattson and Carsten Culmsee

INTRODUCTION

Among neurodegenerative conditions associated with aging, stroke is the major cause of disability and death worldwide. The reduced blood supply that underlies a stroke results in degeneration and death of neurons because of a drastic reduction in their access to oxygen and glucose. Risk factors for stroke include hyperlipidemia (Gorelick et al., 1997), hypertension (Corvol et al., 1997), hyperhomocysteinemia (Perry et al., 1995), and high calorie intake (Bronner et al., 1995). Each of these risk factors has both genetic and environmental components. In addition, there are dominantly inherited disorders that result in stroke. A recent example, in which the genetic defect was identified, is a disorder called cerebral autosomal dominant arteriopathy with subcortical infarcts and leukoencephalopathy (CADASIL). CADASIL results from mutations in the gene encoding Notch-3 and is characterized by multiple cerebral ischemic events occurring during middle age (*see* Penix and Lanska, 1999, for review).

The biochemical and cellular events that lead to ischemic neuronal degeneration are beginning to be understood (*see* Dirnagl et al., 1999, for review). Levels of ATP in neurons are rapidly decreased following the onset of ischemia, resulting in impairment of membrane ion-motive ATPases, which in turn leads directly to membrane depolarization and activation of the N-methyl-D-aspartate (NMDA) subtype of synaptic glutamate receptor and voltage-dependent calcium channels. Levels of reactive oxygen species including superoxide, hydrogen peroxide, hydroxyl radical, and peroxynitrite increase, particularly during the reperfusion phase. Mitochondrial dysfunction ensues as the result of oxidative stress, energy failure, and disruption of cellular calcium homeostasis, and results in further production of free radicals, which damage cellular proteins, DNA, and membrane lipids. Particularly devastating for neurons is membrane lipid peroxidation, which results in the generation of toxic aldehydes such as 4-hydroxynonenal that impair the function of membrane ion-motive ATPases and glucose and glutamate transporters, and thereby amplify disruption of cellular calcium homeostasis (Mattson, 1998). The events just described have been documented in various cell-culture and animal models of stroke and, in many cases, cause–effect relationships have been established. For example, intervention studies have demonstrated the efficacy of glutamate receptor antagonists, calcium-stabilizing agents, and antioxidants in reducing ischemic damage to neurons (Fisher and Bogousslavsky, 1998).

From: *Pathogenesis of Neurodegenerative Disorders* Edited by: M. P. Mattson © Humana Press Inc., Totowa, NJ

In order to study effectively the molecular and cellular mechanisms responsible for neuronal dysfunction and death after a stroke, and to test potential therapeutic treatments, several different animal and cell-culture models have been established that mimic aspects of stroke in humans. In vivo models include the transient global forebrain ischemia model, in which the entire blood supply to the brain is transiently interrupted, and the focal cerebral ischemia model, in which the middle cerebral artery is occluded, resulting in damage to cerebral cortex and striatum in that hemisphere (Ginsberg and Butso, 1989; Mhairi-Macrae, 1992). Either transient or permanent occlusion of the middle cerebral artery can be performed in the focal model, with transient occlusion being generally accepted as the model that most closely replicates stroke in human patients. Studies of cultured dissociated neurons or brain slices have also proven valuable in elucidating mechanisms of ischemic injury and neuroprotection, as such preparations can be manipulated and monitored at levels of resolution not yet possible in vivo. Ischemia is mimicked in such in vitro systems by subjecting the preparations to hypoxia, glucose deprivation, excitatory amino acids, and/or oxidative insults that reproduce certain aspects of the environment neurons encounter following stroke in vivo.

Neurons are postmitotic, and therefore largely irreplaceable once they die (but *see* Svendsen and Smith, 1999, for recent evidence of the presence of neuronal precursor cells in the adult brain). The major focus of stroke research has therefore been to identify the mechanisms of ischemic neuronal injury and death, with the goal of improving functional outcome. Neurons are more vulnerable than most types of cells to ischemic injury because they rely on a constant supply of glucose as an energy source, and because they are excitable and express high levels of receptors for excitatory transmitters, which are concentrated in synapses. Synapses, which are often located at a relatively large distance from the cell body, are sites where the neurodegenerative process may be initiated in ischemic stroke and other neurodegenerative disorders (Mattson et al., 1998; Duan et al., 1999). In addition to its devastating effects on neurons, cerebral ischemia profoundly affects glial cells including astrocytes, microglia, and oligodendrocytes. Each type of glial cell serves important functions in modulating neuronal physiology and survival, and it is therefore important to understand their responses to cerebral ischemia, and how such responses affect neurons.

Focal cerebral ischemia results in characteristic histopathological changes that manifest as a necrotic core of tissue at the center of the infarcted cortex in which all cells die rapidly, and a surrounding "penumbral" region of variable size in which glial cells survive and neurons die over an extended time period of days to weeks (Fig. 1). It is the neurons in the penumbral region of the ischemic territory that can be saved by therapeutic intervention after a stroke in animal models. Biochemical and molecular analyses of brains of rats and mice subjected to focal cerebral ischemia have provided evidence that neurons in the penumbral region undergo a form of programmed cell death called apoptosis. The focus of this chapter will therefore be mainly on the molecular and cellular mechanisms underlying neuronal apoptosis in stroke, with considerations of both the events that trigger and execute neuronal apoptosis, and signaling mechanisms that may protect neurons against ischemic death.

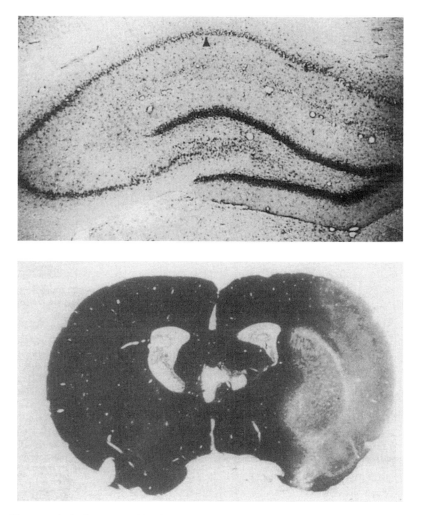

Fig. 1. Characteristic features of brain damage resulting from transient focal cerebral ischemia and transient global forebrain ischemia in rats. The upper micrograph shows a cresyl violet-stained coronal section of hippocampus from a rat that had been subjected to transient global forebrain ischemia (four-vessel occlusion model) and then allowed to live for 3 d. Note selective loss of neurons in region CA1 with preservation of neurons in CA3 and dentate gyrus. The lower micrograph shows an image of a TTC-stained coronal section from the brain of a rat that had been subjected to middle cerebral artery occlusion for 1 h followed by reperfusion for 24 h. Note greatly reduced staining of cortical and striatal tissue in the right (ischemic) hemisphere.

CHARACTERISTICS OF NEURONAL APOPTOSIS

Neurons undergoing apoptosis exhibit cell-body shrinkage, formation of cell-surface "blebs," and nuclear chromatin condensation and DNA fragmentation (Mattson et al., 2000). During the process of apoptosis the plasma membrane and organelles such as mitochondria and endoplasmic reticulum (ER) remain intact. However, fragmentation of neurites (axons and dendrites) occurs very early in the apoptotic process. In contrast to the ballooning and rapid disintegration of the neurites that occur in neurons undergoing necrosis, the neurite fragmentation that occurs during apoptosis manifests little or no swelling. Microglia are activated in response to neurons undergoing apoptosis and phagocytose the dying neurons, resulting in removal of the apoptotic neurons without adverse effects on neighboring cells. Therefore, neurons dying by apoptosis are typically observed in isolation, with adjacent cells being unaffected. Several biochemical changes that may distinguish neurons undergoing apoptosis have been documented including: a rapid increase in prostate apoptosis response-4 (Par-4) protein levels (Guo et al., 1998; Duan et al., 1999); translocation from the cytoplasm to the mitochondria of one or more members of the Bcl-2 protein family (Putcha et al., 1999); mitochondrial membrane depolarization and release of cytochrome c into the cytoplasm (Neame et al., 1998); activation of members of the caspase family of cysteine proteases (Chan and Mattson, 1999); flipping of phosphatidylserine from the inner leaflet of the plasma membrane to the cell surface (a signal for engulfment by microglia); and nuclear DNA fragmentation (Kruman et al., 1997; Mattson et al., 1998). Neurons undergoing necrosis typically do not manifest the aforementioned changes, although in certain experimental models neurons may exhibit both apoptotic and necrotic features (Nicotera and Lipton, 1999). Documentation of the morphological and biochemical changes just described can provide strong evidence that the mode of cell death is apoptosis. However, confirmation of apoptosis requires the demonstration that blockade of key steps in the cell-death process (caspase activation, macromolecular synthesis, and mitochondrial dysfunction) prevent the cell death.

MOLECULAR BIOLOGY OF NEURONAL APOPTOSIS

Complex cascades of gene expression occur in neurons and glial cells within the ischemic penumbra. It has been shown that some of the genes induced encode proteins that induce or enhance cell death, whereas others encode anti-apoptotic proteins (Table 1). Many of the proteins encoded by the injury-responsive genes exert their effects on shared signaling pathways that determine whether or not the neuron dies. For example, some cytokines (e.g., tumor necrosis factor [TNF]) and neurotrophic factors (e.g., nerve growth factor [NGF]) prevent neuronal apoptosis by activating the transcription factor NF-κB resulting in increased expression of anti-apoptotic proteins (Barger et al., 1995; Mattson et al., 2000a; Taglialatela et al., 1997). On the other hand, the pro-apoptotic protein prostate apoptosis response-4 (Par-4) may promote apoptosis by suppressing NF-κB activation (Camandola and Mattson, 2000). Additional pro-apoptotic gene products that have been identified in recent years include members of the caspase family of cysteine proteases such as interleukin-1β converting enzyme (ICE or caspase-1) and caspase-3 (Chan and Mattson, 1999). Caspases cleave proteins after aspartic acid residues and require the presence of the active cysteine in the middle of a conserved QACRG sequence. Some caspases activate themselves or other caspases, and thereby amplify proteolytic cascades. A remarkably large number of proteins are cleaved by caspases, and data are emerging as to if and how cleavage of a particular substrate promotes apoptosis (Chan and Mattson, 1999). Examples include cytoskeletal proteins, such as actin and spectrin, and integral membrane proteins such as glutamate receptor subunits (Chan et al., 1999a; Glazner et al., 2000) and presenilins (Kim et al., 1997;

Table 1
Examples of Proteins
That Promote Neuronal Apoptosis,
and Proteins That Can Prevent Neuronal Death
in Experimental Models of Stroke

Pro-apoptotic proteins	Anti-apoptotic proteins
Bax, Bad	Bcl-2, Bcl-XL
Par-4	XIAP, NAIP
Caspases 3, 8, and 9	PKCzeta, PKA
P53	NF-κB
	MnSOD
	Calmodulin, calbindin

Loetscher et al., 1997) and kinases such as protein kinase C (PKC) (Haussermann et al., 1999). Because of the complex array of caspase-mediated changes that occurs in neurons undergoing apoptosis, it is unlikely that cleavage of a single substrate is sufficient to effect the cell-death process.

Bcl-2 is the mammalian homolog of the *Caenorhabditis elegans* anti-death protein Ced-9 (Kroemer et al., 1997), and is one member of a family of related proteins that either promote or prevent apoptosis. Anti-apoptotic family members include Bcl-2, Bcl-XL, and Mcl-1, whereas pro-apoptotic members include Bax, Bad, and Bak. Overexpression of Bcl-2 in cultured sympathetic neurons or PC12 cells prevents apoptosis induced by NGF withdrawal (Guo et al., 1997), and also protects cultured neurons against death induced by amyloid β-peptide (Aβ) and oxidative insults (Furukawa et al., 1997a; Kruman et al., 1997). Bcl-2 is expressed throughout the developing nervous system, but its expression declines in the adult. Environmental factors affecting Bcl-2 expression are largely unknown, although recent findings suggest that certain cytokines and neurotrophic factors can stimulate Bcl-2 production (Camandola and Mattson, 2000).

OXIDATIVE STRESS AND PERTURBED CALCIUM HOMEOSTASIS IN NEURONAL APOPTOSIS IN STROKE

The evidence that various reactive oxygen species (ROS) play a role in ischemia-induced neuronal apoptosis is very strong. In rodents subjected to transient global forebrain ischemia, or focal ischemia, levels of oxidative damage to proteins, lipids, and DNA have been documented in the vulnerable neuronal populations (Carney et al., 1996; Chan, 1996; Hayashi et al., 1999). Particularly damaging to neurons is membrane lipid peroxidation because of its adverse effects on the function of membrane ion-motive ATPases and glucose and glutamate transporters (Fig. 2). The mechanism whereby lipid peroxidation disrupts neuronal ion homeostasis and induces apoptosis involves production of a toxic aldehyde called 4-hydroxynonenal, which covalently modifies membrane transporter proteins (Na^+/K^+-ATPase, glucose transporter, and glutamate transporter) and impairs their function (Kruman et al., 1997; Mark et al., 1997a, b; Keller et al., 1997). A major site of ROS production is mitochondria that produce high levels of superoxide anion radical, which serves as a precursor to other

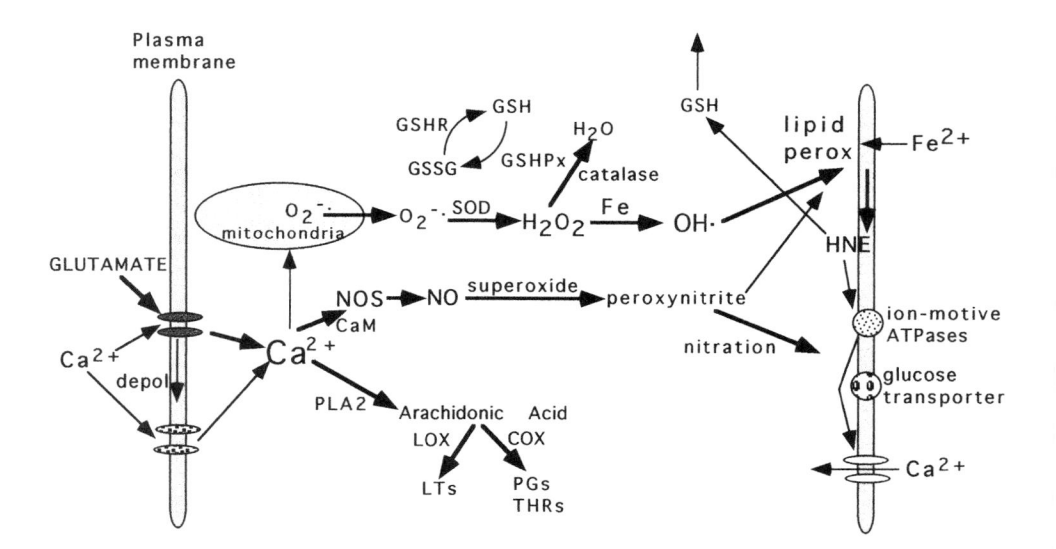

Fig. 2. Mechanisms for production and removal of reactive oxygen species in neurons, and the consequences of membrane lipid peroxidation for cellular ion homeostasis and energy metabolism. *See* text for discussion.

potentially destructive ROS including hydrogen peroxide, hydroxyl radical, and peroxynitrite. Such mitochondria-derived ROS are pivotal for neuronal apoptosis in many settings including ischemia, because overexpression of mitochondrial Mn-SOD can prevent apoptosis induced by an array of insults including cerebral ischemia (Keller et al., 1998). In addition, the anti-apoptotic property of Bcl-2 involves its ability to associate with mitochondria and stabilize their function, and to suppress accumulation of ROS and membrane lipid peroxidation (Hockenbery et al., 1993; Bruce-Keller et al., 1997; Kruman et al., 1997). Moreover, overexpression of Bcl-2 in cortical neurons via virus-mediated transfection reduces focal ischemic injury in vivo (Linnik et al., 1995), and transgenic mice overexpressing Bcl-2 exhibit increased resistance of neurons to ischemic injury (Kitagawa et al., 1998). That ROS production is required for ischemic neuronal injury is supported by data showing that antioxidants such as vitamin E, estrogens, and glutathione can protect cultured neurons from apoptosis induced by insults relevant to ischemic brain injury including glucose deprivation, and exposure to glutamate and Fe^{2+} (Goodman et al., 1996; Mark et al. 1997a). Beneficial effects of antioxidants in animal models of stroke have also been documented. For example, vitamin E protected CA1 neurons against transient global ischemia (Tagami et al., 1999), and uric acid (a scavenger of peroxynitrite and hydroxyl radical) reduced cortical infarct size after transient focal ischemia (Yu et al., 1998).

Disruption of cellular ion homeostasis, particularly increased levels of intracellular calcium, appears to be central to the cell-death process in ischemic stroke (*see* Mattson and Mark, 1996 for review). Calcium serves important functions in many fundamental physiological processes including regulation of neurite outgrowth and synaptogenesis, neurotransmitter release, and synaptic plasticity. Experimental studies have clearly shown, however, that excessive levels of intracellular calcium mediate death of cultured hippocampal and cortical neurons following exposure to hypoxia, hypoglycemia, and glutamate. Mitochondrial calcium overload has also been linked to neuronal apoptosis in some systems, and may play a pivotal role in mitochondrial membrane-permeability transition and release of cytochrome c (Kluck et al., 1997; Budd, 1998; Kruman and Mattson, 1999). In addition, calcium release from endoplasmic reticulum (ER) stores appears to make an important contribution to neuronal calcium overload in several different neurodegenerative conditions including stroke (see Mattson et al., 2000b for review). Accordingly, drugs that block calcium release from ER, including dantrolene (Tasker et al., 1998) and xestospongin (Mattson et al., 2000b), protect neurons against injury in experimental stroke models.

EVIDENCE FOR NEURONAL APOPTOSIS
IN STROKE PATIENTS AND EXPERIMENTAL MODELS

Owing to the relative lack of postmortem brain tissue from stroke patients (at post-stroke time-points that correspond to the time window when neuronal death is occurring), there is very little information available concerning the mode of neuronal death. However, in the few studies that have been performed the data do suggest a contribution of neuronal apoptosis to the brain damage. In one study, a DNA end-labeling technique was employed to identify neurons with DNA damage in parallel analyses of brain tissues from stroke patients and rats subjected to transient global forebrain ischemia (Guglielmo et al., 1998). The latter study revealed a similar temporal profile of appearance of neurons with DNA damage in both stroke patients and in the rat model. Analyses of postmortem brain tissue from patients that suffered severe brain ischemia as the result of cardiac arrest showed that levels of the DNA damage-responsive proteins PARP and Ku80 were increased, particularly in regions that also contained neurons with DNA damage including CA1 hippocampal neurons and neurons in deep layers of cerebral cortex (Love et al., 1998). In another study, a group of stroke patients was followed prospectively for 90 d with clinical evaluation, radiological assessment, and measurements of levels of apoptosis-related proteins in cerebrospinal fluid (CSF) (Tarkowski et al., 1999). Interestingly, levels of soluble Bcl-2 were decreased during the first 3 d following stroke onset. The latter findings suggest that levels of anti-apoptotic proteins may be reduced following stroke and may thereby contribute to apoptotic death of neurons.

Neurons in the penumbral regions of a focal ischemic stroke that die relatively slowly over periods of days to weeks exhibit DNA strand breaks consistent with apoptosis (Linnik et al., 1993). Overexpression of Bcl-2 in neurons in the cerebral cortex (via virus-mediated transfection) increases their resistance to ischemia-induced death (Linnik et al., 1995). The time-course of DNA fragmentation in tissue samples from hippocampus of rats following transient global forebrain ischemia is delayed for several days following the ischemic insult (Fig. 3). Mitochondrial alterations consistent with apoptosis have been documented in neurons after focal cerebral ischemia (Murakami et al., 1998), providing further support for a major contribution of apoptosis to the pathophysiology of stroke. In addition, cytochrome c is released from the mitochondria in cortical tissue following transient focal ischemia in rats (Fujimura et al., 1998). That such mitochondrial alterations are central to the cell-death process is supported by data showing that overexpression of Mn-SOD in transgenic mice results in reduced infarct size following middle cerebral artery occlusion–reperfusion (Keller et al.,

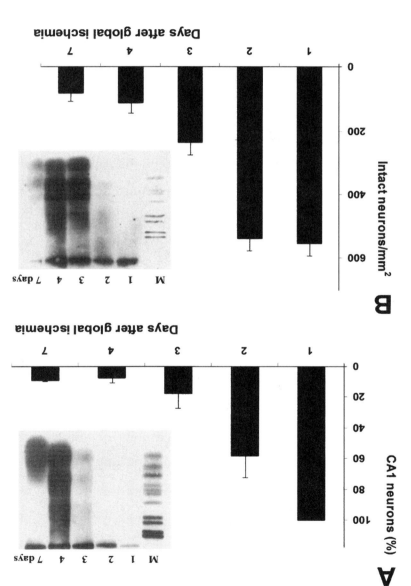

Fig. 3. Delayed apoptosis occurs after cerebral ischemia. Time courses of neuronal damage and associated DNA fragmentation in hippocampus and striatum after transient forebrain ischemia. Rat brains were removed 1, 2, 3, 4, and 7 d after 10 min of global ischemia induced by the occlusion of both common carotid arteries and imposition of hypotension. Neuronal damage in the CA1 subfield of the hippocampus (**A**) and the striatum (**B**) was quantified in 4 rats/time point by celestine blue and acid fuchsin staining. Data are presented as mean +/– SD. For the detection of DNA fragmentation the DNA was extracted from the brain tissue and digoxigenin end-labeled. The blots presented demonstrate the time-course of DNA fragmentation in the hippocampus (A) and the striatum (B) at 1, 2, 3, 4, and 7 d after global ischemia. Note the correlation between delayed neuronal cell death induced by ischemia and the sustained detection of a DNA ladder typical for apoptosis in each brain region. Modified from Zhu et al. (1998).

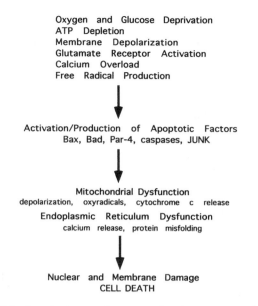

Oxygen and Glucose Deprivation
ATP Depletion
Membrane Depolarization
Glutamate Receptor Activation
Calcium Overload
Free Radical Production

Activation/Production of Apoptotic Factors
Bax, Bad, Par-4, caspases, JUNK

Mitochondrial Dysfunction
depolarization, oxyradicals, cytochrome c release
Endoplasmic Reticulum Dysfunction
calcium release, protein misfolding

Nuclear and Membrane Damage
CELL DEATH

Fig. 4. Sequence of biochemical cascades occurring in neurons subjected to ischemia.

1998), and administration of cyclosporin A (an inhibitor of mitochondrial membrane-permeability transition) also reduces infarct size in a similar ischemia model (Matsumoto et al., 1999). Interestingly, mutations in the *presenilin-1* (*PS1*) gene that cause Alzheimer's disease (AD) promote neuronal apoptosis, and increase vulnerability of cortical neurons to focal ischemia-induced cell death (Mattson et al., 2000c).

Molecular and biochemical cascades known to participate in apoptosis are activated in neurons following cerebral ischemia (Fig. 4). For example, caspase-8 and caspase-3 are activated in neurons in the ischemic penumbra prior to their death (Asahi et al., 1997; Velier et al., 1999). Active caspase-8 was located mainly in pyramidal neurons of layer V of cerebral cortex, whereas active caspase-3 was observed in neurons in layers II and III beginning 24 h following the onset of ischemia. Intraventricular delivery of caspase inhibitors prior to, or 6 h after reperfusion, resulted in a significant decrease in cerebral infarct size (Hara et al., 1997; Endres et al., 1998), and ischemic brain injury was decreased in mice lacking caspase-1 (Schielke et al., 1998). Cellular patterns of gene expression are altered in cells in the ischemic penumbra following focal cerebral ischemia. Levels of the pro-apoptotic receptor Fas and TNF-related apoptosis inducing ligand (TRAIL) increase in vulnerable brain regions following focal ischemia (Martin-Villalba et al., 1999).

Cell shrinkage and DNA fragmentation occur in hippocampal CA1 neurons with a delayed time-course following transient global forebrain ischemia in rodents (MacManus et al., 1993; Nitatori et al., 1995). Both single-strand-breaks and double-strand-breaks occur between 48 and 72 h postreperfusion in the rat model (Jin et al., 1999). Expression of several apoptosis-related genes such as *p53* and *Bax* occurs in hippocampal neurons after transient global forebrain ischemia in rodents (Antonawich et al., 1998; McGahan et al., 1998), and expression of the DNA damage-sensing protein GADD-45 is increased in hippocampal neurons (Hou et al., 1997). However, some neurons may die by necrosis, as indicated by electron microscope analyses, which reveal evidence of necrotic death of CA1 neurons, but no evidence of typical apoptotic morphology, following transient global ischemia (Colbourne et al., 1999). Nevertheless, much of the ischemic neuronal death is prevented or delayed by administration of caspase inhibitors. For example, intraventricular administration of the pseudosubstrate caspase inhibitor N-tosyl-L-phenylalanyl chloromethyl ketone reduced damage to CA1 neurons following transient global ischemia in rats (Hara et al., 1998) and overexpression of X chromosome-linked inhibitor of apoptosis protein (XIAP) prevented caspase-3 activation, death of CA1 neurons, and spatial memory deficits (Xu et al., 1999). Moreover, transgenic mice overexpressing Bcl-2 exhibit increased resistance of hippocampal neurons to transient global forebrain ischemia (Kitagawa et al., 1998).

Additional data concerning the mechanisms of neuronal injury and death in stroke have come from cell-culture studies. Cultured hippocampal or cortical neurons can be exposed to conditions such as glucose deprivation, hypoxia, and/or glutamate that mimic certain aspects of the environment of neurons after ischemic stroke. Morphological changes consistent with apoptosis, particularly cell shrinkage and nuclear DNA fragmentation, occur in cultured rodent-brain neurons following exposure to hypoxia (Bossenmeyer et al., 1998), glucose deprivation (Cheng and Mattson, 1991; Kalda et al., 1998), and glutamate (Ankarcrona et al., 1995) and oxidative insults (Kruman et al., 1997). Mitochondrial alterations known to occur in cells undergoing apoptosis, including a decrease in mitochondrial transmembrane potential (Mattson et al., 1993a, 1995; White and Reynolds, 1996; Bruce-Keller et al., 1999a) and release of cytochrome c (Hortelano et al., 1997), have been documented in several of the cell-culture models. Treatment of cultured neurons with cyclosporin A suppresses cell death induced by ischemia-relevant insults (Keller et al., 1998). Cyclosporin A also protects hippocampal neurons against hypoglycemic injury in vivo (Friberg et al., 1998). Exposure of cultured hippocampal and cortical neurons to excitotoxic and metabolic insults results in activation of caspase-3, and caspase inhibitors protect the neurons against death in such models (Mattson et al., 1998), demonstrating a key role for caspase activation in the cell-death process.

Proteins and lipid mediators that may play key roles in executing the neuronal death process following ischemic injury are being identified in studies of cell culture and in vivo stroke models. The pro-apoptotic protein Par-4 is induced in cultured hippocampal neurons following exposure to glutamate (Duan et al., 1999) and oxidative insults (Chan et al., 1999b). Studies using antisense technology and expression of a dominant-negative form of Par-4 showed that Par-4 is a key link in the chain of events leading to mitochondrial dysfunction, caspase activation, and nuclear apoptosis (Guo et al., 1998; Duan et al., 1999; Chan et al., 1999b). Par-4 protein levels increase in cortical cells in the ischemic penumbral region within 2 h of reperfusion in a rat model of focal ischemic brain injury (Culmsee and Mattson, 2000). Infusion of a Par-4 antisense oligodeoxynucleotide into the lateral ventricle immediately prior to ischemia resulted in a significant decrease in infarct volume, suggesting that Par-4 production is required for the death of many neurons in this model (Culmsee and Mattson, 2001). Hypoxia/reoxygenation in cultured forebrain neurons induces cJun, Jun B, Jun D, c-Fos, JNK1, and JNK3 in association with delayed neuronal apoptosis (Chihab

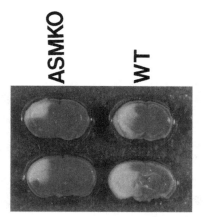

Fig. 5. Focal ischemic brain injury is decreased in mice lacking acidic sphingomyelinase. TTC-stained coronal brain sections, at two different rostral-caudal levels, from a wild-type mouse (WT) and a mouse lacking acidic sphingomyelinase (ASMKO) 24 h after middle cerebral artery occlusion-reperfusion. Note decreased infarct size (white tissue) in both the cortex and striatum of the ASMase-deficient mouse. Modified from Yu et al. (2000).

et al., 1998). Expression/activation of these proteins suggests a role for the transcription factor AP-1 in neuronal apoptosis, although its precise role remains to be established.

Cerebral ischemia results in the production of several different lipid-derived molecules that may promote neuronal death in several different ways. One pathway induced by cerebral ischemia involves cleavage of membrane sphingomyelin by acidic sphingomyelinase (ASMase) resulting in generation of the lipid mediator ceramide. Transient focal cerebral ischemia in mice induces large increases in ASMase activity, ceramide levels, and production of inflammatory cytokines (Yu et al., 2000). The extent of brain-tissue damage is decreased (Fig. 5) and behavioral outcome improved in mice lacking ASMase, and cultured neurons lacking ASMase exhibit decreased vulnerability to excitotoxicity and hypoxia. Moreover, treatment with a drug that inhibits ASMase activation and ceramide production (D-609) reduces focal ischemic brain injury and improves behavioral outcome (Yu et al., 2000). Mice lacking phospholipase-A2 (PLA2), an enzyme required for production of leukotrienes and prostaglandins exhibit decreased brain damage after focal cerebral ischemia, strongly suggesting an important role for one or more lipid mediators generated by PLA2 in ischemic neuronal injury (Bonventre et al., 1997). Platelet-activating factor (PAF) is generated in brain cells in response to cerebral ischemia, and may contribute to neuronal damage because a PAF receptor antagonist significantly decreases infarct size in a focal ischemia model (Aspey et al., 1997). Another lipid-derived compound that is implicated in neuronal apoptosis is lysophosphatidic acid (LPA). Cell-culture studies have shown that LPA can induce neuronal apoptosis, which is mediated by mitochondrial dysfunction and caspase activation (Holtsberg et al., 1998). In addition to lipid-derived signals generated by the activity of enzymes such as PLA2 and acidic sphingomyelinase, neurotoxic lipid-peroxidation products generated by nonenzymatic processes, most notably lipid peroxidation, may contribute to ischemic neuronal apoptosis. Although 4-hydroxynonenal is the most widely studied of such lipid-peroxidation products (Mattson, 1998), additional lipid-derived neurotoxic compounds are being identified including acrolein (Uchida, 1999).

NEUROPROTECTIVE SIGNAL TRANSDUCTION PATHWAYS IN CEREBRAL ISCHEMIA

Studies performed during the past 12 yr have clearly shown that cerebral ischemia results in the activation of several different inter- and intracellular signaling cascades that serve the function of protecting neurons against dysfunction and death (*see* Mattson and Furukawa, 1996; Mattson and Lindvall, 1997 for review). The first group of such neuroprotective signals that was shown to protect neurons against insults relevant to the pathogenesis of stroke was neurotrophic factors. Cell-culture studies showed that basic fibroblast growth factor (bFGF) can protect neurons against ischemia-relevant insults including glutamate toxicity (Mattson et al., 1989) and glucose deprivation (Cheng and Mattson, 1991). Subsequent studies demonstrated that several other neurotrophic factors including nerve growth factor (NGF), brain-derived neurotrophic factor (BDNF), insulin-like growth factors (IGFs), and platelet-derived growth factor (PDGF) protect cultured hippocampal and cortical neurons against injury induced by glucose deprivation and hypoxia (Cheng and Mattson, 1991, 1992, 1994, 1995; Mattson et al., 1993a). A second group of neuroprotective signals is cytokines. For example, tumor necrosis factor-α (TNF-α) protects cultured hippocampal neurons against excitotoxic, metabolic, and oxidative insults relevant to stroke (Cheng et al., 1994; Barger et al., 1995; Mattson et al., 1997a). Two other cytokines that have been demonstrated to protect neurons against ischemia-relevant insults are transforming growth factor-β (TGF-β) (Prehn et al., 1993) and the secreted form of amyloid precursor protein (sAPP) (Mattson et al., 1993b).

In vivo studies have shown that several different neurotrophic factors and cytokines serve a neuroprotective function after stroke. Studies of rodent models of focal and transient global ischemia have documented increases in the expression of several different neurotrophic factors and cytokines in response to ischemia (*see* Mattson and Lindvall, 1997, for review). Levels of bFGF, TNF, and BDNF are increased in cerebral cortex and hippocampus following transient global forebrain ischemia (Endoh et al., 1994; Uno et al., 1997) and transient focal ischemia (Speliotes et al., 1996; Botchkina et al., 1997). Infarct size was significantly decreased in rats administered bFGF (via intravenous infusion) (Fisher et al., 1995). Intraventricular administration of NGF (Shigeno et al., 1991), sAPPα (Smith-Swintosky et al., 1994), and TGF-β (Henrich-Noack et al., 1996) reduced damage to CA1 hippocampal neurons following transient global forebrain ischemia. Overexpression of NGF, effected by transfection of cells with a plasmid containing a c-fos-responsive promoter, resulted in decreased brain damage following permanent occlusion of the middle cerebral artery in rats (Guegan et al., 1998). Intraventricular infusion of BDNF after cerebral ischemia protected CA1 neurons against delayed death in the four-vessel occlusion rat model of transient global forebrain ischemia (Kiprianova et al., 1999). The latter study further showed that BDNF suppresses activation of astrocytes and microglia as indicated by a decrease in levels of inducible nitric oxide synthase (NOS) production. Administration of TGF-β reduces infarct size in a mouse focal ischemia model (Prehn et al., 1993), and injection of glial cell-derived neurotrophic factor (GDNF) into the dorsal hippocampus protects CA1 neurons against ischemic injury (Miyazaki et al., 1999).

Blockade of specific neurotrophic factor and cytokine signaling pathways, using gene targeting and pharmacological approaches, have provided insight into the roles for these injury-induced factors in the pathogenesis of stroke. Data obtained in studies of TNF receptor knockout mice demonstrated that ischemia-induced TNF serves a neuroprotective function following focal ischemia-reperfusion (Bruce et al., 1996; Gary et al., 1998). The drug clenbuterol has been shown to increase NGF production in brain cells and, when administered to adult rats, clenbuterol reduces infarct volume following permanent focal ischemia; administration of NGF antisense oligonucleotides abolished the protective effect of clenbuterol indicating a key role for endogenous NGF in the neuroprotective mechanism (Culmsee et al., 1999a).

The specific mechanisms whereby neurotrophic factors and cytokines protect neurons against ischemic injury are being elucidated (*see* Mattson and Furukawa, 1996, for review). The signal transduction mechanism begins with activation of membrane receptors, which, in the cases of neurotrophins (NGF, BDNF, NT-3 and NT-4/5), bFGF, and IGFs, possess intrinsic tyrosine kinase activity. Binding of neurotrophins results in receptor dimerization and transautophosphorylation of cytoplasmic of cytoplasmic domains that, in turn, initiates cascades of kinase activation. Some of the kinases (e.g., mitogen-activated protein kinase [MAP]) activate transcription factors. Genes regulated by neurotrophic factors include those encoding antioxidant enzymes such as superoxide dismutases (SOD) and glutathione peroxidase (Cheng and Mattson, 1995; Mattson et al., 1995; 1997a), calcium-binding proteins such as calbindin (Cheng et al., 1994), and excitatory amino acid receptors (Mattson et al., 1993c; Cheng and Mattson, 1995). Neurotrophic factors and cytokines may also induce expression of anti-apoptotic members of the Bcl-2 family (Camandola and Mattson, 2000).

TNF can protect neurons against apoptosis by a signaling pathway involving activation of the transcription factor NF-κB (Fig. 6, and *see* Mattson et al., 2000a, for review). NF-κB is comprised of three subunits, two of which (p50 and p65) form the activate transcription factor dimer, with an additional subunit called IκB serving as an inhibitory subunit. NF-κB can be activated by many different stimuli including neurotrophic factors, neurotransmitters, and oxidative stress. Activation may require phosphorylation of IκB, which causes it to dissociate from the other two subunits, which then translocate to the nucleus where they bind to specific DNA sequences present in the enhancer region of κB-responsive genes. The mechanism whereby NF-κB activation increases resistance of neurons to apoptosis (Barger et al., 1995; Mattson et al., 1997a) involves increased expression of genes encoding manganese SOD (Bruce et al., 1996; Mattson et al., 1997a; Yu et al., 1999a, b) and Bcl-2 (Tamatani et al., 1999; Camandola and Mattson, 2000). Cerebral ischemia induces activation of NF-κB activity in neurons, which apparently represents a neuroprotective response because neuronal damage is increased when NF-κB activity is blocked using a decoy DNA approach (Mattson et al., 1997a; Yu et al., 1999a, b). It should be noted, however, that NF-κB activation in microglia may enhance the production of neurotoxic agents in this type of immune cell, which may contribute to ischemic neuronal injury (*see* Mattson et al., 2000a, for review). Another example of NF-κB-mediated neuroprotection comes from studies of

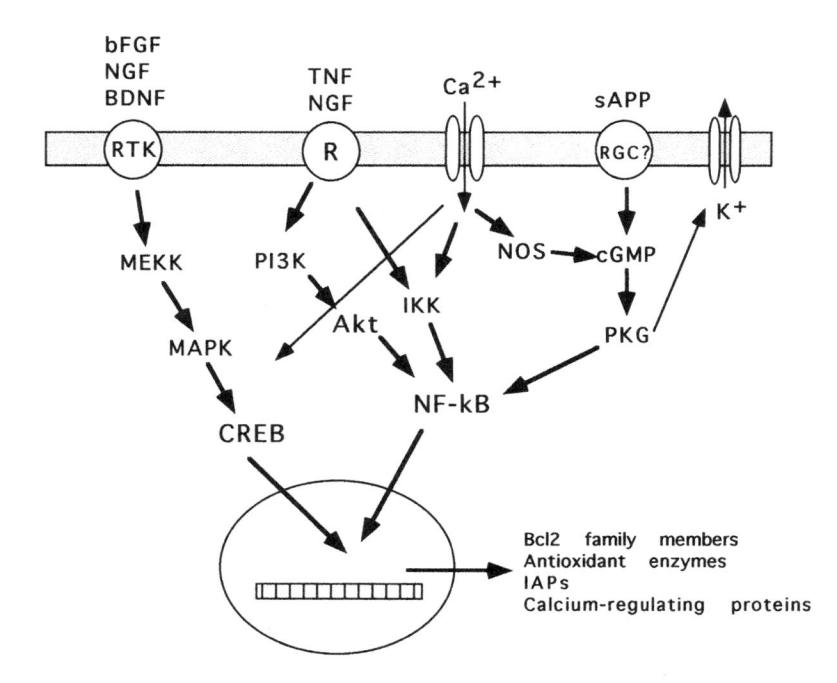

Fig. 6. Examples of different signaling pathways that can protect neurons against ischemic injury. BDNF, brain-derived neurotrophic factor; bFGF, basic fibroblast growth factor; cGMP, cyclic guanosine monophosphate; CREB, cyclic AMP response binding protein; IKK, IκB kinase; MAPK, mitogen-activated protein kinase; NGF, nerve growth factor; NOS, nitric oxide synthase; RTK, receptor tyrosine kinase; sAPP, secreted form of amyloid precursor protein; TNF, tumor necrosis factor.

the signal transduction mechanism of sAPPα. sAPPα stimulates production of cyclic GMP (cGMP) and activation of cGMP-dependent protein kinase (Barger and Mattson, 1996). The cGMP pathway may protect cells via both a transcription-independent pathway involving dephosphorylation (activation) of potassium channels (Furukawa et al., 1996), and a transcription-dependent pathway involving NF-κB (Barger and Mattson, 1996). Collectively, the available data suggest that neurotrophic factors and some cytokines can protect neurons against ischemic cell death by activating signaling pathways that result in suppression of oxyradical production and stabilization of cellular calcium homeostasis.

In addition to neuroprotective signaling pathways activated by intercellular signals, ischemia may activate anti-apoptotic signaling mechanisms within cells. One example involves actin filaments. Elevation of intracellular calcium levels, as occurs in neurons subjected to ischemia, activates a protein called gelsolin that cleaves actin filaments. Actin depolymerization results in reduced calcium influx through NMDA receptors and voltage-dependent calcium channels (Furukawa et al., 1995, 1997b) by a mechanism that likely involves interactions of actin-associated proteins with membrane calcium channels. Experiments performed in gelsolin knockout mice have provided evidence that this actin-based signaling pathway serves a neuroprotective role following excitotoxic and ischemic insults (Furukawa et al., 1997b; Endres et al., 1999).

Ischemia also induces expression of "heat-shock" proteins, many of which serve as chaperones that control protein folding and trafficking, and removal of damaged proteins. Levels of several such stress-responsive proteins including heat-shock protein-70 (HSP-70) and glucose-regulated protein-78 (GRP-78) are increased in neurons following cerebral ischemia (Wang et al., 1993; Wagstaff et al., 1996). Experimental studies have shown that HSP-70 and GRP-78 play pivotal roles in determining whether or not a neuron survives exposure to excitotoxic, oxidative, and metabolic insults (Lowenstein et al., 1991; Yu et al., 1999a; Yu and Mattson, 1999). Mice overexpressing HSP-70 exhibit reduced neuronal damage following permanent focal ischemia compared to wild-type mice, suggesting a key role for HSP-70 in preserving neurons following stroke (Plumier et al., 1997). Stress proteins may play a central role in the well-established phenomenon of "ischemic preconditioning" in which rodents subjected to mild ischemia are more resistant to a subsequent ischemic stroke (Barone et al., 1998).

INVOLVEMENT OF GLIAL CELLS IN ISCHEMIC STROKE

Astrocytes, oligodendrocytes, and microglia are the three major types of glial cells present in the brain. Quite striking alterations in each type of glial cell have been documented in experimental stroke models.

Astrocytes possess gap junctions, which are channels that connect the cytoplasm of one cell with that of a neighboring cell and provide electronic coupling that serves the purpose of rapid propagation of signals through cell networks. Studies of primary hippocampal neurons growing on a monolayer of coupled astrocytes have provided evidence that gap junction communication between astrocytes can serve a neuroprotective function. Uncoupling astrocyte junctions result in enhanced generation of intracellular peroxides, impairment of mitochondrial function, and cell death after exposure of the cultures to oxidative insults (Blanc et al., 1998). Increases in intracellular calcium levels in neurons were induced by oxidative insults were greatly enhanced when astrocytic gap junctions were uncoupled. Uncoupling of gap junctions also greatly increased the vulnerability of neurons in hippocampal slices to excitotoxin-induced injury (Blanc et al., 1998). Finally, GA exacerbated kainate- and $FeSO_4$-induced injury to pyramidal neurons in organotypic hippocampal slice cultures. The data suggest that interastrocytic gap-junctional communication decreases neuronal vulnerability to oxidative injury by a mechanism involving stabilization of cellular calcium homeostasis and dissipation of oxidative stress.

IMPLICATIONS FOR TREATMENT OF STROKE

The mechanisms of neuronal degeneration and neuroprotective signaling pathways involved in ischemic stroke suggest several different approaches aimed at reducing neuronal death and improving outcome in stroke patients. Antioxidants that have proven effective in animal stroke models, such as vitamin E and uric acid, merit trials in stroke patients and, indeed, preliminary trials of vitamin E do suggest beneficial effects (Leppala et al., 2000). Calcium influx or release blockers provide another category of potential therapeutic agents for stroke. Glutamate receptor antagonists and blockers of voltage-dependent calcium channels (Kobayashi and Mori, 1998) and inhibitors of calcium release from endoplasmic reticulum (Mattson et al., 2000b) have proven effective in one or more animal stroke models and are in development for clinical trials in stroke patients.

Glucocorticoids and estrogens are two types of steroid hormones that are increasingly recognized as potent modulators of neuronal vulnerability in various experimental models of neurodegenerative disorders including stroke. Cerebral ischemia is a severe stress and, as such, there is a large increase in glucocorticoid production that occurs after a stroke. Studies in which production of glucocorticoids is blocked by removal of the adrenal glands or by administration of metyrapone, an inhibitor of glucocorticoid synthesis, have shown that glucocorticoids enhance ischemic neuronal death (Smith-Swintosky et al., 1996). Epidemiological studies and data from animal stroke models suggest gender differences in stroke outcome, which are likely related to the neuroprotective properties of estrogens (Alkayed et al., 1998). Estrogens have been shown to protect neurons against excitotoxic, oxidative, and metabolic insults (Goodman et al., 1996; Mattson et al., 1997b). Infarct size after focal ischemia is significantly decreased in rats and mice administered estrogens (Culmsee et al., 1999b; Fukuda et al., 2000), and estrogen also protects CA1 hippocampal neurons against transient global ischemia (Sudo et al., 1997).

A very effective prophylactic method for reducing neuronal injury in models of various neurodegenerative conditions, including stroke, was recently demonstrated. Maintenance of adult rats and mice on a dietary restriction regimen (reduced calorie intake with maintained nutrition) increased resistance of neurons to ischemia and related excitotoxic and oxidative insults, and improved behavioral outcome (Bruce-Keller et al., 1999b; Duan and Mattson, 1999; Yu and Mattson, 1999). The latter studies provided evidence that the mechanism whereby dietary restriction benefits neurons involves upregulation of stress proteins. Dietary restriction is known to extend lifespan and reduce incidence of various age-related diseases including cardiovascular disease, diabetes, and cancer (Sohal and Weindruch, 1996). Based on experimental and epidemiological data, we expect that dietary restriction will prove to be an effective approach for decreasing the incidence of stroke, and for improving stroke outcome.

CONCLUSIONS

Cell-culture and animal models of ischemic brain injury have proven invaluable in elucidating the cellular and molecular underpinnings of neuronal cell death in stroke. A working model of the cascade of events that leads to neuronal apoptosis following stroke is shown in Fig. 4. Oxyradical production, perturbed calcium homeostasis, caspase activation, and mitochondrial dysfunction appear to be important mediators of neuronal apoptosis following stroke. Evolving technologies for monitoring and manipulating reactive oxygen species, intracellular calcium levels, mitochondrial function, and caspase activity provide the opportunity to establish cause-effect relationships in many different paradigms of apoptosis relevant to the pathogenesis of stroke. A variety of signaling pathways that prevent neuronal apoptosis are being identified in cell-culture studies. Such pathways often involve induction of genes that encode proteins that have antioxidant functions or that modulate cellular calcium homeostasis. Such studies are identifying compounds and dietary manipulations with anti-apoptotic actions that may prove beneficial in preventing neuronal death and improving outcome following stroke in humans.

REFERENCES

Alkayed, N. J., Harukuni, I., Kimes, A. S., London, E. D., Traystman, R. J., and Hurn, P. D. (1998) Gender-linked brain injury in experimental stroke. *Stroke* **29,** 159–165.

Ankarcrona. M., Dypbukt, J. M., Bonfoco, E., Zhivotovsky, B., Orrenius, S., Lipton, S. A., and Nicotera, P. (1995) Glutamate-induced neuronal death: a succession of necrosis or apoptosis depending on mitochondrial function. *Neuron* **15,** 961–973.

Antonawich, F. J., Krajewski, S., Reed, J. C., and Davis, J. N. (1998) Bcl-x(l) Bax interaction after transient global ischemia. *J. Cereb. Blood Flow Metab.* **18,** 882–886.

Asahi, M., Hoshimaru, M., Uemura, Y., Tokime, T., Kojima, M., Ohtsuka, T., et al. (1997) Expression of interleukin-1 beta converting enzyme gene family and bcl-2 gene family in the rat brain following permanent occlusion of the middle cerebral artery. *J. Cereb. Blood Flow Metab.* **17,** 11–18.

Aspey, B. S., Alp, M. S., Patel, Y., and Harrison, M. J. (1997) Effects of combined glutamate and platelet-activating factor inhibition on the outcome of focal cerebral ischemia— an initial screening study. *Metab. Brain Dis.* **12,** 237–249.

Barger, S. W., Horster, D., Furukawa, K., Goodman, Y., Krieglstein, J., and Mattson, M. P. (1995) Tumor necrosis factors a and b protect neurons against amyloid β-peptide toxicity: evidence for involvement of a κB-binding factor and attenuation of peroxide and Ca2+ accumulation. *Proc. Natl. Acad. Sci. USA* **92,** 9328–9332.

Barger, S. W. and Mattson, M. P. (1996) Induction of neuroprotective κB-dependent transcription by secreted forms of the Alzheimer's β-amyloid precursor. *Mol. Brain Res.* **40,** 116–126.

Barone, F. C., White, R. F., Spera, P. A., Ellison, J., Currie, R. W., Wang, X., and Feuerstein, G. Z. (1998) Ischemic preconditioning and brain tolerance: temporal histological and functional outcomes, protein synthesis requirement, and interleukin-1 receptor antagonist and early gene expression. *Stroke* **29,** 1937–1950.

Blanc, E. M., Bruce-Keller, A. J., and Mattson, M. P. (1998) Astrocytic gap junctional communication decreases neuronal vulnerability to oxidative stress-induced disruption of Ca^{2+} homeostasis and cell death. *J. Neurochem.* **70,** 958–970.

Bonventre, J. V., Huang, Z., Taheri, M. R., O'Leary, E., Li, E., Moskowitz, M. A., and Sapirstein, A. (1997) Reduced fertility and post-ischemic brain injury in mice deficient in cytosolic phospholipase A2. *Nature* **390,** 622–625.

Bossenmeyer, C., Chihab, R., Muller, S., Schroeder, H., and Daval, J. L. (1998) Hypoxia/ reoxygenation induces apoptosis through biphasic induction of protein synthesis in cultured rat brain neurons. *Brain Res.* **787,** 107–116.

Botchkina, G. I., Meistrell, M. E., Botchkina, I. L., and Tracey, K. J. (1997) Expression of TNF and TNF receptors (p55 and p75) in the rat brain after focal cerebral ischemia. *Mol. Med.* **3,** 765–781.

Bronner, L. L., Kanter, D. S., and Manson, J. E. (1995) Primary prevention of stroke. *N. Engl. J. Med.* **333,** 1392–1400.

Bruce, A. J., Boling, W., Kindy, M. S., Peschon, J., Kraemer, P. J., Carpenter, M. K., et al. (1996) Altered neuronal and microglial responses to brain injury in mice lacking TNF receptors. *Nature Med.* **2,** 788–794.

Bruce-Keller, A. J., Begley, J. G., Fu, W., Butterfield, D. A., Bredesen, D. E., Hutchins, J. B., et al. (1997) Bcl-2 protects isolated plasma and mitochondrial membranes against lipid peroxidation induced by hydrogen peroxide and amyloid b-peptide. *J. Neurochem.* **70,** 31–39.

Bruce-Keller, A. J., Geddes, J. W., Knapp, P. E., McFall, R. W., Keller, J. N., Holtsberg, F. W., et al. (1999a) Anti-death properties of TNF against metabolic poisoning: mitochondrial stabilization by MnSOD. *J. Neuroimmunol.* **93,** 53–71.

Bruce-Keller, A. J., Umberger, G., McFall, R., and Mattson, M. P. (1999b) Food restriction reduces brain damage and improves behavioral outcome following excitotoxic and metabolic insults. *Ann. Neurol.* **45,** 8–15.

Budd, S. L. (1998) Mechanisms of neuronal damage in brain hypoxia/ischemia: focus on the role of mitochondrial calcium accumulation. *Pharmacol. Ther.* **80,** 203–229.

Camandola, S. and Mattson, M. P. (2000) The pro-apoptotic action of Par-4 involves inhibition of NF-κB activity and suppression of Bcl-2 expression. *J. Neurosci. Res.* **61,** 134–139.

Carney, J. M., Hall, N. C., Cheng, M., Wu, J., and Butterfield, D. A. (1996) Protein and lipid oxidation following ischemia/reperfusion injury, the role of polyamines: an electron paramagnetic resonance analysis. *Adv. Neurol.* **71,** 259–268.

Chan, P. H. (1996) Role of oxidants in ischemic brain damage. *Stroke* **27,** 1124–1129.

Chan, S. L. and Mattson, M. P. (1999) Caspase and calpain substrates: roles in synaptic plasticity and cell death. *J. Neurosci. Res.* **58,** 167–190.

Chan, S. L., Griffin, W. S., and Mattson, M. P. (1999a) Evidence for caspase-mediated cleavage of AMPA receptor subunits in neuronal apoptosis and Alzheimer's disease. *J. Neurosci. Res.* **57,** 315–323.

Chan, S. L., Tammariello, S. P., Estus, S., and Mattson, M. P. (1999b) Par-4 mediates trophic factor withdrawal-induced apoptosis of hippocampal neurons: actions prior to mitochondrial dysfunction and caspase activation. *J. Neurochem.* **73,** 502–512.

Cheng, B., Christakos, S., and Mattson, M. P. (1994) Tumor necrosis factors protect neurons against excitotoxic/metabolic insults and promote maintenance of calcium homeostasis. *Neuron* **12,** 139–153.

Cheng, B. and Mattson. M. P. (1991) NGF and bFGF protect rat and human central neurons against hypoglycemic damage by stabilizing calcium homeostasis. *Neuron* **7,** 1031–1041.

Cheng, B. and Mattson, M. P. (1992) IGF-I and IGF-II protect cultured hippocampal and septal neurons against calcium-mediated hypoglycemic damage. *J. Neurosci.* **12,** 1558–1566.

Cheng, B. and Mattson, M. P. (1994) NT-3 and BDNF protect CNS neurons against metabolic/excitotoxic insults. *Brain Res.* **640,** 56–67.

Cheng, B. and Mattson, M. P. (1995) PDGFs protect hippocampal neurons against energy deprivation and oxidative injury: evidence for induction of antioxidant pathways. *J. Neurosci.* **15,** 7095–7104.

Chihab, R., Ferry, C., Koziel, V., Monin, P., and Daval, J. L. (1998) Sequential activation of activator protein-1-related transcription factors and JNK protein kinases may contribute to apoptotic death induced by transient hypoxia in developing brain neurons. *Mol. Brain. Res.* **63,** 105–120.

Colbourne. F., Sutherland, G. R., and Auer, R. N. (1999) Electron microscopic evidence against apoptosis as the mechanism of neuronal death in global ischemia. *J. Neurosci.* **19,** 4200–4210.

Corvol, P., Soubrier, F., and Jeunemaitre, X. (1997) Molecular genetics of the renin-angiotensin-aldosterone system in human hypertension. *Pathol. Biol.* **45,** 229–239.

Culmsee, C. and Mattson, M. P. (2001) Participation of Par-4 in neuronal death after focal cerebral ischemia-reperfusion in mice. *J. Cereb. Blood Flow Metab.*, in press.

Culmsee, C., Semkova, I., and Krieglstein, J. (1999a) NGF mediates the neuroprotective effect of the beta2-adrenoceptor agonist clenbuterol in vitro and in vivo: evidence from an NGF antisense study. *Neurochem. Int.* **35,** 47–57.

Culmsee, C., Vedder, H., Ravati, A., Junker, V., Otto, D., Ahlemeyer, B., et al. (1999b) Neuroprotection by estrogens in a mouse model of focal cerebral ischemia and in cultured neurons: evidence for a receptor-independent anti-oxidative mechanism. *J. Cereb. Blood Flow Metab.* **19,** 1263–1269.

Dirnagl, U., Iadecola, C., and Moskowitz, M. A. (1999) Pathobiology of ischaemic stroke: an integrated view. *Trends Neurosci.* **22,** 391–397.

Duan, W. and Mattson, M. P. (1999) Dietary restriction and 2-deoxyglucose administration improve behavioral outcome and reduce degeneration of dopaminergic neurons in models of Parkinson's disease. *J. Neurosci. Res.* **57,** 185–206.

Duan, W., Rangnekar, V. M., and Mattson, M. P. (1999) Prostate apoptosis response-4 production in synaptic compartments following apoptotic and excitotoxic insults: evidence for a pivotal role in mitochondrial dysfunction and neuronal degeneration. *J. Neurochem.* **72,** 2312–2322

Endoh, M., Pulsinelli, W. A., and Wagner, J. A. (1994) Transient global ischemia induces dynamic changes in the expression of bFGF and the FGF receptor. *Mol. Brain Res.* **22,** 76–88.

Endres, M., Fink, K., Zhu, J., Stagliano, N. E., Bondada, V., Geddes, J. W., et al. (1999) Neuroprotective effects of gelsolin during murine stroke. *J. Clin. Invest.* **103,** 347–354.

Endres, M., Namura, S., Shimizu-Sasamata, M., Waeber, C., Zhang, L., Gomez-Isla, T., et al. (1998) Attenuation of delayed neuronal death after mild focal ischemia in mice by inhibition of the caspase family. *J. Cereb. Blood Flow Metab.* **18,** 238–247.

Fisher, M. and Bogousslavsky, J. (1998) Further evolution toward effective therapy for acute ischemic stroke. *JAMA* **279,** 1298–1303.

Fisher, M., Meadows, M. E., Do, T., Weise, J., Trubetskoy, V., Charette, M., and Finklestein, S. P. (1995) Delayed treatment with intravenous basic fibroblast growth factor reduces infarct size following permanent focal cerebral ischemia in rats. *J. Cereb. Blood Flow Metab.* **15,** 953–959.

Friberg, H., Ferrand-Drake, M., Bengtsson, F., Halestrap, A. P., and Wieloch, T. (1998) Cyclosporin A, but not FK 506, protects mitochondria and neurons against hypoglycemic damage and implicates the mitochondrial permeability transition in cell death. *J. Neurosci.* **18,** 5151–5159.

Fujimura, M., Morita-Fujimura, Y., Murakami, K., Kawase, M., and Chan, P. H. (1998) Cytosolic redistribution of cytochrome c after transient focal cerebral ischemia in rats. *J. Cereb. Blood Flow Metab.* **18,** 1239–1247.

Fukuda, K., Yao, H., Ibayashi, S., Nakahara, T., Uchimura, H., Fujishima, M., and Hall, E. D. (2000) Ovariectomy exacerbates and estrogen replacement attenuates photothrombotic focal ischemic brain injury in rats. *Stroke* **31,** 155–160.

Furukawa, K., Barger, S. W., Blalock, E., and Mattson, M. P. (1996) Activation of K$^+$ channels and suppression of neuronal activity by secreted β-amyloid precursor protein. *Nature* **379,** 74–78.

Furukawa, K., Estus, S., Fu, W., and Mattson, M. P. (1997a) Neuroprotective action of cycloheximide involves induction of Bcl-2 and antioxidant pathways. *J. Cell Biol.* **136,** 1137–1150.

Furukawa, K., Fu, W., Witke, W., Kwiatkowski, D. J., and Mattson, M. P. (1997b) The actin-severing protein gelsolin modulates calcium channel and NMDA receptor activities and vulnerability to excitotoxicity in hippocampal neurons. *J. Neurosci.* **17,** 8178–8186.

Furukawa, K., Smith-Swintosky, V. L., and Mattson, M. P. (1995) Evidence that actin depolymerization protects hippocampal neurons against excitotoxicity by stabilizing [Ca^{2+}]$_i$. *Exp. Neurol.* **133,** 153–163.

Gary, D. S., Bruce-Keller, A. J., Kindy, M. S., and Mattson, M. P. (1998) Ischemic and excitotoxic brain injury is enhanced in mice lacking the p55 tumor necrosis factor receptor. *J. Cereb. Blood Flow Metab.* **18,** 1283–1287.

Ginsberg, M. D. and Busto, R. (1989) Rodent models of cerebral ischemia. *Stroke* **20,** 1627–1642.

Glazner, G. W., Chan, S. L., and Mattson, M. P. (2000) Caspase-mediated degradation of AMPA receptor subunits: a mechanism for preventing excitotoxic necrosis and ensuring apoptosis. *J. Neurosci.* **20,** 3641–3649.

Goodman, Y., Bruce, A. J., Cheng, B., Mattson, M. P. (1996) Estrogens attenuate and corticosterone exacerbates excitotoxicity, oxidative injury, and amyloid beta-peptide toxicity in hippocampal neurons. *J. Neurochem.* **66,** 1836–1844.

Gorelick, P. B., Schneck, M., Berglund, L. F., Feinberg, W., and Goldstone, J. (1997) Status of lipids as a risk factor for stroke. *Neuroepidemiology* **16,** 107–115.

Guegan, C., Onteniente, B., Makiura, Y., Merad-Boudia, M., Ceballos-Picot, I., and Sola, B. (1998) Reduction of cortical infarction and impairment of apoptosis in NGF-transgenic mice subjected to permanent focal ischemia. *Mol. Brain Res.* **55,** 133–140.

Guglielmo, M. A., Chan, P. T., Cortez, S., Stopa, E. G., McMillan, P., Johanson, C. E., et al. (1998) The temporal profile and morphologic features of neuronal death in human stroke resemble those observed in experimental forebrain ischemia: the potential role of apoptosis. *Neurol. Res.* **20,** 283–296.

Guo, Q., Fu, W., Xie, J., Luo, H., Sells, S. F., Geddes, J. W., et al. (1998) Par-4 is a novel mediator of neuronal degeneration associated with the pathogenesis of Alzheimer's disease. *Nature Med.* **4,** 957–962.

Guo, Q., Sopher, B. L., Pham, D. G., Furukawa, K., Robinson, N., Martin, G. M., and Mattson, M. P. (1997) Alzheimer's presenilin mutation sensitizes neural cells to apoptosis induced by trophic factor withdrawal and amyloid β-peptide: involvement of calcium and oxyradicals. *J. Neurosci.* **17,** 4212–4222.

Hara, H., Friedlander, R. M., Gagliardini, V., Ayata, C., Fink, K., Huang, Z., et al. (1997) Inhibition of interleukin 1beta converting enzyme family proteases reduces ischemic and excitotoxic neuronal damage. *Proc. Natl. Acad. Sci. USA* **94,** 2007–2012.

Hara, A., Niwa, M., Nakashima, M., Iwai, T., Uematsu, T., Yoshimi, N., and Mori, H. (1998) Protective effect of apoptosis-inhibitory agent, N-tosyl-L-phenylalanyl chloromethyl ketone against ischemia-induced hippocampal neuronal damage. *J. Cereb. Blood Flow Metab.* **18,** 819–823.

Haussermann, S., Kittstein, W., Rincke, G., Johannes, F. J., Marks, F., and Gschwendt, M. (1999) Proteolytic cleavage of protein kinase Cmu upon induction of apoptosis in U937 cells. *FEBS Lett.* **462,** 442–446.

Hayashi, T., Sakurai, M., Itoyama, Y., and Abe, K. (1999) Oxidative damage and breakage of DNA in rat brain after transient MCA occlusion. *Brain Res.* **832,** 159–163.

Henrich-Noack, P., Prehn, J. H., and Krieglstein, J. (1996) TGF-β1 protects hippocampal neurons against degeneration caused by transient global ischemia. Dose-response relationship and potential neuroprotective mechanisms. *Stroke* **27,** 1609–1614.

Hockenbery, D., Oltvai, Z. N., Yin, X. M., Milliman, C. L., and Korsmeyer, S. H. (1993) Bcl-2 functions in an antioxidant pathway to prevent apoptosis. *Cell* **75,** 241–251.

Holtsberg, F. W., Steiner, M. R., Keller, J. N., Mark, R. J., Mattson, M. P., Steiner, S. M. (1998) Lysophosphatidic acid induces necrosis and apoptosis in hippocampal neurons. *J. Neurochem.* **70,** 66–76.

Hortelano, S., Dallaporta, B., Zamzami, N., Hirsch, T., Susin, S. A., Marzo, I., et al. (1997) Nitric oxide induces apoptosis via triggering mitochondrial permeability transition. *FEBS Lett.* **410,** 373–377.

Hou, S. T., Tu, Y., Buchan, A. M., Huang, Z., Preston, E., Rasquinha, I., et al. (1997) Increases in DNA lesions and the DNA damage indicator Gadd45 following transient cerebral ischemia. *Biochem. Cell Biol.* **75,** 383–392.

Jin, K., Chen, J., Nagayama, T., Chen, M., Sinclair, J., Graham, S. H., and Simon, R. P. (1999) In situ detection of neuronal DNA strand breaks using the Klenow fragment of DNA polymerase I reveals different mechanisms of neuron death after global cerebral ischemia. *J. Neurochem.* **72,** 1204–1214.

Kalda, A., Eriste, E., Vassiljev, V., and Zharkovsky, A. (1998) Medium transitory oxygen-glucose deprivation induced both apoptosis and necrosis in cerebellar granule cells. *Neurosci. Lett.* **240,** 21–24.

Keller, J. N., Kindy, M. S., Holtsberg, F. W., St. Clair, D. K., Yen, H. C., Germeyer, A., et al. (1998) Mitochondrial manganese superoxide dismutase prevents neural apoptosis and reduces ischemic brain injury: suppression of peroxynitrite production, lipid peroxidation, and mitochondrial dysfunction. *J. Neurosci.* **18,** 687–697.

Keller, J. N., Pang, Z., Geddes, J. W., Begley, J. G., Germeyer, A., Waeg, G., and Mattson, M. P. (1997) Impairment of glucose and glutamate transport and induction of mitochondrial oxidative stress and dysfunction in synaptosomes by amyloid β-peptide: role of the lipid peroxidation product 4-hydroxynonenal. *J. Neurochem.* **69,** 273–284.

Kim, T. W., Pettingell, W. H., Jung, Y. K., Kovacs, D. M., and Tanzi, R. E. (1997) Alternative cleavage of Alzheimer-associated presenilins during apoptosis by a caspase-3 family protease. *Science* **277,** 373–376.

Kiprianova, I., Freiman, T. M., Desiderato, S., Schwab, S., Galmbacher, R., Gillardon, F., and Spranger, M. (1999) Brain-derived neurotrophic factor prevents neuronal death and glial activation after global ischemia in the rat. *J. Neurosci. Res.* **56,** 21–27.

Kitagawa, K., Matsumoto, M., Tsujimoto, Y., Ohtsuki, T., Kuwabara, K., Matsushita, K., et al. (1998) Amelioration of hippocampal neuronal damage after global ischemia by neuronal overexpression of BCL-2 in transgenic mice. *Stroke* **29,** 2616–2621.

Kluck, R. M., Bossy-Wetzel, E., Green, D. R., and Newmeyer, D. D. (1997) The release of cytochrome c from mitochondria: a primary site for Bcl-2 regulation of apoptosis. *Science* **275,** 1132–1136.

Kobayashi, T. and Mori, Y. (1998) Ca2+ channel antagonists and neuroprotection from cerebral ischemia. *Eur. J. Pharmacol.* **363,** 1–15.

Kroemer, G., Zamzami, N., and Susin, S. A. (1997) Mitochondrial control of apoptosis. *Immunol. Today* **18,** 44–51.

Kruman, I., Bruce-Keller, A. J., Bredesen, D. E., Waeg, G., and Mattson, M. P. (1997) Evidence that 4-hydroxynonenal mediates oxidative stress-induced neuronal apoptosis. *J. Neurosci.* **17,** 5089–5100.

Kruman, I. and Mattson, M. P. (1999) Pivotal role of mitochondrial calcium uptake in neural cell apoptosis and necrosis. *J. Neurochem.* **72,** 529–540.

Leppala, J. M., Virtamo. J., Fogelholm, R., Huttunen, J. K., Albanes, D., Taylor, P. R., and Heinonen, O. P. (2000) Controlled trial of alpha-tocopherol and beta-carotene supplements on stroke incidence and mortality in male smokers. *Arterioscler. Thromb. Vasc. Biol.* **20,** 230–235.

Linnik, M. D., Zahos, P., Geschwind, M. D., and Federoff, H. J. (1995) Expression of bcl-2 from a defective herpes simplex virus-1 vector limits neuronal death in focal cerebral ischemia. *Stroke* **26,** 1670–1674.

Linnik, M. D., Zobrist, R. H., and Hatfield, M. D. (1993) Evidence supporting a role for programmed cell death in focal cerebral ischemia in rats. *Stroke* **24,** 2002–2008.

Loetscher, H., Deuschle, U., Brockhaus, M., Reinhardt, D., Nelboeck, P., Mous, J., et al. (1997) Presenilins are processed by caspase-type proteases. *J. Biol. Chem.* **272,** 20,655–20,659.

Love, S., Barber, R., and Wilcock, G. K. (1998) Apoptosis and expression of DNA repair proteins in ischaemic brain injury in man. *Neuroreport* **9,** 955–959.

Lowenstein, D. H., Chan, P., and Miles, M. (1991) The stress protein response in cultured neurons: characterization and evidence for a protective role in excitotoxicity. *Neuron* **7,** 1053–1060.

MacManus, J. P., Buchan, A. M., Hill, I. E., Rasquinha, I., and Preston, E. (1993) Global ischemia can cause DNA fragmentation indicative of apoptosis in rat brain. *Neurosci. Lett.* **164,** 89–92.

Mark, R. J., Lovell, M. A., Markesbery, W. R., Uchida, K., and Mattson, M. P. (1997a) A role for 4-hydroxynonenal, an aldehydic product of lipid peroxidation, in disruption of ion homeostasis and neuronal death induced by amyloid β-peptide. *J. Neurochem.* **68,** 255–264.

Mark, R. J., Pang, Z., Geddes, J. W., Uchida, K., and Mattson, M. P. (1997b) Amyloid β-peptide impairs glucose uptake in hippocampal and cortical neurons: involvement of membrane lipid peroxidation. *J. Neurosci.* **17,** 1046–1054.

Martin-Villalba, A., Herr, I., Jeremias, I., Hahne, M., Brandt, R., Vogel, J., et al. (1999) CD95 ligand (Fas-L/APO-1L) and tumor necrosis factor-related apoptosis-inducing ligand mediate ischemia-induced apoptosis in neurons. *J. Neurosci.* **19**, 3809–3817.

Matsumoto, S., Friberg, H., Ferrand-Drake, M., and Wieloch, T. (1999) Blockade of the mitochondrial permeability transition pore diminishes infarct size in the rat after transient middle cerebral artery occlusion. *J. Cereb. Blood Flow Metab.* **19**, 736–741.

Mattson, M. P. (1998) Modification of ion homeostasis by lipid peroxidation: roles in neuronal degeneration and adaptive plasticity. *Trends Neurosci.* **21**, 53–57.

Mattson, M. P., Chan, S. L., LaFerla, F. M., Leissring, M., Shepel, P. N., and Geiger, J. D. (2000b) Calcium signaling in the ER: its role in neuronal plasticity and neurodegenerative disorders. *Trends Neurosci.* **23**, 222–229.

Mattson, M. P., Cheng, B., Culwell, A., Esch, F., Lieberburg, I., and Rydel, R. E. (1993b) Evidence for excitoprotective and intraneuronal calcium-regulating roles for secreted forms of β-amyloid precursor protein. *Neuron* **10**, 243–254.

Mattson, M. P., Culmsee, C., Yu, Z., and Camandola, S. (2000a) Roles of NF-κB in neuronal survival and plasticity. *J. Neurochem.* **74**, 443–456.

Mattson, M. P., Estus, S., and Rangnekar, V., eds. (2000) *Programmed Cell Death,* Vol. 1 (Adv. Cell Aging Gerontology) JAI Press, Greenwich, CT, in press.

Mattson, M. P. and Furukawa, K. (1996) Programmed cell life: anti-apoptotic signaling and therapeutic strategies for neurodegenerative disorders. *Restorative Neurol. Neurosci.* **9**, 191–205.

Mattson, M. P., Goodman, Y., Luo, H., Fu, W., and Furukawa, K. (1997a) Activation of NF-κB protects hippocampal neurons against oxidative stress-induced apoptosis: evidence for induction of Mn-SOD and suppression of peroxynitrite production and protein tyrosine nitration. *J. Neurosci. Res.* **49**, 681–697.

Mattson, M. P., Keller, J. N., and Begley, J. G. (1998) Evidence for synaptic apoptosis. *Exp. Neurol.* **153**, 35–48.

Mattson, M. P., Kumar, K., Cheng, B., Wang, H., and Michaelis, E. K. (1993c) Basic FGF regulates the expression of a functional 71 kDa NMDA receptor protein that mediates calcium influx and neurotoxicity in cultured hippocampal neurons. *J. Neurosci.* **13**, 4575–4588.

Mattson, M. P. and Lindvall, O. (1997) Neurotrophic factor and cytokine signaling in the aging brain, in *The Aging Brain* (Mattson, M. P. and Geddes, J. W., eds.), JAI Press, Greenwich CT, *Adv. Cell Aging Gerontol.* **2**, 299–345.

Mattson, M. P., Lovell, M. A, Furukawa, K., and Markesbery, W. R (1995) Neurotrophic factors attenuate glutamate-induced accumulation of peroxides, elevation of $[Ca^{2+}]_i$ and neurotoxicity, and increase antioxidant enzyme activities in hippocampal neurons. *J. Neurochem.* **65**, 1740–1751.

Mattson, M. P. and Mark, R. J. (1996) Excitotoxicity and excitoprotection in vitro. *Adv. Neurol.* **71**, 1–37.

Mattson, M. P., Murrain, M., Guthrie, P. B., and Kater, S. B (1989) Fibroblast growth factor and glutamate: opposing actions in the generation and degeneration of hippocampal neuroarchitecture. *J. Neurosci.* **9**, 3728–3740.

Mattson, M. P., Robinson, N., and Guo, Q. (1997b) Estrogens stabilize mitochondrial function and protect neural cells against the pro-apoptotic action of mutant presenilin-1. *Neuroreport* **8**, 3817–3821.

Mattson, M. P., Zhang, Y., and Bose, S. (1993a) Growth factors prevent mitochondrial dysfunction, loss of calcium homeostasis and cell injury, but not ATP depletion in hippocampal neurons deprived of glucose. *Exp. Neurol.* **121**, 1–13.

Mattson, M. P., Zhu H., Yu, J., and Kindy, M. S. (2000c) Presenilin-1 mutation increases neuronal vulnerability to focal ischemia in vivo, and to hypoxia and glucose deprivation in cell culture: involvement of perturbed calcium homeostasis. *J. Neurosci.* **20**, 1358–1364.

McGahan, L., Hakim, A. M., and Robertson, G. S. (1998) Hippocampal Myc and p53 expression following transient global ischemia. *Mol. Brain Res.* **56,** 133–145.

Mhairi-Macrae, I. (1992) New models of focal cerebral ischaemia. *Br. J. Clin. Pharmacol.* **34,** 302–308.

Miyazaki, H., Okuma, Y., Fujii, Y., Nagashima, K., and Nomura, Y. (1999) Glial cell line-derived neurotrophic factor protects against delayed neuronal death after transient forebrain ischemia in rats. *Neuroscience* **89,** 643–647.

Murakami, K., Kondo, T., Kawase, M., Li, Y., Sato, S., Chen, S. F., and Chan, P. H. (1998) Mitochondrial susceptibility to oxidative stress exacerbates cerebral infarction that follows permanent focal cerebral ischemia in mutant mice with manganese superoxide dismutase deficiency. *J. Neurosci.* **18,** 205–213.

Neame, S. J., Rubin, L. L., and Philpott, K. L. (1998) Blocking cytochrome c activity within intact neurons inhibits apoptosis. *J. Cell Biol.* **142,** 1583–1593.

Nicotera, P. and Lipton, S. A. (1999) Excitotoxins in neuronal apoptosis and necrosis. *J. Cereb Blood Flow Metab.* **19,** 583–591.

Nitatori, T., Sato, N., Waguri, S., Karasawa, Y., Araki, H., Shibanai, K., et al. (1995) Delayed neuronal death in the CA1 pyramidal cell layer of the gerbil hippocampus following transient ischemia is apoptosis. *J. Neurosci.* **15,** 1001–1011.

Penix, L. and Lanska, D. (1999) Cerebrovascular disease, in *Genetic Aberrancies in Neurodegenerative Disorders* (Mattson, M. P., ed.), JAI Press, Greenwich, CT.

Perry, I. J., Refsum, H., Morris, R. W., Ebrahim, S. B., Ueland, P. M., and Shaper, A. G. (1995) Prospective study of serum total homocysteine concentration and risk of stroke in middle-aged British men. *Lancet* **346,** 1395–1398.

Plumier, J. C., Krueger, A. M., Currie, R. W., Kontoyiannis, D., Kollias, G., and Pagoulatos, G. N. (1997) Transgenic mice expressing the human inducible Hsp70 have hippocampal neurons resistant to ischemic injury. *Cell Stress Chaperones* **2,** 162–167.

Prehn, J. H., Backhauss, C., and Krieglstein, J. (1993) Transforming growth factor-β1 prevents glutamate neurotoxicity in rat neocortical cultures and protects mouse neocortex from ischemic injury in vivo. *J. Cereb. Blood Flow Metab.* **13,** 521–525.

Putcha, G. V., Deshmukh, M., and Johnson, E. M., Jr. (1999) BAX translocation is a critical event in neuronal apoptosis: regulation by neuroprotectants, BCL-2, and caspases. *J. Neurosci.* **19,** 7476–7485.

Schielke, G. P, Yang, G. Y., Shivers, B. D., and Betz, A. L. (1998) Reduced ischemic brain injury in interleukin-1 beta converting enzyme-deficient mice. *J. Cereb. Blood Flow Metab.* **18,** 180–185.

Shigeno, T., Mima, T., Takakura, K., Graham, D. I., Kato, G., Hashimoto, Y., and Furukawa, S. (1991) Amelioration of delayed neuronal death in the hippocampus by nerve growth factor. *J Neurosci.* **11,** 2914–2919.

Smith-Swintosky, V. L., Pettigrew, L. C., Craddock, S. D., Culwell, A. R., Rydel, R. E., and Mattson, M. P. (1994) Secreted forms of β-amyloid precursor protein protect against ischemic brain injury. *J. Neurochem.* **63,** 781–784.

Smith-Swintosky, V. L., Pettigrew, L. C., Sapolsky, R. M., Phares, C., Craddock, S. D., Brooke, S. M., and Mattson, M. P. (1996) Metyrapone, an inhibitor of glucocorticoid production, reduces brain injury induced by focal and global ischemia and seizures. *J. Cereb. Blood Flow Metab.* **16,** 585–598.

Sohal, R. S. and Weindruch, R. (1996) Oxidative stress, caloric restriction, and aging. *Science* **273,** 59–63.

Speliotes, E. K., Caday, C. G., Do, T., Weise, J., Kowall, N. W., and Finklestein, S. P. (1996) Increased expression of basic fibroblast growth factor (bFGF) following focal cerebral infarction in the rat. *Mol. Brain Res.* **39,** 31–42.

Sudo, S., Wen, T. C., Desaki, J., Matsuda, S., Tanaka, J., Arai, T., et al. (1997) Beta-estradiol protects hippocampal CA1 neurons against transient forebrain ischemia in gerbil. *Neurosci. Res.* **29,** 345–354.

Svendsen, C. N. and Smith, A. G. (1999) New prospects for human stem-cell therapy in the nervous system. *Trends Neurosci.* **22,** 357–364.

Tagami, M., Ikeda, K., Yamagata, K., Nara, Y., Fujino, H., Kubota, A., et al. (1999) Vitamin E prevents apoptosis in hippocampal neurons caused by cerebral ischemia and reperfusion in stroke-prone spontaneously hypertensive rats. *Lab Invest.* **79,** 609–615.

Taglialatela, G., Robinson, R., and Perez-Polo, J. R. (1997) Inhibition of nuclear factor κB (NFκB) activity induces nerve growth factor-resistant apoptosis in PC12 cells. *J. Neurosci. Res.* **47,** 155–162.

Tamatani, M., Che, Y. H., Matsuzaki, H., Ogawa, S., Okado, H., Miyake, S., et al. (1999) Tumor necrosis factor induces Bcl-2 and Bcl-x expression through NF-κB activation in primary hippocampal neurons. *J. Biol. Chem.* **274,** 8531–8538.

Tarkowski, E., Rosengren, L., Blomstrand, C., Jensen, C., Ekholm, S., and Tarkowski, A. (1999) Intrathecal expression of proteins regulating apoptosis in acute stroke. *Stroke* **30,** 321–327.

Tasker, R. C., Sahota, S. K., Cotter, F. E., and Williams, S. R. (1998) Early postischemic dantrolene-induced amelioration of poly(ADP-ribose) polymerase-related bioenergetic failure in neonatal rat brain slices. *J. Cereb. Blood Flow Metab.* **18,** 1346–1356.

Uchida, K. (1999) Current status of acrolein as a lipid peroxidation product. *Trends Cardiovasc. Med.* **9,** 109–113.

Uno, H., Matsuyama, T., Akita, H., Nishimura, H., and Sugita, M. (1997) Induction of tumor necrosis factor-alpha in the mouse hippocampus following transient forebrain ischemia. *J. Cereb. Blood Flow Metab.* **17,** 491–499.

Velier, J. J., Ellison, J. A., Kikly, K. K., Spera, P. A., Barone, F. C., and Feuerstein, G. Z. (1999) Caspase-8 and caspase-3 are expressed by different populations of cortical neurons undergoing delayed cell death after focal stroke in the rat. *J. Neurosci.* **19,** 5932–5941.

Wagstaff, M. J., Collaco-Moraes, Y., Aspey, B. S., Coffin, R. S., Harrison, M. J., Latchman, D. S., and de Belleroche, J. S. (1996) Focal cerebral ischaemia increases the levels of several classes of heat shock proteins and their corresponding mRNAs. *Mol. Brain Res.* **42,** 236–244.

Wang, S., Longo, F. M., Chen, J., Butman, M., Graham, S. H., Haglid, K. G., and Sharp, F. R. (1993) Induction of glucose regulated protein (grp78) and inducible heat shock protein (hsp70). mRNAs in rat brain after kainic acid seizures and focal ischemia. *Neurochem. Int.* **23,** 575–582.

White, J. R. and Reynolds, I. J. (1996) Mitochondrial depolarization in glutamate stimulated neurons: an early signal specific to excitotoxin exposure. *J. Neurosci.* **16,** 5688–5697.

Xu, D., Bureau, Y., McIntyre, D. C., Nicholson, D. W., Liston, P., Zhu, Y., et al. (1999) Attenuation of ischemia-induced cellular and behavioral deficits by X chromosome-linked inhibitor of apoptosis protein overexpression in the rat hippocampus. *J. Neurosci.* **19,** 5026–5033.

Yu, Z. F., Bruce-Keller, A. J., Goodman, Y., and Mattson, M. P. (1998) Uric acid protects neurons against excitotoxic and metabolic insults in cell culture, and against focal ischemic brain injury in vivo. *J. Neurosci. Res.* **53,** 613–625.

Yu, Z. F., Luo, H., Fu, W., and Mattson, M. P. (1999a) The endoplasmic reticulum stress-responsive protein GRP78 protects neurons against excitotoxicity and apoptosis: suppression of oxidative stress and stabilization of calcium homeostasis. *Exp. Neurol.* **155,** 302–314.

Yu, Z. F. and Mattson, M. P. (1999) Dietary restriction and 2-deoxyglucose administration reduce focal ischemic brain damage and improve behavioral outcome: evidence for a preconditioning mechanism. *J. Neurosci. Res.* **57,** 830–839.

Yu, Z. F., Nikolova-Karakashian, M., Zhou, D., Cheng, G., Schuchman, E. H., and Mattson, M. P. (2000) Pivotal role for acidic sphingomyelinase in cerebral ischemia-induced ceramide and cytokine production, and neuronal death. *J. Mol. Neurosci.* **87,** 85–97.

Yu, Z. F., Zhou, D., Bruce-Keller, A. J., Kindy, M. S., and Mattson, M. P. (1999b) Lack of the p50 subunit of NF-κB increases the vulnerability of hippocampal neurons to excitotoxic injury. *J. Neurosci.* **19,** 8856–8865.

Zhu, Y., Culmsee, C., Semkova, I., and Krieglstein, J. (1998) Stimulation of beta2-adreno-ceptors inhibits apoptosis in rat brain tissue after transient forebrain ischemia. *J. Cereb. Blood Flow Metab.* **18,** 1032–1039.

11
Spinal-Cord Injury

Isabel Klusman and Martin E. Schwab

INTRODUCTION

It has been known for more than a century that axons of the peripheral nervous system (PNS) have the capacity to regenerate after injury, while axons of the central nervous system (CNS) show no regenerative growth. For many decades this fact was accepted as a "law of nature" and research in this field was very limited. Lately, however, it has become clear that this inability can be overcome by adding growth-promoting neurotrophic factors, blocking inhibitory factors, or bridging the lesion area with growth-permissive tissue. The mechanisms responsible for the inability of CNS axons to regenerate are being unravelled at a rapid pace. This knowledge, in parallel with new concepts on neuronal death and survival, and the realization that the inflammatory processes probably also play important roles in the processes following injury, have led to an increased understanding of the pathophysiology of spinal-cord lesions. Unravelling of the cascade of events that take place after an injury to the CNS will lead to new therapeutic approaches that can improve the situation for spinal-cord and brain-injured patients. The following chapter will describe spinal-cord injuries in general, review the cellular mechanisms of neurodegeneration seen after such an injury, as far as they are known, and describe the regenerative processes after injury.

SPINAL-CORD INJURY

The spinal cord is one of the seven anatomical regions of the adult CNS. It receives sensory information from the skin, joints, and muscles of the trunk and limbs and contains the motor neurons responsible for voluntary and reflex movements. The spinal cord contains ascending and descending fiber tracts for communication between the brain and the rest of the body. After a spinal-cord injury, disruption of these pathways often leads to a dramatic loss of function. In experimental spinal-cord injury models, the descending corticospinal and rubrospinal tract and the ascending fibers of the dorsal funiculus have been studied in great detail.

The initial, mechanical insult triggers a wave of molecular and cellular changes that occur within minutes and persist for days to weeks. These processes enlarge the primary lesion and are therefore referred to as secondary tissue damage. Understanding these secondary processes will help to develop strategies to minimize spinal cord tissue

From: *Pathogenesis of Neurodegenerative Disorders* Edited by: M. P. Mattson © Humana Press Inc., Totowa, NJ

Fig. 1. The dying lioness: a detail from Ashurbanipal's lion-hunt series of Assyrian stone bas-reliefs (about 645 BC).

loss, thus preserving functions and creating better conditions for injured axons to regenerate.

Historical Remarks

One of the oldest surviving medical documents, an Egyptian papyrus dating from the year 1600 BC, precisely describes the principal signs of spinal-cord injury "...there is paralysis of the arms, legs and sphincters, with bad prognosis..." (Breasted, 1930). Another document dates from the seventh century BC and depicts scenes from Ashurbanipal's lion-hunt: a dying lioness is depicted with paralyzed legs dragging behind as an arrow has penetrated her back (Fig. 1). Today, traumatic spinal cord lesions in humans are usually due to motor-vehicle accidents, falls, penetrating injuries, or sports accidents. The injuries predominantly affect young, active male members of our society. Over the centuries, the outcome of damage to the spinal cord has remained the same: full or partial paralysis and loss of sensation below the level of the injury.

NATURE OF THE SPINAL-CORD INJURY

Owing to the complex structure and segmental organization of the spinal cord, the effects of an injury are heterogeneous and variable. Depending on the region and severity of the trauma, the clinical deficits range from paralysis of the lower extremities to total loss of movement and an inability to breathe independently as a consequence of paralysis of the diaphragm. Over the last few decades, numerous animal models have been developed to study the different stages of injury and to examine possible strategies to overcome the devastating consequences of such a CNS lesion.

Human Injury

Depending on the level of the injury, patients can be classified clinically as paraplegic or quadriplegic and as either complete or incomplete according to the amount of spared function. Most frequently, dislocation of the vertebrae or bone fragments mechanically damage axons and causes the typical contusions of the spinal cord. Only in a few cases, mostly shot-gun injuries, is the cord completely transected. The initial mechanical insult, in turn, triggers a complex cascade of secondary tissue damage that enlarges the primary lesion.

Severe damage to the spinal cord produces a state of spinal shock characterized by a flaccid paralysis of the extremities below the level of the injury. The pathophysiology underlying this shock is unclear but the syndrome is transient.

High-resolution magnetic resonance imaging (MRI) has become the main tool to monitor lesion size and the changes seen after injury. However, to correlate the picture observed in human patients to the histology and the pathophysiological development of the lesion, animal models have been developed.

Clinical Considerations

As a result of injuries to the spinal cord in Western countries, about 30 people per million inhabitants become paraplegic or quadriplegic each year. Spinal-cord injury mostly strikes the young and healthy suddenly and with devastating outcome. For many years there was no way of limiting the disability besides stabilizing the cord to prevent additional destruction, treating infections, and prescribing rehabilitative therapy. Promising results achieved over the last couple of years give new hope to spinal cord-injured people. Treatment strategies can be divided into two different approaches. The first approach is directed at intervening with the acute phase and thus limiting tissue damage, while the second approach deals with the stabilization of the status of the patient and encompasses rehabilitation aimed at training of reflexes and remaining circuits to provide optimal living conditions. A third, future approach, is directed at the induction of axonal regeneration. In animal models of spinal-cord injury, axonal regeneration has already been achieved using different strategies (for reviews, *see* Schwab and Bartholdi, 1996; Olson, 1997; Stichel and Müller, 1998).

Neuroprotective Treatments

Although preventive efforts to reduce the overall incidence of spinal-cord injury are important, it is vital to minimize secondary injury once the initial trauma has occurred. These secondary processes include local ischemia and edema formation, changes in calcium and excitatory amino acids levels, production of free radicals followed by necrosis, and inflammatory processes (for review, *see* Schwab and Bartholdi, 1996). Recent efforts have focused on understanding the biochemical basis of secondary injury and developing pharmacologic agents to intervene in the progression of neurologic deterioration. A large number of pharmacological substances have been tested in experimental models of spinal-cord injury, but owing to the lack of standardized evaluation techniques, so far the data have often been conflicting. There are, however, a few compounds that have been shown to have neuroprotective properties.

One of the first drugs that was able to limit spinal-cord damage, probably by reducing edema and inflammation and by scavenging free radicals, was the synthetic steroid methylprednisolone. Very high doses of the drug administered at early time points after lesion were shown to exert a protective effect in both various animal models (Young and Flamm, 1982; Anderson et al., 1982) and patients (Bracken et al., 1997; Nesathurai, 1998). The exact mode of action remains hypothetical to date. Three multicenter randomized clinical trials (National Acute Spinal Cord Injury Studies, NASCIS) have been carried out in patients with acute spinal-cord injury. The second study showed that high-dose treatment within 8 h after injury led to a significantly improved neurological recovery of the patients (Bracken et al., 1990). The early time of treatment (within 8 h of injury) was of crucial importance because application at a later time point was not effective. In the NASCIS III study (Bracken et al., 1997), the treatment strategy could be specified to a higher degree: patients receiving methylprednisolone within 3 h of injury should be maintained on the treatment regimen for 24 h, whereas patients treated between 3 and 8 h should remain on steroid therapy for 48 h.

Naloxone, an opioid receptor antagonist, given at high doses after injury also showed an improvement in neurological function in the NASCIS II trial (Bracken and Holford, 1993), although earlier trials showed conflicting results (Bracken et al., 1990).

Contradictory results also exist with regard to the ganglioside GM-1. The Maryland GM-1 Ganglioside Study reported enhanced neurological recovery after treatment with GM-1 ganglioside in a small group of patients (Geisler, 1993, 1998; Geisler et al., 1993); the results of a larger placebo-controlled multicenter study are expected to be published soon.

A number of other agents including vitamin E as an antioxidant, dimethyl sulfoxide (DMSO), N-methyl-D-aspartate (NMDA), and AMPA/kainate antagonists and calcium channel blockers have been tested in different models of spinal-cord injury (for review, *see* Tator, 1995). The results from most of these studies, however, have been conflicting. Future application of neuroprotective treatments will probably include a combination of several drugs delivered at different time-points after injury, whereby each drug will specifically protect distinct cell populations.

Animal Models of Spinal-Cord Injury

The field of spinal-cord injury research has expanded tremendously over the last decade, but because there is no "stereotypical" clinical spinal-cord injury, there is not a single, "ideal" experimental animal model. To obtain a useful model, the experimental design should be reproducible and as similar to the human situation as possible. Considering the many facets in human injuries and the fact that certain aspects of the lesion are difficult to imitate, this assignment is not easy. Various experimental injury models ranging from weight-drop contusion injuries to complete or partial scissors transections at different spinal-cord levels are being used. Already at the beginning of our century (Allen, 1911), the weight-drop injury was applied to dogs, and today the model is still used in a number of other animal species ranging from monkey to cat and rat (Basso et al., 1996). However, the weight-drop injury compresses the cord posteriorly, whereas most patients show anterior compression (Schwab and Bartholdi, 1996). Another dissimilarity between the clinical injury and many animal lesions, i.e., all section lesions, is that the latter usually use an open laminectomy, whereas most human injuries occur in a closed vertebral canal. Microsurgical transections, hemisections, or complete transections are generally applied to study the reaction of specific axonal pathways. In these models, it is of importance to lesion the axons of a given tract in a defined, and, if possible, complete way to study their regenerative capacity after specific treatments.

Despite the diversity of spinal-cord injury models, results achieved using different paradigms have led to a greatly improved understanding of the pathophysiology of spinal-cord lesion as summarized in the next section.

PATHOLOGICAL CHANGES IN RESPONSE TO TRAUMA

Primary tissue damage that is immediate and irreversible is caused by the mechanical trauma. During the following minutes to hours, a much larger amount of tissue is lost. It is therefore important to clarify this secondary tissue destruction and to develop methods that can intervene in the degenerative processes that take place during this phase. The additional destruction is a result of a cascade of cellular, vascular, and biochemical events following the initial insult. The pathological changes comprise posttraumatic ischemia, edema, hemorrhagic necrosis, release of excitatory amino acids, Ca^{2+} overload, inflammation, uncontrolled transmitter release, membrane lipid peroxidation, and activation of the arachidonic-acid cascade (Tator, 1995; Schwab and Bartholdi, 1996; Springer et al., 1997). These secondary processes can be divided into three separate phases; an acute, a subacute, and a late phase.

Acute Phase

The acute phase is dominated by ischemia and edema. First alterations are seen in the gray matter within 30 min to 3 h with a concomitant spread to the white matter. First signs of morphological abnormality are the changes seen in the vasculature that lead to hemorrhages and thrombosis. Swelling and necrosis of neurons can be seen early (from 3 h on) in the ischemic areas surrounding the primary lesion. A common phenomenon of spinal-cord injury is the swelling of the cord and the concomitant compression of the tissue due to the restricting dura and spinal canal. The axons can be disrupted directly by the injury but can also suffer indirectly due to ischemia and swelling. A typical feature of degenerating axons is the bud-like retraction bulb that forms at the distal end of the injured fiber stump. The isolating myelin sheath surrounding the axons often shows swelling or even complete fragmentation. Astrocytes and oligodendrocytes can be abolished completely in the case of a large trauma. Microglia become activated within minutes; together with infiltrating macrophages they play an important role in the inflammatory reaction that is characteristic for the subacute phase. This reaction starts with an early flash-like production of mRNA for the pro-inflammatory cytokines interleukin-1 (IL-1), IL-6, and tumor necrosis factor (TNF) and the chemokines macrophage inflammatory protein (MIP)-1α and β (Bartholdi and Schwab, 1997; Wang et al., 1997; McTigue et al., 1998; Streit et al., 1998). The meninges are very resistant to trauma and ischemic events, and their main role is to contain the swollen cord.

Breakdown of the blood–brain barrier (BBB) after injury can be due to mechanical damage of the vessels that allows influx of blood cells and plasma proteins into the parenchyma. The release of reactive molecules (Azbill et al., 1997) like reactive oxygen species, bradykinins, histamines, and nitric oxide (NO) can also contribute to the enhanced permeability of the BBB.

Subacute Phase

The subacute phase is marked by a reactive response of CNS microglia and astrocytes (reactive gliosis), and a massive infiltration of cells from the periphery. Microglial cells become activated (specific cell-surface markers are expressed) and can transform into brain macrophages. These activated microglia could release a number of mediators and factors, and possibly also free radicals that are able to damage membranes and other components of healthy cells and thus lead to their death. Scar tissue is built up of reactive astrocytes, which are larger and have more processes than quiescent astrocytes. The reactive astrocytes show an increased expression of glial fibrillary acidic protein (GFAP) with a peak response at 14 d.

The inflammatory response seen after CNS injury is similar to the reactions taking place in other organs of the body; however, it is delayed in onset, and inflammatory cells can often still be observed in the injured region over long time periods. This delay leads to a prolonged presence of damaged tissue surrounding the lesion, which could add additional complications for regeneration of injured axons. The slow removal of damaged tissue from the CNS could be explained by the fact that oligodendrocytes in contrast to Schwann cells do not have phagocytic properties. Moreover, recruitment of blood-borne macrophages to a peripheral injury is quicker than to a central insult (for review, *see* Perry et al., 1995).

Two main waves of cellular infiltration are observed after a CNS lesion (Blight, 1992; Dusart and Schwab, 1994; Popovich et al., 1997; Carlson et al., 1998). The first wave consists of polymorphonuclear granulocytes that start to infiltrate the spinal-cord tissue within the first few hours after injury, reaching a maximum at 24 h and disappearing by 3 d. There seems to be a correlation between the number of infiltrating neutrophils and the degree of hemorrhage at the lesion site. The second wave of cells that enter the lesion site from the periphery belong to the macrophage/monocyte cell lineage, and their main function seems to be to phagocytose cell debris. It was hypothesized that the macrophages could be responsible for the neurodegenerative processes taking place shortly after the injury owing to their ability to act cytotoxically in vitro. The amount of total tissue loss in the rat, however, already reaches its maximum at 6–12 h following injury, a time period that is correlated to ischemic necrosis but not to macrophage invasion that peaks at 4 d. The role of the inflammatory cells that invade the injured tissue when most of the tissue necrosis has already taken place, is still poorly understood (for review, *see* Schwab and Bartholdi, 1996; Schwartz et al., 1999). Small numbers of lymphocytes also appear at the lesion site in the first week after injury (Schnell at al., 1997; Popovich et al., 1997); their role is unknown. Apart from the immune cells, other cells from the periphery have also been observed to invade the spinal cord. Schwann cells that migrate into the lesioned area are thought to be beneficial for axonal regeneration due to their ability to produce neurotrophic factors and the capacity to replace the lost myelin sheaths. To a certain degree, fibroblasts also infiltrate into the injured spinal cord, where their proliferation rate can be influenced by the production of basic fibroblast growth factor (bFGF) that has been reported to be upregulated after injury (Follesa et al., 1994). The roles of these cells are mainly seen in the context of scar formation.

Late Phase

A typical feature of the late phase is the formation of fluid-filled cysts surrounded by activated astrocytes that form a scar-tissue interface with the intact spinal-cord tissue. The cavities are often connected to the central canal and are therefore filled with cerebrospinal fluid (CSF). In human patients at very late stages, secondary, very elongated cavities mostly in the rostral direction can form (syringomyelia). These cavities can lead to complications and further neurologic deterioration of the patient accompanied by increased spasticity and pain. The late phase of injury is also characterized by Wallerian degeneration of lesioned fibers and loss of myelin in the white matter. At later time points after injury, sprouting of fibers occurs in the cavity region, often in association with Schwann cells. Distinct differences are observed in the amount of remyelination of demyelinated axons; in general, small lesions show a higher amount of remyelinated fibers.

Because a large amount of additional damage arises after the initial injury, finding ways to block the processes that are responsible for this degeneration could lead to useful interventions in treating spinal-cord injury. Methylprednisolone is currently the only widely used drug that is being successfully used in acute spinal-cord injured patients.

SPROUTING AND REGENERATION OF LESIONED AXONS

Morphological features of axonal injury, axonal degeneration, and the frequently observed spontaneous, transient sprouting response were already described at the beginning of this century (Ramon y Cajal, 1928). These first studies hypothesized that CNS neurons, in contrast to the neurons of the PNS, either lack the intrinsic capacity to regenerate after axotomy or did not grow due to a lack of attractive or trophic factors (Tello, 1911; Ramon y Cajal, 1928). Current investigations suggest that central neurons may be less capable to regenerate due to extrinsic mechanisms, although intrinsic neuronal factors certainly also play an important role.

Axonal Regeneration and Sprouting in Amphibians and Fish

In contrast to the situation in the CNS of higher mammals and birds, a lesion of the CNS in certain amphibians and fish leads to functional regeneration of many of the injured nerve fibers (Lurie and Selzer, 1991). Many studies have employed the lamprey, an ancestral fish and living fossil belonging to the class of Cyclostomes. A complete transection of the spinal cord in these fish results in full functional recovery within a few months. Many axons sprout and regenerate over long distances into the caudal spinal cord, where they form functional synapses. The situation in the goldfish is similar in that after spinal-cord transection, functional recovery is observed and certain populations of lesioned nerve fibers regenerate. In salamanders, a comparable situation is observed, as is the case of the larval tadpole stage in the frog. Regeneration of fibers is, however, poor in the frog spinal cord after metamorphosis (but is retained in the optic nerve) (Lang et al., 1995). In lower vertebrates, growth processes remain activated throughout life, probably as the whole organism keeps growing. Scar reactions seem to be different, myelin phagocytosis is very rapid, and both fish and frog myelin contains fewer factors inhibitory for neurite outgrowth than myelin from mammals (Bastmeyer et al., 1991; Sivron and Schwartz, 1994).

Responses of PNS Axons

After injury of a peripheral nerve, the axon proximal to the lesion site undergoes degeneration up to the previous node of Ranvier. The myelin sheath surrounding the damaged distal nerves is broken down and is phagocytised by infiltrating and resident macrophages as well as by activated Schwann cells. Schwann cells play a crucial role in axonal growth by producing trophic factors (e.g., brain-derived neurotrophic factor [BDNF] and cilary neurotrophic factor [CNTF]). These cells also express a number of adhesion molecules (L1, neuronal cell adhesion molecule [NCAM]) that have been shown to be involved in neurite outgrowth (Seilheimer and Schachner, 1988). Lesioned CNS axons can even grow into Schwann cell grafts implanted into the cavity of the injured spinal cord (Paino et al., 1994). At the proximal stumps, within 1 d of injury, bud-like swellings contain cytoskeletal components, mitochondria, and elements of the smooth endoplasmic reticulum. These structures differentiate into growth cones that can then initiate regeneration.

In the PNS, it is therefore evident that rapid breakdown of the axon and the myelin sheath is followed by a fast invasion of macrophages and an activation of Schwann cells that remove debris and optimize the tissue for regenerative processes. In the CNS, the cells responsible for the removal of the damaged material react much more slowly and less efficiently than in the PNS (Castano et al., 1996).

Responses of Spinal Cord Axons of Higher Vertebrates

Both intrinsic neuronal as well as extrinsic factors contribute to the absence of axonal long-distance regrowth after injury. The observations made by Tello (1911) argued that the lack of trophic support might be the reason for the poor nerve growth in the adult CNS: Tello described the regenerative growth of cortical neurites into transplants of peripheral nerves. Inspired by the differences of regenerative properties of CNS vs PNS nerves, many studies were aimed at elucidating the role of external factors that influence axonal growth. In the 1980s, Aguayo and co-workers initiated a series of transplantation experiments, using either CNS or PNS tissue as a graft (Richardson et al., 1980; Aguayo et al., 1981; David and Aguayo, 1981). They concluded that the microenvironment surrounding outgrowing neurites determines the regenerative capacities of the axons, whereby Schwann cells proved to be highly growth-support-ive. Subsequently, in a series of in vitro experiments, it became apparent that CNS tissue, in particular CNS white matter and oligodendrocytes, actively inhibited neurite outgrowth (Schwab and Thoenen, 1985; Schwab and Caroni, 1988; Savio and Schwab, 1989). Myelin proteins of molecular weight 35 and 250 kDa were found to be respon-sible for much of the inhibitory activity of the CNS white matter (Caroni and Schwab, 1988a). These results were strengthened by the observation that after a lesion of the CNS in neonatal animals extensive rearrangements of the remaining fiber tracts were seen (for reviews see Kolb and Whishaw, 1998; Kaas et al., 1999). These phenomena are limited to a critical period that usually ends shortly after birth and is correlated in time with CNS myelination in the different parts of the CNS (Keirstead et al., 1992; Kapfhammer and Schwab, 1994a). In order to investigate the role of myelin on the sprouting and regenerative capacity of neurons, myelination was suppressed by X-irradiation. This treatment selectively kills oligodendrocyte precursor cells and prevents their expression of myelin-associated neurite growth inhibitors. Indeed, long-distance axonal regeneration was seen in myelin-free spinal cord within 2–3 wk after lesion (Savio and Schwab, 1990).

It is also clear, however, that intrinsic neuronal factors play a role in the process of regeneration of spinal-cord nerve fibers after an insult. Intrinsic factors, including a set of growth-related genes (e.g., growth-associated protein [GAP-43]) are upregulated as a spontaneous reaction of many neurons to injury, but their increase is transitory, if regeneration cannot occur. Overexpression of GAP-43 in neurons of transgenic mice leads to enhanced sprouting in spinal cord, brain, and in the periphery (Aigner et al., 1995).

FACTORS AND SUBSTRATES
THAT INFLUENCE REGENERATION AFTER SPINAL-CORD INJURY

Neurotrophic factors, inhibitory molecules, cells, or tissues that provide structural support and scar tissue are some of the important extrinsic factors that positively or negatively influence the regenerative capacity of outgrowing CNS neurites.

Growth-Promoting Factors

The neurotrophins are a group of factors regulating neuronal differentiation, axonal outgrowth, synapse function, and neuronal survival in the nervous system, and recently nerve growth factor (NGF) has also been implicated to have cell death-promoting properties (Frade et al., 1996; for reviews see Barde, 1994; Chao et al., 1998).

NGF, BDNF, and neurotrophin 3 and 4 (NT-3 and NT-4) are, however, also produced and transported in the adult CNS. It was therefore obvious to study their effects on injured adult neurons. Neuronal survival and axonal growth, i.e., enhanced sprouting and regeneration, have been observed by several investigators after administration of neurotrophic factors to the injured spinal cord (Schnell et al., 1994; Giehl and Tetzlaff, 1996; Bregman et al., 1997; Grill et al., 1997; Junger and Varon, 1997; Jakeman et al., 1998).

Several other neurotrophic factors have also been shown to have survival effects for young and adult neurons projecting to the spinal cord. Corticospinal neurons survive an axotomy close to the cell body if supplied with either CNTF or glia-cell line-derived neurotrophic factor (GDNF) (Dale et al., 1995; Junger and Varon, 1997). Leukemia inhibitory factor (LIF) enhances axonal growth after spinal-cord injury in adult rats, although probably indirectly via NT-3 (Blesch et al., 1999).

A third class of growth factors that have been associated mainly with survival of injured nerve fibers in the nervous system are the fibroblast growth factors (Teng et al., 1998). All these findings probably just mark the beginning of a promising field of research and intervention, as many growth-promoting factors and guidance molecules remain to be discovered and studied in their effects on the lesioned spinal-cord tissue.

Growth-Inhibiting Factors

In addition to the growth promoting factors that influence nerve regeneration, there are also factors that inhibit axonal growth in the CNS. One major difference between

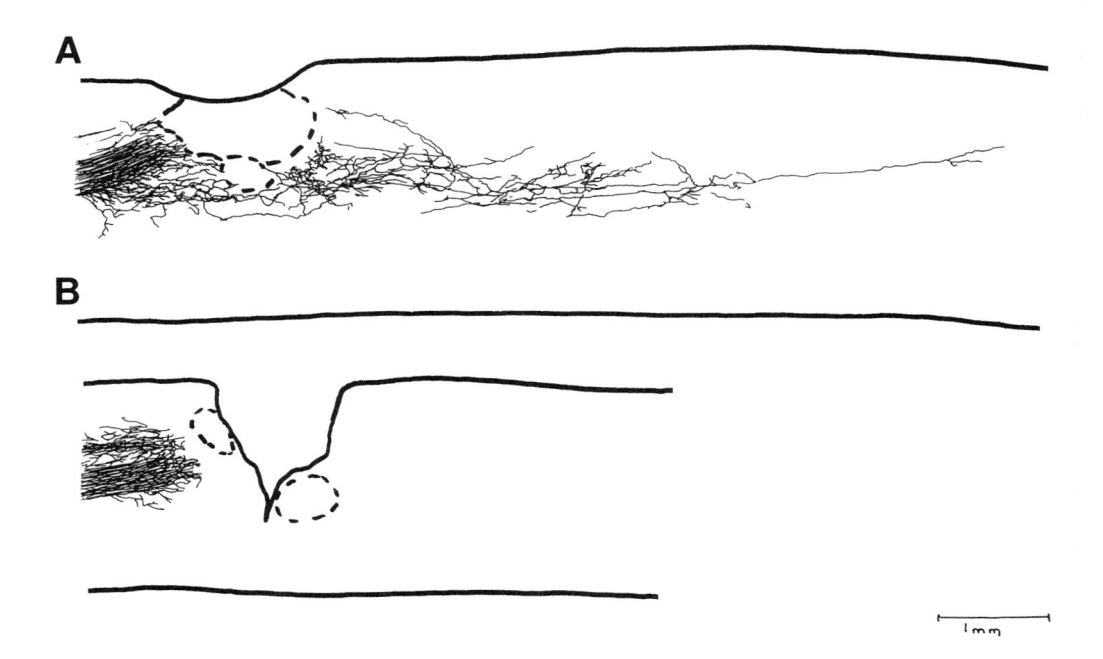

Fig. 2. Camera lucida reconstructions of consecutive series of parasagittal sections (50 μm) of the thoracic and high lumbar spinal cord of a representative rIN-1 Fab-treated animal (**A**) and a comparable control (**B**). rIN-1-treated animals exhibited long-distance regeneration for up to more than 9 mm with axon branching profusely into the gray-matter areas, whereas control animals showed no regenerative growth.

the CNS and the PNS indeed lies in the fact that the Schwann cells that myelinate PNS axons do not produce the same strong nerve-growth inhibition that is exerted by oligodendrocytes. CNS oligodendrocytes contain inhibitory proteins that suppress outgrowth of several types of neurons in vitro (Caroni and Schwab, 1988a; Spillmann et al., 1998). The potent inhibitory protein NI-250, now known as Nogo-A, has been characterized and its cDNA was cloned (Chen et al., 2000). Antibodies were raised that neutralize the inhibitory action of this protein in vitro (monoclonal antibody IN-1; Caroni and Schwab, 1988b; Spillmann et al., 1997) and enable axonal regeneration after spinal-cord injury in adult rats (rIN-1 Fab; Fig. 2) (Schnell and Schwab, 1990; Schnell et al., 1994; Bregman et al., 1995; Raineteau et al., 1999). Antibody administration also enhances structural plasticity of unlesioned nerve fibers in the CNS and leads to an increased functional recovery of the animals (Fig. 3) (Thallmair et al., 1998; Z'Graggen et al., 1998). Immunological suppression of myelin development also leads to an enhanced axonal regeneration after injury (Keirstead et al., 1995).

Apart from Nogo-A, an inhibitory effect of the myelin-associated glycoprotein MAG has also been reported: a MAG substrate inhibits axonal growth and sprouting of cultured adult dorsal root ganglia, and the MAG knockout mouse shows enhanced regeneration in the PNS (Schafer et al., 1996). When Schwann cells are induced to overexpress MAG by retroviral infection, axonal outgrowth and neurite branching are greatly inhibited (Shen et al., 1998). This suggests that MAG not only affects axonal regeneration but may also play a role in the control of axonal branching. Another group

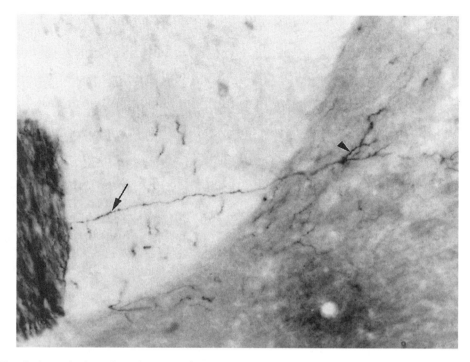

Fig. 3. An unlesioned corticospinal fiber crosses into the area of the degenerated, contralateral corticospinal tract (arrow) and branches into the denervated region (arrow head) after treatment of a cervically injured rat with mAb IN-1 (Thallmair et al., 1998).

of molecules that are neuronal repellents are the collapsin/semaphorin subfamily members (for reviews, *see* Tessier-Lavigne and Goodman, 1996; Pasterkamp et al., 1998); their potential role in peripheral and central regenerative processes, however, are unknown at present.

Lesion Scar

After injury to the CNS, astrocytes become hypertrophic and form scar tissue that walls off the CNS against the debris zone and the forming cysts. Axons of the CNS seem to be inhibited by the scar tissue and for many years the lesion scar was seen as a physical barrier for regenerating axons. Current theories suggest that the regenerative failure of axons is not solely due to a mechanical obstruction. The expression of inhibitory molecules, e.g., extracellular matrix molecules, by reactive astrocytes seems to be correlated with the reduction of neurite outgrowth following CNS injury in this region (McKeon et al., 1991; for review see Fitch and Silver, 1997). Different types of proteoglycans, in particular, chondroitin sulfate proteoglycans, have been reported to be upregulated following trauma to the adult CNS in vivo (Levine, 1994; Fitch and Silver, 1997). The expression of certain proteoglycans is co-localized with strongly positive glial fibrillary acid protein (GFAP) astrocytes (McKeon et al., 1991). Moreover, it was shown that in vitro scar tissue of neonatal animals supports axonal growth, whereas that of adult animals had an inhibitory effect (Rudge and Silver, 1990). At least three different proteoglycans—phosphacan, neurocan, and NG2—are expressed in adult CNS scar tissue. Different members of the tenascin gene family are expressed both in the developing and in the lesioned CNS and exert both attractant as well as repellent properties for axonal growth (for reviews, *see* Faissner, 1997; Fawcett and Asher, 1999).

The role of this upregulation of inhibitory signals still remains to be unraveled. In fact, it may be a part of the wound-healing process, and the consequences for regeneration may be an unwelcome side effect.

Transplants of Peripheral and Fetal Nerve Tissue, Embryonic Stem Cells, Schwann Cells, Olfactory Nerve Ensheathing Cells, and Artificial Bridges

Understanding and manipulating the endogenous growth-promoting and growth-inhibiting factors of the CNS is important. In the field of axonal regeneration the possibilities of changing the environment of the regenerating axons by tissue grafts has also been studied extensively. Spinal axons can regenerate into a peripheral nerve graft placed in the spinal-cord lesion site (Aguayo et al., 1981, 1984). Bridging the lesion site with peripheral nerve tissue showed axonal growth into the graft and back into the CNS environment where the axonal growth stopped, however, after a short distance (David and Aguayo, 1981). Spinal-cord transection and directed implantation of peripheral nerve bridges can lead to recovery of coordinated gait in the transplanted rats and thus suggests the formation of functionally relevant neuronal communication across the lesion (Cheng et al., 1997).

Fetal nerve tissue, usually E14-E17 rat tissue, used as a graft in spinal-cord injury paradigms showed that transplants can rescue axotomized central neurons (Bregman and Reier, 1986; Tessler et al., 1988). Axonal growth occurs to some extent into and through these transplants and can be enhanced by addition of neurotrophic factors (Bregman et al., 1997). Multipotent neural progenitors or stem cells can integrate into the host tissue. In the future, such cells may be able to replace dysfunctional neurons and glia of the injured host tissue (for review, *see* Snyder and Macklis, 1995–1996). Inducing the stem cells to produce neurotrophic factors or specific antibodies could lead to new therapeutic approaches.

Experiments were also done transplanting Schwann cells into the spinal-cord lesion site as a supportive substrate for injured nerve fibers. Many of the important experiments with Schwann cell implants after CNS injury were done in the lab of Richard and Mary Bunge. It was shown that highly purified populations of Schwann cells within a guidance channel grafted into lesioned adult rat spinal cord stimulate growth of axons into the graft. Moreover, these regenerating axons were myelinated or ensheathed by the grafted Schwann cells (Paino et al., 1994; Xu et al., 1995). The Schwann cells in these experiments were obtained from rat or human peripheral nerve explants.

The most efficient regeneration-promoting cell implants currently available are purified olfactory nerve ensheathing cells (Li et al., 1997; Ramon-Cueto and Nieto-Sampedro, 1994; Ramon-Cueto et al., 1998). One difference between Schwann cells and ensheathing glia is that the latter can migrate down the white matter tracts, preparing a highly growth-promoting territory and taking axons with them, whereas Schwann cells remain where they are transplanted.

Many other materials, including collagen, hydrogels, or nitrocellulose filters containing astrocytes or laminin, have been used to try to artificially bridge the gap of the lesion. In general, growth of axons into artificial bridges was poor and the implants did not integrate well with the spinal-cord tissue.

FUNCTIONAL ASPECTS AFTER SPINAL-CORD INJURY

All the progress described above, although clearly heading toward new therapies, did not yet have practical consequences for spinal cord-injured patients today. Patients have to learn to cope with the function remaining after the injury. Functional recovery after CNS injury probably depends largely on remyelination and the spontaneous reorganization (plasticity) of undamaged neural pathways that takes place to a limited extent after a lesion. Recent studies have observed enhanced functional recovery after experimental interventions, and work with spinal-injured humans demonstrates that training can improve functional locomotor abilities due to stabilization and possibly reorganization of the still-existing nerve fibers.

Sparing of Function

The neural systems responsible for locomotion are largely located at the spinal level and are referred to as the central pattern generators. Central pattern generators seem to be present in all vertebrates and are independent of sensory input. The principal role of spinal locomotor central pattern generators (CPGs) is to provide oscillatory motor commands to individual muscles and to control the exact timing of those commands thus leading to efficient, coordinated locomotor behavior (for review, *see* Sigvardt and Miller, 1998). The CPG can be activated in spinal cords completely disconnected from the brain, demonstrating their independence of sensory input. Initiation and control of many functions can be observed in lesions showing only a small percentage of descending intact fibers, and incomplete lesions of the lower thoracic spinal cord can lead to preservation of hindlimb stepping and locomotion in different species including man (Eidelberg et al., 1981; Brustein and Rossignol, 1998). There seems to be an adaptation of local spinal-cord circuits to a smaller number of remaining axons.

Locomotion training on a treadmill, aims at mobilizing the patient and retaining an overall good constitution (Fig. 4). An important element in the treadmill training for both spinal-injured cats and humans is the provision of adequate locomotion-related sensory input, which can physiologically modulate the spinal locomotor circuitry and thus improve the functional recovery. Similar to the situation seen in experimental spinal-cord-injury models, human patients can be trained to show an improved locomotor function after a spinal-cord injury (Dietz et al., 1998). The patients show a return of rhythmic locomotor muscle activation patterns best explained by the activation of a spinal pattern generator. Immobilized spinal cord-lesioned rats showed poor motor recovery, whereas rats with a similar lesion and free mobility showed a marked recovery (Little et al., 1991). These results suggest that disuse of spinal circuits due to injury could lead to a loss of function in the undamaged spinal circuit.

Spontaneous Sprouting after Injury

After injury to the spinal cord, spontaneous sprouting of spared fibers can be expected to take place. This sprouting could lead to plastic rearrangements of the circuitry that could have beneficial effects on the central control of the spinal networks. Sprouting could, however, also lead to connections that generate neurogenic pain. Unilateral corticospinal tract lesions in perinatal animals lead to sprouting of fibers

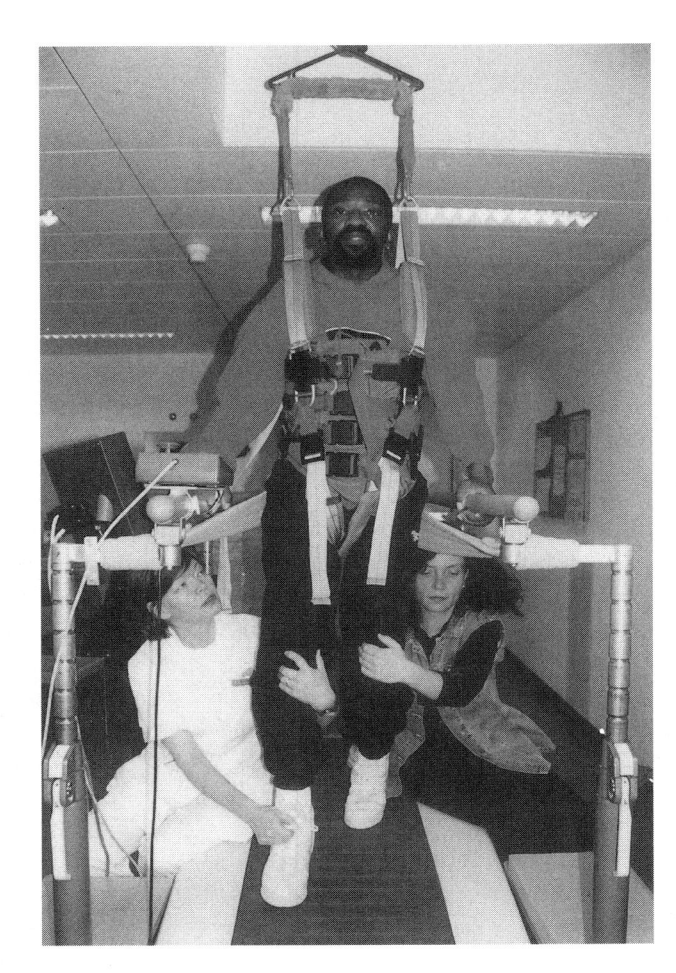

Fig. 4. A spinal cord-injured patient is trained on a treadmill with the help of physiotherapists and a weight-supporting device. The goal is to train the spinal circuits for locomotion and to stabilize connections to remaining descending fibers. (Photo, Gerry Colombo, Paraplegic Center, Balgrist, University Hospital, Zurich).

from the intact tract across the midline into the denervated areas of the brainstem and spinal cord (Kalil and Reh, 1982). Similar lesions in adult animals do not show spared fibers sprouting across the midline. In human patients with large cortical lesions, a comparable situation is present. Young children with a damaged sensory-motor cortex show a large degree of recovery of the motor capacity of the denervated side, which may at least partially be due to projections coming from spared fibers of the intact side, as shown by the occasional occurrence of mirror movements, especially of the hands (Cao et al., 1994). This phenomenon does not occur in adult cortex-injured patients, suggesting that the immature brain is able to reorganize in response to injury and that this reorganization is clearly age-dependent (Kennard-principle; Kennard, 1936). The results suggest the existence of regulatory factors, such as myelin-associated neurite outgrowth inhibiting proteins. Indeed, the growth associated protein GAP-43 is downregulated during postnatal development in parallel to myelin formation in the brain and spinal cord (Kapfhammer and Schwab, 1994b).

Fig. 5. An injured rat treated with the mAb IN-1 shows an improvement of behavioral function elevated in different tests, e.g., the rope-climbing test. (Photo, Roland Schob, Scientific Photographer, Brain Research Institute, Zurich, Switzerland).

Improvement of Function Under Experimental Influences

Neurotrophins are promising candidates for stimulating local sprouting and regeneration of injured fibers. Indeed, NT-3 has been reported to not only promote axonal growth of corticospinal fibers, but also to enhance functional recovery of spinal cord-injured rats (Grill et al., 1997; Houweling et al., 1998). Functional recovery was also observed after administration of an antibody that neutralizes the myelin-associated neurite growth inhibitors (mAb IN-1). Using different lesion paradigms, adult spinal cord-injured rats showed regeneration and plastic responses as well as a significant behavioral recovery (Fig. 5) (Bregman et al., 1995; Thallmair et al., 1998; Z'Graggen et al., 1998). A combination of NT-3 and mAb IN-1 treatment after spinal-cord lesion even showed a slight improvement of functional recovery after a 2-wk delay between lesion and treatment (von Meyenburg et al., 1998). Continuous intrathecal administration of bFGR during a 1-wk period following moderate spinal-cord contusion in adult rats significantly reduced the extent of spinal cord-tissue damage and improved recovery of coordinated hindlimb movements (Rabchevsky et al., 1999).

Fetal spinal-cord transplants at the site of neonatal spinal-cord injury resulted in enhanced recovery of several parameters of locomotion, probably induced by anatomical plasticity elicited by the transplants (Kunkel-Bagden and Bregman, 1990). Fetal grafts were also placed into the spinal cord of rats 10 d after a contusion injury; again, a recovery in motor behavior was observed (Stokes and Reier, 1992).

The improvements seen after experimental interventions can possibly be enhanced by treadmill training, a combination that may represent the strategy for future therapies.

CONCLUSIONS

The field of spinal cord-injury research has made large progress over the last decade, and many researchers believe that basic science is slowly getting closer to clinical applications. New therapies may include a combination of interventions that suppress endogenous growth-inhibitory components of the adult CNS and stimulate neuronal survival and neurite outgrowth. By manipulating the scar and the inflammatory reactions after injury, or by bridging the lesion site with growth-permissive material, the damaged tissue might be remodeled to form an optimal substrate for regenerating nerve fibers. Although the number of regenerating fibers in experimental spinal cord-injury models is small at present, an impressive degree of functional recovery can be observed for many functions. For spinal cord-injured patients, even a modest gain of function would improve their quality of life to a large degree.

REFERENCES

Aguayo, A. J., David, S., and Bray, G. M. (1981) Influences of the glial environment on the elongation of axons after injury: transplantation studies in adult rodents. *J. Exp. Biol.* **95,** 231–240.

Aguayo, A. J., Bjorklund, A., Stenevi, U., and Carlstedt, T. (1984) Fetal mesencephalic neurons survive and extend long axons across peripheral nervous system grafts inserted into the adult rat striatum. *Neurosci. Lett.* **45,** 53–58.

Aigner, L., Arber, S., Kapfhammer, J. P., Laux, T., Schneider, C., Botteri, F., et al. (1995) Overexpression of the neural growth-associated protein GAP-43 induces nerve sprouting in the adult nervous system of transgenic mice. *Cell* **83,** 269–278.

Allen, A. R. (1911) Surgery of experimental lesions of spinal cord equivalent to crush injury of fracture dislocation. Preliminary report. *J. Am. Med. Assoc.* **57,** 878–880.

Anderson, D. K., Means, E. D., Waters, T. R., and Green, E. S. (1982) Microvascular perfusion and metabolism in injured spinal cord after methylprednisolone treatment. *J. Neurosurg.* **56,** 106–113.

Azbill, R. D., Mu, X., Bruce-Keller, A. J., Mattson, M. P., and Springer, J. E. (1997) Impaired mitochondrial function, oxidative stress and altered antioxidant enzyme activities following traumatic spinal cord injury. *Brain Res.* **765,** 283–290.

Barde, Y. A. (1994) Neurotrophic factors: an evolutionary perspective. *J. Neurobiol.* **25,** 1329–1333.

Bartholdi, D. and Schwab, M. E. (1997) Expression of pro-inflammatory cytokine and chemokine mRNA upon experimental spinal cord injury in mouse: an in situ hybridization study. *Eur. J. Neurosci.* **9,** 1422–1438.

Bastmeyer, M., Beckmann, M., Schwab, M. E., and Stuermer, C. A. (1991) Growth of regenerating goldfish axons is inhibited by rat oligodendrocytes and CNS myelin but not but not by goldfish optic nerve tract oligodendrocytelike cells and fish CNS myelin. *J. Neurosci.* **11,** 626–640.

Basso, D. M., Beattie, M. S., Bresnahan, J. C., Anderson, D. K., Faden, A. I., Gruner, J. A., et al. (1996) MASCIS evaluation of open field locomotor scores: effects of experience and teamwork on reliability. Multicenter Animal Spinal Cord Injury Study. *J. Neurotrauma* **13,** 343–359.

Blesch, A., Uy, H. S., Grill, R. J., Cheng, J. G., Patterson, P. H., and Tuszynski, M. H. (1999) Leukemia inhibitory factor augments neurotrophin expression and corticospinal axon growth after adult CNS injury. *J. Neurosci.* **19,** 3556–3566.

Blight, A. R. (1992) Macrophages and inflammatory damage in spinal cord injury. *J. Neurotrauma* **9(Suppl. 1),** S83–S91.

Bracken, M. B., Shepard, M. J., Collins, W. F., Holford, T. R., Young, W., Baskin, D. S., et al. (1990) A randomized, controlled trial of methylprednisolone or naloxone in the treatment of acute spinal-cord injury. Results of the Second National Acute Spinal Cord Injury Study. *N. Engl. J. Med.* **322,** 1405–1411.

Bracken, M. B. and Holford, T. R. (1993) Effects of timing of methylprednisolone or naloxone administration on recovery of segmental and long-tract neurological function in NASCIS 2. *J. Neurosurg.* **79,** 500–507.

Bracken, M. B., Shepard, M. J., Holford, T. R., Leo-Summers, L., Aldrich, E. F., Fazl, M., et al. (1997) Administration of methylprednisolone for 24 or 48 hours or tirilazad mesylate for 48 hours in the treatment of acute spinal cord injury. Results of the Third National Acute Spinal Cord Injury Randomized Controlled Trial. National Acute Spinal Cord Injury Study. *JAMA* **277,** 1597–1604.

Breasted, J. H. (1930) *The Edwin Smith Surgical Papyrus*. University of Chicago Press, Chicago, IL.

Bregman, B. S., Kunkel-Bagden, E., Schnell, L., Dai, H. N., Gao, D., and Schwab, M. E. (1995) Recovery from spinal cord injury mediated by antibodies to neurite growth inhibitors. *Nature* **378,** 498–501.

Bregman, B. S., McAtee, M., Dai, H. N., and Kuhn, P. L. (1997) Neurotrophic factors increase axonal growth after spinal cord injury and transplantation in the adult rat. *Exp. Neurol.* **148,** 475–494.

Bregman, B. S. and Reier, P. J. (1986) Neural tissue transplants rescue axotomized rubrospinal cells from retrograde death. *J. Comp. Neurol.* **244,** 86–95.

Brustein, E. and Rossignol, S. (1998) Recovery of locomotion after ventral and ventrolateral spinal lesions in the cat. I. Deficits and adaptive mechanisms. *J. Neurophysiol.* **80,** 1245–1267.

Cao, Y., Vikingstad, E. M., Huttenlocher, P. R., Towle, V. L., and Levin, D. N. (1994) Functional magnetic resonance studies of the reorganization of the human hand sensorimotor area after unilateral brain injury in the perinatal period. *Proc. Natl. Acad. Sci. USA* **91,** 9612–9616.

Carlson, S. L., Parrish, M. E., Springer, J. E., Doty, K., and Dossett, L. (1998) Acute inflammatory response in spinal cord following impact injury. *Exp. Neurol.* **151,** 77–88.

Caroni, P. and Schwab, M. E. (1988a) Two membrane protein fractions from rat central myelin with inhibitory properties for neurite growth and fibroblast spreading. *J. Cell Biol.* **106,** 1281–1288.

Caroni, P. and Schwab, M. E. (1988b) Antibody against myelin-associated inhibitor of neurite growth neutralizes nonpermissive substrate properties of CNS white matter. *Neuron* **1,** 85–96.

Castano, A., Bell, M. D., and Perry, V. H. (1996) Unusual aspects of inflammation in the nervous system: Wallerian degeneration. *Neurobiol. Aging* **17,** 745–751.

Chao, M., Casaccia-Bonnefil, P., Carter, B., Chittka, A., Kong, H., and Yoon, S. O. (1998) Neurotrophin receptors: mediators of life and death. *Brain Res. Brain Res. Rev.* **26,** 295–301.

Chen, M., Huber, A. B., van der Haar, M., Frank, M., Schnell, L., Spillmann, A. A., Christ, F., and Schwab, M. E. (2000) Nogo-A is a myelin-associated neurite outgrowth inhibitor and an antigen for monoclonal antibody IN-1. *Nature* **403,** 434–439.

Cheng, H., Almstrom, S., Gimenez-Llort, L., Chang, R., Ove Ogren, S., Hoffer, B., and Olson, L. (1997) Gait analysis of adult paraplegic rats after spinal cord repair. *Exp. Neurol.* **148,** 544–557.

Dale, S. M., Kuang, R. Z., Wei, X., and Varon, S. (1995) Corticospinal motor neurons in the adult rat: degeneration after intracortical axotomy and protection by ciliary neurotrophic factor (CNTF). *Exp. Neurol.* **135,** 67–73

David, S. and Aguayo, A. J. (1981) Axonal elongation into peripheral nervous system "bridges" after central nervous system injury in adult rats. *Science* **214,** 931–933.

Dietz, V., Wirz, M., Curt, A., and Colombo, G. (1998) Locomotor pattern in paraplegic patients: training effects and recovery of spinal cord function. *Spinal Cord* **36,** 380–390.

Dusart, I. and Schwab, M. E. (1994) Secondary cell death and the inflammatory reaction after dorsal hemisection of the rat spinal cord. *Eur. J. Neurosci.* **6,** 712–724.

Eidelberg, E., Story, J. L., Walden, J. G., and Meyer, B. L. (1981) Anatomical correlates of return of locomotor function after partial spinal cord lesions in cats. *Exp. Brain Res.* **42,** 81–88.

Faissner, A. (1997) The tenascin gene family in axon growth and guidance. *Cell Tissue Res.* **290,** 331–341.

Fawcett, J. W. and Asher, R. A. (1999) The glial scar and central nervous system repair. *Brain Res. Bull.* **49,** 377–391.

Fitch, M. T. and Silver, J. (1997) Activated macrophages and the blood-brain barrier: inflammation after CNS injury leads to increases in putative inhibitory molecules. *Exp. Neurol.* **148,** 587–603.

Follesa, P., Wrathall, J. R., and Mocchetti, I. (1994) Increased basic fibroblast growth factor mRNA following contusive spinal cord injury. *Brain Res. Mol. Brain Res.* **22,** 1–8.

Frade, J. M., Rodriguez-Tebar, A., and Barde, Y. A. (1996) Induction of cell death by endogenous nerve growth factor through its p75 receptor. *Nature* **383,** 166–168.

Geisler, F. H. (1993) GM-1 ganglioside and motor recovery following human spinal cord injury. *J. Emerg. Med.* **11(Suppl. 1),** 49–55.

Geisler, F. H., Dorsey, F. C., and Coleman, W. P. (1993) Past and current clinical studies with GM-1 ganglioside in acute spinal cord injury. *Ann. Emerg. Med.* **22,** 1041–1047.

Geisler, F. H. (1998) Clinical trials of pharmacotherapy for spinal cord injury. *Ann. NY Acad. Sci.* **845,** 374–381.

Giehl, K. M. and Tetzlaff, W. (1996) BDNF and NT-3, but not NGF, prevent axotomy-induced death of rat corticospinal neurons in vivo. *Eur. J. Neurosci.* **8,** 1167–1175.

Grill, R., Murai, K., Blesch, A., Gage, F. H., and Tuszynski, M. H. (1997) Cellular delivery of neurotrophin-3 promotes corticospinal axonal growth and partial functional recovery after spinal cord injury. *J. Neurosci.* **17,** 5560–5572.

Houweling, D. A., Lankhorst, A. J., Gispen, W. H., Bar, P. R., and Joosten, E. A. (1998) Collagen containing neurotrophin-3 (NT-3) attracts regrowing injured corticospinal axons in the adult rat spinal cord and promotes partial functional recovery. *Exp. Neurol.* **153,** 49–59.

Jakeman, L. B., Wei, P., Guan, Z., and Stokes, B. T. (1998) Brain-derived neurotrophic factor stimulates hindlimb stepping and sprouting of cholinergic fibers after spinal cord injury. *Exp. Neurol.* **154,** 170–184.

Junger, H. and Varon, S. (1997) Neurotrophin-4 (NT-4) and glial cell line-derived neurotrophic factor (GDNF) promote the survival of corticospinal motor neurons of neonatal rats in vitro. *Brain Res.* **762,** 56–60.

Kaas, J. H., Florence, S. L., and Jain, N. (1999) Subcortical contributions to massive cortical reorganizations. *Neuron* **22,** 657–660.

Kalil, K. and Reh, T. (1982) A light and electron microscopic study of regrowing pyramidal tract fibers. *J. Comp. Neurol.* **211,** 265–275.

Kapfhammer, J. P. and Schwab, M. E. (1994a) Inverse patterns of myelination and GAP-43 expression in the adult CNS: neurite growth inhibitors as regulators of neuronal plasticity? *J. Comp. Neurol.* **340,** 194–206.

Kapfhammer, J. P. and Schwab, M. E. (1994b) Increased expression of the growth-associated protein GAP-43 in the myelin-free rat spinal cord. *Eur. J. Neurosci.* **6,** 403–411.

Keirstead, H. S., Hasan, S. J., Muir, G. D., and Steeves, J. D. (1992) Suppression of the onset of myelination extends the permissive period for the functional repair of embryonic spinal cord. *Proc. Natl. Acad. Sci. USA* **89,** 11,664–11,668.

Keirstead, H. S., Dyer, J. K., Sholomenko, G. N., McGraw, J., Delaney, K. R., and Steeves, J. D. (1995) Axonal regeneration and physiological activity following transection and immunological disruption of myelin within the hatchling chick spinal cord. *J. Neurosci.* **15,** 6963–6974.

Kennard, M. A. (1936) Age and other factors in motor recovery from precentral lesions in monkeys. *Am. J. Physiol.* **115,** 138–146.

Kolb, B. and Whishaw, I. Q. (1998) Brain plasticity and behavior. *Annu. Rev. Psychol.* **49,** 43–64.

Kunkel-Bagden, E. and Bregman, B. S. (1990) Spinal cord transplants enhance the recovery of locomotor function after spinal cord injury at birth. *Exp. Brain Res.* **81,** 25–34.

Lang, D. M., Rubin, B. P., Schwab, M. E., and Stuermer, C. A. (1995) CNS myelin and oligo-dendrocytes of the Xenopus spinal cord—but not optic nerve—are nonpermissive for axon growth. *J. Neurosci.* **15,** 99–109.

Levine, J. M. (1994) Increased expression of the NG2 chondroitin-sulfate proteoglycan after brain injury. *J. Neurosci.* **14,** 4716–4730.

Li, Y., Field, P. M., and Raisman, G. (1997) Repair of adult rat corticospinal tract by transplants of olfactory ensheathing cells. *Science* **277,** 2000–2002.

Little, J. W., Harris, R. M., and Lerner, S. J. (1991) Immobilization impairs recovery after spinal cord injury. *Arch. Phys. Med. Rehabil.* **72,** 408–412.

Lurie, D. I. and Selzer, M. E. (1991) Axonal regeneration in the adult lamprey spinal cord. *J. Comp. Neurol.* **306,** 409–416.

McKeon, R. J., Schreiber, R. C., Rudge, J. S., and Silver, J. (1991) Reduction of neurite outgrowth in a model of glial scarring following CNS injury is correlated with the expression of inhibitory molecules on reactive astrocytes. *J. Neurosci.* **11,** 3398–3411.

McTigue, D. M., Tani, M., Krivacic, K., Chernosky, A., Kelner, G. S., Maciejewski, D., et al. (1998) Selective chemokine mRNA accumulation in the rat spinal cord after contusion injury. *J. Neurosci. Res.* **53,** 368–376.

Nesathurai, S. (1998) Steroids and spinal cord injury: revisiting the NASCIS 2 and NASCIS 3 trials. *J. Trauma* **45,** 1088–1093.

Olson, L. (1997) Regeneration in the adult central nervous system: experimental repair strategies. *Nat. Med.* **3,** 1329–1335.

Paino, C. L., Fernandez-Valle, C., Bates, M. L., and Bunge, M. B. (1994) Regrowth of axons in lesioned adult rat spinal cord: promotion by implants of cultured Schwann cells. *J. Neurocytol.* **23,** 433–452.

Pasterkamp, R. J., De Winter, F., Giger, R. J., and Verhaagen, J. (1998) Role for semaphorin III and its receptor neuropilin-1 in neuronal regeneration and scar formation? *Prog. Brain Res.* **117,** 151–170.

Perry, V. H., Bell, M. D., Brown, H. C., and Matyszak, M. K. (1995) Inflammation in the nervous system. *Curr. Opin. Neurobiol.* **5,** 636–641.

Popovich, P. G., Wei, P., and Stokes, B. T. (1997) Cellular inflammatory response after spinal cord injury in Sprague-Dawley and Lewis rats. *J. Comp. Neurol.* **377,** 443–464.

Rabchevsky, A. G., Fugaccia, I., Fletcher-Turner, A., Blades, D. A., Mattson, M. P., and Scheff, S. W. (1999) Basic fibroblast growth factor (bFGF) enhances tissue sparing and functional recovery following moderate spinal cord injury. *J. Neurotrauma* **16,** 817–830.

Raineteau, O., Z'Graggen, W. J., Thallmair, M., and Schwab, M. E. (1999) Sprouting and regeneration after pyramidotomy and blockade of the myelin-associated neurite growth inhibitors NI 35/250 in adult rats. *Eur. J. Neurosci.* **11,** 1486–1490.

Ramon-Cueto, A. and Nieto-Sampedro, M. (1994) Regeneration into the spinal cord of transected dorsal root axons is promoted by ensheathing glia transplants. *Exp. Neurol.* **127,** 232–244.

Ramon-Cueto, A., Plant, G. W., Avila, J., and Bunge, M. B. (1998) Long-distance axonal regeneration in the transected adult rat spinal cord is promoted by olfactory ensheathing glia transplants. *J. Neurosci.* **18,** 3803–3815.

Ramon y Cajal, S. (1928) *Degeneration and Regeneration of the Nervous System.* Oxford University Press, Oxford, UK.

Richardson, P. M., McGuinness, U. M., and Aguayo, A. J. (1980) Axons from CNS neurons regenerate into PNS grafts. *Nature* **284,** 264–265.

Rudge, J. S. and Silver, J. (1990) Inhibition of neurite outgrowth on astroglial scars in vitro. *J. Neurosci.* **10,** 3594–3603.

Savio, T. and Schwab, M. E. (1989) Rat CNS white matter, but not gray matter, is nonpermissive for neuronal cell adhesion and fiber outgrowth. *J. Neurosci.* **9,** 1126–1133.

Savio, T. and Schwab, M. E. (1990) Lesioned corticospinal tract axons regenerate in myelin-free rat spinal cord. *Proc. Natl. Acad. Sci. USA* **87,** 4130–4133.

Schafer, M., Fruttiger, M., Montag, D., Schachner, M., and Martini, R. (1996) Disruption of the gene for the myelin-associated glycoprotein improves axonal regrowth along myelin in C57BL/Wlds mice. *Neuron* **16,** 1107–1113.

Schnell, L. and Schwab, M. E. (1990) Axonal regeneration in the rat spinal cord produced by an antibody against myelin-associated neurite growth inhibitors. *Nature* **343,** 269–272.

Schnell, L., Schneider, R., Kolbeck, R., Barde, Y. A., and Schwab, M. E. (1994) Neurotrophin-3 enhances sprouting of corticospinal tract during development and after adult spinal cord lesion. *Nature* **367,** 170–173.

Schnell, L., Schneider, R., Berman, M. A., Perry, V. H., and Schwab, M. E. (1997) Lymphocyte recruitment following spinal cord injury in mice is altered by prior viral exposure. *Eur. J. Neurosci.* **9,** 1000–1007.

Schwab, M. E. and Bartholdi, D. (1996) Degeneration and regeneration of axons in the lesioned spinal cord. *Physiol. Rev.* **76,** 319–370.

Schwab, M. E. and Caroni, P. (1988) Oligodendrocytes and CNS myelin are nonpermissive substrates for neurite growth and fibroblast spreading in vitro. *J. Neurosci.* **8,** 2381–2393.

Schwab, M. E. and Thoenen, H. (1985) Dissociated neurons regenerate into sciatic but not optic nerve explants in culture irrespective of neurotrophic factors. *J. Neurosci.* **5,** 2415–2423.

Schwartz, M., Lazarov-Spiegler, O., Rapalino, O., Agranov, I., Velan, G., and Hadani, M. (1999) Potential repair of rat spinal cord injuries using stimulated homologous macrophages. *Neurosurgery* **44,** 1041–1046.

Seilheimer, B. and Schachner, M. (1988) Studies of adhesion molecules mediating interactions between cells of peripheral nervous system indicate a major role for L1 in mediating sensory neuron growth on Schwann cells in culture. *J. Cell Biol.* **107,** 341–351.

Shen, Y. J., DeBellard, M. E., Salzer, J. L., Roder, J., and Filbin, M. T. (1998) Myelin-associated glycoprotein in myelin and expressed by Schwann cells inhibits axonal regeneration and branching. *Mol. Cell Neurosci.* **12,** 79–91.

Sigvardt, K. A. and Miller, W. L. (1998) Analysis and modeling of the locomotor central pattern generator as a network of coupled oscillators. *Ann. NY Acad. Sci.* **860,** 250–265.

Sivron, T. and Schwartz, M. (1994) The enigma of myelin-associated growth inhibitors in spontaneously regenerating nervous systems. *Trends Neurosci.* **17,** 277–281.

Snyder, E. Y. and Macklis, J. D. (1995–1996) Multipotent neural progenitor or stem-like cells may be uniquely suited for therapy for some neurodegenerative conditions. *Clin. Neurosci.* **3,** 310–316.

Spillmann, A. A., Amberger, V. R., and Schwab, M. E. (1997) High molecular weight protein of human central nervous system myelin inhibits neurite outgrowth: an effect which can be neutralized by the monoclonal antibody IN-1. *Eur. J. Neurosci.* **9,** 549–555.

Spillmann, A. A., Bandtlow, C. E., Lottspeich, F., Keller, F., and Schwab, M. E. (1998) Identification and characterization of a bovine neurite growth inhibitor (bNI-220). *J. Biol. Chem.* **273,** 19,283–19,293.

Springer, J. E., Azbill, R. D., Mark. R. J., Begley, J. G., Waeg, G., and Mattson, M. P. (1997) 4-Hydroxynonenal, a lipid peroxidation product, rapidly accumulates following spinal cord injury and inhibits glutamate uptake. *J. Neurochem.* **68,** 2469–2476.

Stichel, C. C. and Müller, H. W. (1998) Experimental strategies to promote axonal regeneration after traumatic central nervous system injury. *Prog. Neurobiol.* **56,** 119–148.

Streit, W. J., Semple-Rowland, S. L., Hurley, S. D., Miller, R. C., Popovich, P. G., and Stokes, B. T. (1998) Cytokine mRNA profiles in contused spinal cord and axotomized facial nucleus suggest a beneficial role for inflammation and gliosis. *Exp. Neurol.* **152,** 74–87.

Stokes, B. T. and Reier, P. J. (1992) Fetal grafts alter chronic behavioral outcome after contusion damage to the adult rat spinal cord. *Exp. Neurol.* **116,** 1–12

Tator, C. H. (1995) Update on the pathophysiology and pathology of acute spinal cord injury. *Brain Pathol.* **5,** 407–413.

Tello, F. (1911) La influencia del neurotropismo en la regeneracion de los centros nerviosos. *Trab. Lab. Invest. Biol.* **9,** 123–159.

Teng, Y. D., Mocchetti, I., and Wrathall, J. R. (1998) Basic and acidic fibroblast growth factors protect spinal motor neurons in vivo after experimental spinal cord injury. *Eur. J. Neurosci.* **10,** 798–802.

Tessier-Lavigne, M. and Goodman, C. S. (1996) The molecular biology of axon guidance. *Science* **274,** 1123–1133.

Tessler, A., Himes, B. T., Houle, J., and Reier, P. J. (1988) Regeneration of adult dorsal root axons into transplants of embryonic spinal cord. *J. Comp. Neurol.* **270,** 537–548.

Thallmair, M., Metz, G. A., Z'Graggen, W. J., Raineteau, O., Kartje, G. L., and Schwab, M. E. (1998) Neurite growth inhibitors restrict plasticity and functional recovery following corticospinal tract lesions. *Nat. Neurosci.* **1,** 124–131.

von Meyenburg, J., Brosamle, C., Metz, G. A., and Schwab, M. E. (1998) Regeneration and sprouting of chronically injured corticospinal tract fibers in adult rats promoted by NT-3 and the mAb IN-1, which neutralizes myelin-associated neurite growth inhibitors. *Exp. Neurol.* **154,** 583–594.

Xu, X. M., Guenard, V., Kleitman, N., and Bunge, M. B. (1995) Axonal regeneration into Schwann cell-seeded guidance channels grafted into transected adult rat spinal cord. *J. Comp. Neurol.* **351,** 145–160.

Young, W. and Flamm, E. S. (1982) Effect of high-dose corticosteroid therapy on blood flow, evoked potentials, and extracellular calcium in experimental spinal injury. *J. Neurosurg.* **57,** 667–673.

Wang, C. X., Olschowka, J. A., and Wrathall, J. R. (1997) Increase of interleukin-1beta mRNA and protein in the spinal cord following experimental traumatic injury in the rat. *Brain Res.* **759,** 190–196.

Z'Graggen, W. J., Metz, G. A., Kartje, G. L., Thallmair, M., and Schwab, M. E. (1998) Functional recovery and enhanced corticofugal plasticity after unilateral pyramidal tract lesion and blockade of myelin-associated neurite growth inhibitors in adult rats. *J. Neurosci.* **18,** 4744–4757.

12

The Pathogenesis of Duchenne Muscular Dystrophy

Edward A. Burton and Kay E. Davies

INTRODUCTION

Duchenne muscular dystrophy (DMD) is a lethal X-linked inherited muscle-wasting disease (Duchenne, 1868; Gowers, 1879; Emery, 1993). It is the most common genetic neuromuscular disease, with an estimated incidence of 1 in 3500 live male births (Emery, 1993).

This review is divided into four major sections. First, we briefly describe the clinical and histopathological features of DMD and its allelic variants. Second, we discuss dystrophin, the product of the *DMD* gene, and its protein-binding partners in muscle. We review the reported mutations of the *DMD* gene and their phenotypic correlations, and we review mutations in the genes encoding associated proteins that result in similar muscle-wasting phenotypes. Third, we examine the various animal models that have allowed investigation of the pathophysiology of DMD. Finally, we assess the various hypotheses and supportive experimental data that attempt to explain the mechanistic link between loss of dystrophin from myogenic cells and muscle degeneration.

DUCHENNE MUSCULAR DYSTROPHY

Clinical Features

Typically, infants are clinically normal at birth. The first symptoms are often related to locomotor problems developing around the age of 3 yr (for example, difficulty running, abnormal gait, or unsteadiness). Examination shows symmetrical, predominantly proximal, lower-limb weakness. Weakness of the knee and hip extensors results in difficulty standing from a lying position—the child assists the weak muscles by using his arms to climb up his body (Gowers' sign; Gowers, 1879). Pseudohypertrophy of the calf muscles (caused by replacement of muscle tissue with adipose and fibrous tissue) is almost invariable and is occasionally found in other muscle groups (Emery, 1993). Early presentation may be atypical, for example, as a hypotonic infant or with delayed early motor milestones.

The disease runs an inexorably progressive course. Deteriorating lower-limb function eventually leads to inability to walk or weight-bear, and the child becomes chairbound. In the majority, this occurs by the age of 12; the median age was 8.5 yr, and the 95th percentile 11.9 yr in one series (Emery, 1993). Progressive upper-limb weakness

From: *Pathogenesis of Neurodegenerative Disorders* Edited by: M. P. Mattson © Humana Press Inc., Totowa, NJ

follows, and contractures often develop in affected areas, particularly the elbows, hips, and knees. Axial weakness eventually makes the patient unable to sit unaided, and results in progressive kyphoscoliosis. Death ensues, usually a result of respiratory complications arising from the combination of kyphoscoliosis and respiratory muscle weakness. The terminal event is often respiratory infection. The majority of patients die before the age of 20 yr; the median was 15.5 yr and the 95th percentile 20.5 yr in one series (Emery, 1993). Involvement of the extraocular and pharyngeal muscles is not seen, although rarely there is mild facial weakness.

Histological Features

DMD muscle initially appears unremarkable, with the exception of occasional eosinophilic hypercontracted fibers (Lotz and Engel, 1987). These are present in fetal samples and more numerous at birth (Emery, 1977). The characteristic findings of muscle-fiber degeneration and regeneration are almost invariably visible at birth and become more prominent with time (*see* Fig. 5) (Bell and Conen, 1968; Bradley et al., 1972). Necrotic fibers are often seen in clusters ("grouped necrosis") and studies of serial transverse sections show that the necrosis is often confined to limited longitudinal portions of the muscle fiber ("segmental necrosis") (Schmalbruch, 1984; Gorospe et al., 1997). The necrotic fibers are often seen to be undergoing phagocytosis, and the endomysium and perimysium are infiltrated with macrophages and CD4$^+$ lymphocytes (Arahata and Engel, 1984; Engel and Arahata, 1986; McDouall et al., 1990). Muscle regeneration is evident, with proliferating myoblasts and nascent myofibers, recognizable by virtue of their centrally placed nuclei, small diameter, and basophilic RNA-rich cytoplasm (Bell and Conen, 1968; Bradley et al., 1972; Schmalbruch, 1984). The eventual failure of muscle regeneration in the context of continuing necrosis results in reduced numbers of myofibers. Muscle tissue becomes replaced by adipose and fibrous tissue, giving rise to the clinical appearance of pseudohypertrophy followed by atrophy (reviewed in Emery, 1993). It has been proposed that the evolution of histopathological abnormalities in DMD be divided into an initial necrotic/regenerative phase and a later fibrotic/degenerative phase to distinguish features directly consequent to dystrophin deficiency from secondary pathological changes (Gorospe et al., 1997).

Involvement of Tissues Other Than Skeletal Muscle

The majority of patients have evidence of cardiac involvement, by electrocardiography or histological examination, although clinically important cardiac dysfunction is less common (reviewed in Emery, 1993). Echocardiographic evidence of ventricular muscle dysfunction has been reported in a high proportion of patients (de Kermadec et al., 1994; Melacini et al., 1996; Sasaki et al., 1998). Clinical cardiac failure (15% patients; Boland et al., 1996) and dilated cardiomyopathy (24% patients; Melacini et al., 1996) are reported to be less common, although these complications may be fatal when present. The picture of cardiac dysfunction in DMD is further complicated by the presence of pulmonary hypertension in some patients with nocturnal respiratory problems (Melacini et al., 1996).

Many patients have mild mental retardation; the mean IQ of DMD patients has been variously reported as 68–89 in different studies. Apart from some degree of intellectual impairment, specific deficits in memory and behavior have been reported (reviewed in Emery, 1993).

Allelic Variants

Becker Muscular Dystrophy

Becker muscular dystrophy is a milder X-linked muscle-wasting disease (Becker and Keiner, 1955) that results from mutations in the same gene as DMD (*see* below). The phenotype has a later clinical onset and slower, more variable progression. Some of the patients have a normal life span and may even remain ambulant until old age. There is a continuous spectrum of disease severity between the severe DMD and mild BMD phenotypes.

X-Linked Cardiomyopathy

Some cases of X-linked dilated cardiomyopathy are caused by mutations in the same gene as DMD. This phenotype involves progressive ventricular wall dysfunction, dilated cardiomyopathy, and cardiac failure in the absence of a skeletal myopathy (Ferlini et al., 1999).

Carriers of DMD Mutations

Uneven X-inactivation and balanced X-autosome translocations are thought responsible for the proportion of female carriers of DMD and BMD mutations that manifest clinical features of the disease. A recent study examined the incidence and characteristics of clinical features in carriers of known *DMD* gene mutations (Hoogerwaard et al., 1999). About one quarter of women had symptoms, ranging from mild proximal weakness or muscle cramps to dilated cardiomyopathy. There was no correlation between the phenotype and genotype. Interestingly, the serum creatine kinase (CK) activity was only raised in approx one-half of the patients studied, and there was no correlation between muscle weakness and raised serum CK.

DYSTROPHIN AND THE DMD GENE

Dystrophin

DMD results from mutations in the gene encoding dystrophin (Monaco et al., 1985, 1986). Dystrophin is a 427 kDa cytoskeletal protein expressed in skeletal muscle, heart, and the brain (Koenig et al., 1987, 1988). In skeletal muscle, dystrophin is localized at the cytoplasmic face of the sarcolemma (Cullen et al., 1990; Zubrzycka Gaarn et al., 1988) and enriched at the myotendinous (Samitt and Bonilla, 1990) and neuromuscular junctions (Byers et al., 1991; Sealock et al., 1991; Watkins et al., 1988). It is a component of the cytoskeleton and is involved in a complex series of molecular interactions at the sarcolemma of muscle fibers (Fig. 1). The N-terminal domain of the protein shows homology to α-actinin; this region has been shown to bind the F-actin component of the subsarcolemmal cytoskeleton (Hemmings et al., 1992; Way et al., 1992) and may play an important role in the organization of the cytoskeleton. The central domain of dystrophin consists of a series of coiled-coil spectrin-like repeats, separated by four proline-rich "hinge" regions predicted to give the molecule a flexible rod-shaped structure (Koenig and Kunkel, 1990; Koenig et al., 1988). A cysteine-rich region separates the rod domain from the carboxyl terminus. The cysteine-rich region contains putative calcium-binding EF-hand domains in addition to a putative zinc-finger ZZ domain that may be involved in protein-protein interactions (Ponting et al., 1996).

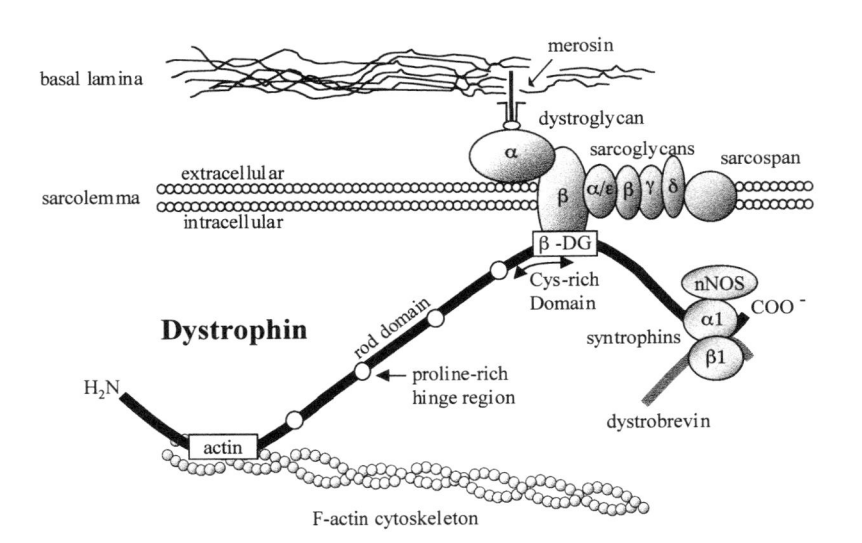

Fig. 1. Dystrophin and associated proteins at the sarcolemma. Dystrophin (black) and components of the dystrophin-associated glycoprotein complex (gray) are depicted schematically at the sarcolemma of muscle fibers. The actin-binding and β-dystroglycan binding domains of dystrophin are indicated by boxes.

In conjunction with the first part of the carboxy-terminus domain, the cysteine-rich domain binds to β-dystroglycan, a transmembrane protein that is a component of a multimeric complex (Campbell and Kahl, 1989; Yoshida and Ozawa, 1990). The carboxy-terminal domain of dystrophin binds to the syntrophins (Adams et al., 1993, 1995; Yang et al., 1994; Ahn and Kunkel, 1995; Ahn et al., 1996) and dystrobrevins (Blake et al., 1996a, 1998; Peters et al., 1997b; Sadoulet Puccio et al., 1997). These interactions are thought to occur through coiled-coil structures in the C-terminus of dystrophin (Blake et al., 1995b). The interaction between components of the dystrophin-associated glycoprotein complex (DAPC) and extracellular matrix (Ibraghimov Beskrovnaya et al., 1992) completes a molecular link between the muscle cytoskeleton and the exterior of the cell.

In the brain, dystrophin is localized to the postsynaptic regions of neurons in the cerebral and cerebellar cortex (Lidov et al., 1990), where it associates with a similar, but distinct, complex of proteins to those found in muscle (Blake et al., 1999). In the heart, dystrophin is localized to the sarcolemma of cardiomyocytes (Tanaka and Ozawa, 1990; Pons et al., 1994a) and the cell membranes of Purkinje fibers (Bies et al., 1992).

Expression of Dystrophin in DMD, BMD, and XL-DCM

Mutations that cause DMD generally abolish production of dystrophin protein (Bonilla et al., 1988; Hoffman et al., 1987, 1988a; Arahata et al., 1989). In contrast, muscle biopsies from patients with BMD often show retained sarcolemmal immunostaining for dystrophin, but immunoblot analysis usually shows an abnormal form or reduced amount of the protein (Arahata et al., 1989). X-linked dilated cardiomyopathy (XL-DCM) results from mutations that abolish the cardiac expression of dystrophin, while retaining the skeletal-muscle expression (Ferlini et al., 1999).

Dystrophin Homologs

There are three types of protein that show homology to dystrophin:

1. Spectrin/α-actinin: These are homologous to the actin-binding and rod domain of dystrophin, and are discussed above.
2. Dystrobrevins: These share homology with the cysteine-rich and C-terminal domain of dystrophin, with which they associate. They are an important intracellular part of the dystrophin-associated protein complex, and are considered later in the section Dystrophin-Associated Glycoprotein Complex (DAPC).
3. Utrophin: Utrophin (previously called dystrophin-related protein; DRP) is the autosomal homolog of dystrophin. It was first identified as a 13 kb fetal muscle transcript by low-stringency screening of a human fetal muscle library using a probe to the 3' end of dystrophin (Love et al., 1989). The gene was localized to human chromosome 6 (Love et al., 1989) and mouse chromosome 10 (Buckle et al., 1990). The 13 kb transcript encodes a 395 kDa protein (Tinsley et al., 1992) that is ubiquitously expressed (Khurana et al., 1991, 1992; Love et al., 1991b; Nguyen et al., 1991, 1992; Blake et al., 1996b).

Examination of the primary sequence of utrophin reveals marked similarity to dystrophin along the length of both proteins (Tinsley et al., 1992). Dystrophin and utrophin share similar functional domains, including the actin-binding and C-terminal domains, which are over 80% identical between the two proteins (Tinsley et al., 1992; Winder et al., 1995a, 1995b; Winder and Kendrick Jones, 1995; Morris et al., 1999). Utrophin has similar C-terminal protein binding partners to dystrophin (James et al., 1995, 1996; Peters et al., 1997a) and associates with a similar complex of transmembrane components as dystrophin (Matsumura et al., 1992; Peters et al., 1997a). In contrast to dystrophin, utrophin is ubiquitously expressed and down-regulated during development. (Khurana et al., 1990, 1992; Love et al., 1991b; Nguyen et al., 1991; Schofield et al., 1993; Rigoletto et al., 1995). In mature skeletal muscle, utrophin is localized to the neuromuscular and myotendinous junctions (Nguyen et al., 1991; Clerk et al., 1993). In embryonic and regenerating muscle, however, utrophin is localized to the sarcolemma (Helliwell et al., 1992; Matsumura et al., 1992; Clerk et al., 1993; Karpati et al., 1993; Pons et al., 1993; Wilson et al., 1994). It is possible that utrophin and dystrophin have different functions that are dictated by their expression patterns rather than by their structure. Indeed, functional redundancy between the two has been demonstrated (Tinsley et al., 1996, 1998; Deconinck et al., 1997b; Deconinck et al., 1997; Rafael et al., 1998; Wakefield et al., 2000; and *see* below).

Dystrophin-related protein 2 (DRP2) is a phylogenetically conserved X-encoded transcript that predicts a protein of 110kDa (Roberts et al., 1996). The inferred translation shows homology to the C-terminal of dystrophin, which shows a much closer relationship to utrophin than to DRP2. The DRP2 transcript is widely distributed within the nervous system (Dixon et al., 1997). DRP2 protein is yet to be demonstrated; consequently, there are no available data concerning its localization or protein-binding partners. Indeed, it is not yet known whether the DRP2 transcript is translated. The gene is mentioned here for completeness.

Organization of the Dystrophin Gene

The gene encoding dystrophin is one of the largest described; it spans approx 2.5 Mb of genomic sequence on the short arm of the X chromosome (Koenig et al., 1987;

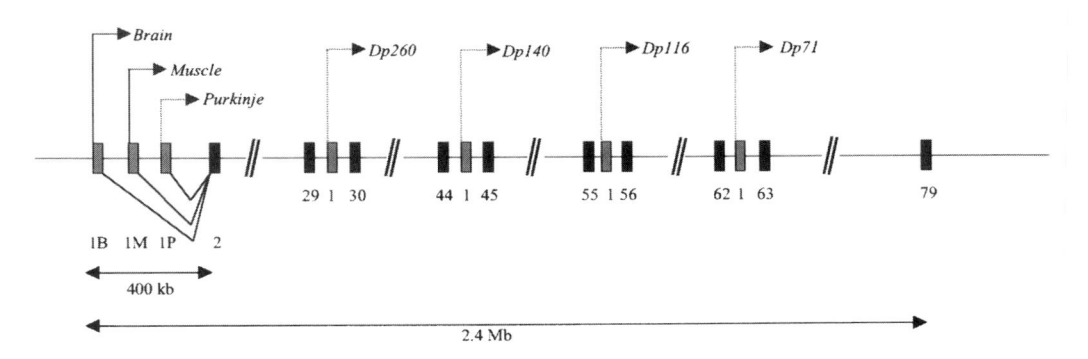

Fig. 2. Organization of the dystrophin gene. A schematic summary of the exon organization of the dystrophin gene is shown. Exons common to more than one transcript are depicted in black, whereas unique 5' exons exclusively associated with particular transcripts and promoters are shown in gray. Arrows mark the positions of promoters in the gene. Solid arrows indicate the positions of promoters that have been characterized; broken arrows show presumptive positions of other promoters. The diagram is not to scale, but an approximate indication of the large size of the gene and its 5' regulatory region is shown.

Coffey et al., 1992; Monaco et al., 1992). The gene consists of 79 exons (Koenig et al., 1987; Roberts et al., 1993) that give rise to a 14 kb transcript (Koenig et al., 1987).

Expression of the full-length transcript is controlled by three independently regulated promoters (Fig. 2) (Nudel et al., 1989; Klamut et al., 1990; Boyce et al., 1991; Makover et al., 1991; Gorecki et al., 1992). These give rise to unique 5' exons that splice into common full-length transcripts at exon 2. The muscle promoter is expressed in skeletal muscle, cerebral glial cells, and cardiomyocytes (Barnea et al., 1990; Chelly et al., 1990a). The cerebral neuronal promoter is expressed in cortical neurons (Barnea et al., 1990; Chelly et al., 1990a). The Purkinje-cell promoter is expressed in cerebellar Purkinje cells, and skeletal muscle (Gorecki et al., 1992; Holder et al., 1996). The three 5' promoters are distributed over a vast genomic interval. The muscle promoter and first exon are separated from exon 2 by 120–150 kb. This large intron contains the Purkinje-cell promoter and first exon, which lie approx 60 kb 3' of the muscle promoter and first exon. The brain promoter, and its associated first exon, lie 200 kb 5' to the muscle promoter (Boyce et al., 1991; Gorecki et al., 1992; Whittaker et al., 1993).

In addition, a number of internal promoters give rise to shorter transcripts that encode truncated C-terminal isoforms of dystrophin (Figs. 2 and 3):

- Dp71 is a 71 kDa protein whose transcription is regulated by a promoter located between exons 62 and 63 (Lederfein et al., 1993). It is expressed in a wide range of tissues, including brain, liver, lung, and kidney (Cox et al., 1994; Greenberg et al., 1994, 1996; Howard et al., 1998, 1999; Sarig et al., 1999).
- Dp116 is a 116 kDa dystrophin C-terminal isoform whose production is regulated by a promoter located between exons 55 and 56. The expression of this protein is limited to Schwann cells of the peripheral nervous system (PNS) and cultured glial cells (Byers et al., 1993).
- Dp140 is a 140 kDa protein localized to the central nervous system (CNS); the promoter responsible for the Dp140 transcript is located in intron 44 (Lidov et al., 1995).

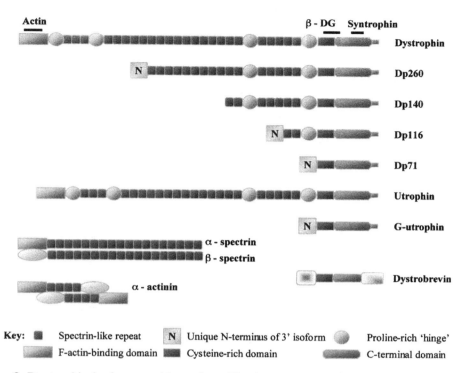

Fig. 3. Dystrophin isoforms and homologs. The important protein domains and functional modules of full-length dystrophin and its 3' isoforms are shown schematically in the upper part of the figure. The positions of the actin, β-dystroglycan, and syntrophin-binding domains of dystrophin are indicated by solid lines. The lower part of the figure depicts the relationship between dystrophin and the utrophin, dystrobrevin, and spectrin groups of proteins.

- Dp260, a 260kDa isoform is produced by transcription from a promoter in intron 29 (D'Souza et al., 1995; Kameya et al., 1997; Rodius et al., 1997). Expression of this isoform is limited to the retina, CNS, and heart.

These C-terminal isoforms of dystrophin share identity with the C-terminus of the full-length protein, and therefore contain the necessary binding sites for interaction with many dystrophin-associated proteins. The functions of these isoforms have not been elucidated; they may be involved in stabilization of DAPC components in the membranes of nonmuscle cells.

Interestingly, utrophin shares many similarities with dystrophin in its genomic organization, suggesting that the two were derived from duplication of a common ancestral gene. Both genes contain multiple short exons, but have a single long exon encoding the 3'UTR (Pearce et al., 1993). The organization of the 5' ends of the genes is similar, in that both have large introns separating the first and second coding exons, and multiple 5' promoters and enhancer elements with alternatively spliced unique 5' exons (Burton et al., 1999). In addition, both genes have 3' isoforms with unique first exons, presumably driven by internal promoters located within 3' introns (Blake et al., 1995a, 1995b; Wilson et al., 1999).

Mutations of the Dystrophin Gene

Many hundreds of different mutations have been described within the gene encoding dystrophin. The high mutation rate (1×10^{-4} genes per generation) reflects in part the large size of the locus. There is some correlation between mutations and the resulting phenotype.

Large Deletions in DMD and BMD

The most common mutations in the dystrophin gene are gross deletions, which account for approx 60–70% of DMD and BMD cases (Monaco et al., 1985; Koenig et al., 1987; Baumbach et al., 1989; Liechti Gallati et al., 1989; Beggs et al., 1990; Chelly et al., 1990b). Following recognition that the dystrophin gene was mutated in both the severe DMD and the milder BMD phenotype, an apparent paradox was noted in that the phenotype did not correlate with the size or position of the deletion. This was resolved by the reading frame hypothesis (Monaco et al., 1988; Koenig et al., 1989). The intron/exon boundaries of the dystrophin gene do not all have identical phase. In a large deletion in which several exons are lost, preservation of the translational reading frame depends on the phase of the splice boundaries of the exons that are brought together by the deletion. Thus, a deletion bringing together two exons whose donor and acceptor boundaries had compatible codon phase would preserve the translational reading from, resuting in the production of an internally truncated, partially functional dystrophin molecule and a BMD phenotype. A deletion causing a juxtaposition exons with incompatible splice boundary phase would result in a translational frame shift, and premature termination of translation. The resulting protein would be nonfunctional and probably unstable, resulting in a DMD phenotype. By knowledge of the splice sequences and which exons were deleted, it was possible to predict the phenotype resulting from large deletions in over 90% of cases (Koenig et al., 1989). Many other reports corroborate the reading frame rule (Baumbach et al., 1989; Liechti Gallati et al., 1989; England et al., 1990; Love et al., 1990, 1991a).

The majority of large deletions occur in two deletion "hot spots" within the *DMD* gene. Deletion cluster region 1 spans exons 44–53 (Beggs et al., 1990; Todoroya et al., 1996); deletion cluster region 2 spans exons 3–20 (Liechti Gallati et al., 1989). The vast majority of large deletions can be detected easily using a multiplex polymerase chain reaction (PCR) test that screens the exons most commonly deleted in this type of mutation (Beggs et al., 1990). It is worth noting that very large deletions within the rod domain may result in a mild phenotype, provided the translational reading frame is perserved. A patient with almost 50% of the coding sequence deleted remained ambulant into later life (England et al., 1990), presumably because a grossly truncated dystrophin molecule retained partial function by virtue of preserved actin-binding and β-dystroglycan-binding domains. The design of truncated dystrophin molecules small enough to deliver in gene therapy vectors has largely been based on such patients (Love et al., 1991a).

There are some exceptions to the reading frame rule, and it is not always possible to predict the phenotype on the basis of knowledge of the genotype. Out-of-frame mutations at the 5' end of the gene have been associated with a mild BMD phenotype, occurring through initiation of protein synthesis from a downstream ATG codon (Malhotra et al., 1988), or by aberrant splicing allowing production of truncated in-frame tran-

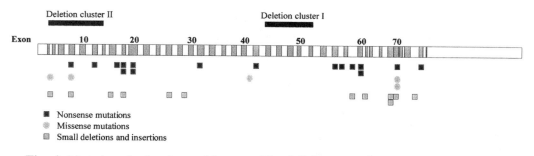

Fig. 4. Mutations in the dystrophin gene. The full-length dystrophin cDNA is illustrated diagrammatically, to scale. The exons of the transcript are numbered; even-numbered exons are shown in gray and odd-numbered exons in white. The positions of the exons affected by the two deletion clusters that are associated with both DMD and BMD (according to the reading frame rule, *see* text) are shown above the schematic cDNA. The positions of small coding sequence mutations associated with DMD are shown below the schematic cDNA.

scripts (Chelly et al., 1990b). In addition, deletions of the C-terminal domain of the protein may result in a severe phenotype regardless of reading frame preservation, implying that this portion of the molecule has crucial functions.

Mutations Other Than Large Deletions Associated with DMD

SMALL DELETIONS AND INSERTIONS

Some smaller intraexonic deletions also give rise to the severe DMD phenotype by causing a translational frameshift. Unlike the large deletions, which cluster in two "hotspots," small deletions and insertions are scattered throughout the gene (Fig. 4) (Prior et al., 1995). The majority of small deletions and insertions that have been described are associated with a severe DMD phenotype; it is unlikely that small in-frame deletions result in a phenotype at all.

POINT MUTATIONS

The vast majority of point mutations that have been described in the coding sequence of the dystrophin gene are nonsense mutations that result in a DMD phenotype (Fig. 4) (Lenk et al., 1993; Roberts et al., 1994). These account for approx 18% of all DMD mutations, or 60–70% of those cases not associated with large deletions (Roberts et al., 1994), and are scattered throughout the gene (Prior et al., 1995). Interestingly, although it might be predicted that such mutations would give rise to a truncated N-terminal fragment, usually no protein is detectable at all. This suggests that abnormally truncated forms of dystrophin are unstable (Hoffman et al., 1987). This mechanism is also responsible for the loss of dystrophin in the *mdx* mouse (*see* below) (Sicinski et al., 1989).

In some instances in biology, the clustering of missense mutations has given useful clues about the position and nature of important functional domains in a particular protein. This has not proved to be the case with dystrophin. Several missense mutations have been described in the dystrophin gene. There is no clear clustering of these mutations (Fig. 4), and the ones reported were associated with loss of protein stability and undetectable levels, rather than loss of discrete functional domains in the presence of normal amounts of protein (Roberts et al., 1994; Prior et al., 1995).

Exceptions to this observation are provided by the following cases: mutation of a conserved leucine to arginine in the actin-binding domain (amino acid 54) of dystrophin gave rise to a DMD phenotype in the presence of normal amounts of protein (Prior et al., 1993). A missense mutation, D3335H, in the β-dystroglycan-binding domain of dystrophin gave rise to a severe phenotype, but normal localization and amounts of dystrophin and associated proteins (Goldberg et al., 1998). Substitution of a conserved cysteine at position 3340, also within the β-dystroglycan-binding domain, resulted in a severe phenotype with reduced but detectable levels of dystrophin on muscle biopsy (Lenk et al., 1996).

Mutations Other Than Large Deletions Associated with BMD

Although a similar proportion of DMD and BMD cases are associated with large deletions of the dystrophin gene, small coding sequence mutations are relatively under-represented in BMD patients (Prior et al., 1995). There are, however, an increasing number of splice-site mutations being described in BMD patients. The results of these mutations do not always adhere to the reading frame rule. It appears that mutation of consensus splice sites can have unpredictable effects on splicing that allow multiple exons to be skipped, and the reading frame to be preserved (Chelly et al., 1990b; Wilton et al., 1993; Hagiwara et al., 1994; Bartolo et al., 1996; Shiga et al., 1997). In addition, there are reports of mutations affecting the promoter regions of the dystrophin gene that result in reduced levels of dystrophin in skeletal muscle, and a mild BMD phenotype (Bushby et al., 1991). It appears that this type of mutation, however, is relatively uncommon (Tubiello et al., 1995).

Mutations Associated with XL-DCM

Several mutations have been described, including rearrangements in the muscle-specific promoter (with retained skeletal muscle expression driven from the Purkinje-cell promoter) and mutations of elements that affect transcriptional regulation or splicing in the heart (reviewed in Ferlini et al., 1999). The common feature of many of these mutations is that they result in tissue-specific loss of dystrophin from the heart.

Mental Retardation

There appears to be little correlation between the type of dystrophin mutation and the presence or severity of mental retardation in DMD patients (Hodgson et al., 1992; Rapaport et al., 1992; Lenk et al., 1993, 1996; Nicholson et al., 1993; Bushby et al., 1995; Tubiello et al., 1995; North et al., 1996). There is a trend towards an association between 3' gene deletions and mental retardation (Bushby et al., 1995), but no specific part of the gene or the protein has been implicated in its pathogenesis. The same mutations can produce quite different phenotypes in different patients, and one series identified a group of mentally normal patients whose combined deletions covered almost the entire gene (Hodgson et al., 1992).

Dystrophin-Associated Glycoprotein Complex (DAPC) (Fig. 1)

The protein/glycoprotein complex that interacts with the C-terminal and cysteine-rich domains of dystrophin consists of a number of separate species:

- α- and β-dystroglycan (DG) are derived from a single gene by posttranslational processing of a precursor protein. The C-terminal of β-dystroglycan directly interacts with the cysteine-rich and C-terminal domains of dystrophin (Campbell and Kahl, 1989). β-DG is a type I integral membrane protein with a single transmembrane helix. The N-terminal of β-DG interacts with the C-terminal of α-DG. α-DG is entirely extracellular in location. Binding of its N-terminal to components of the extracellular matrix completes a mechanical link between the F-actin cytoskeleton and the basal lamina of muscle, via dystrophin and β-DG (Montanaro et al., 1999). Dystroglycans are expressed in many nonmuscle tissues, where their functions are not clear. Although they associate with utrophin and possibly dystrophin isoforms in nonmuscle tissue, the rest of the protein complex appears different to that seen in muscle tissue (Durbeej and Campbell, 1999).

- There are five members of the sarcoglycan (SG) family (α-, β-, γ-, δ-, ε-), each encoded by separate genes (Roberds et al., 1994; Lim et al., 1995; Noguchi et al., 1995; Jung et al., 1996; Nigro et al., 1996b; Ettinger et al., 1997). α- and ε-SG are type I transmembrane proteins and are closely related (McNally et al., 1998). Each has an intracellular C-terminal domain, a single transmembrane helix, and an extracellular N-terminal domain. β-, γ-, and δ- SG are type II integral membrane proteins with single transmembrane helices. The sarcoglycans form a subcomplex with sarcospan (*see* below) that associates with the dystroglycan subcomplex. The nature of the interaction between the two subcomplexes is not clear at present; it appears to be weaker than the interactions holding the SGs and DGs together in their respective subcomplexes (Crosbie et al., 1999). The function of the SG complex is not yet known; a role in stabilization of the DG complex has been postulated (*see* below) (Crosbie et al., 1999). It is emerging that α- and ε-sarcoglycan may associate with separate complexes containing β-, γ-, and δ-SG in skeletal muscle, and may thus give rise to SG complexes with discrete functional roles (Liu and Engvall, 1999). It also appears that ε-SG replaces α-SG in smooth muscle, and that the SG complex in smooth muscle contains only β, δ- and ε-SG (Straub et al., 1999).
- Sarcospan is a recently described tetraspan integral membrane protein that associates with the SG subcomplex (Crosbie et al., 1997). Its correct targeting to the plasma membrane is dependent on other components of the SG complex (Crosbie et al., 1999), but its function is unknown.
- Filamin 2 is a recently described member of the filamin protein family. It associates with γ- and δ-SG as part of the DAPC. Filamin family proteins have been shown to have diverse functions, including signal-transduction cascades associated with cell migration, adhesion, differentiation, and force transduction. In addition, filamin proteins have been implicated in the maintenance of membrane integrity in response to force (Thompson et al., 2000).
- The three syntrophins, α1-, β1-, and β2-, are encoded by separate genes that give rise to highly related proteins (Adams et al., 1993, 1995; Ahn et al., 1996). The syntrophins are intracellular proteins that are located at the cytoplasmic face of the plasma membrane. Each dystrophin-associated protein complex contains two syntrophin molecules (Yoshida and Ozawa, 1990; Ervasti and Campbell, 1991; Sadoulet Puccio et al., 1997), one interacting directly with dystrophin and the other with dystrobrevin (Peters et al., 1997a; Sadoulet Puccio et al., 1997). Syntrophins are enriched at the neuromuscular junction, and there is evidence of differential association of syntrophin types with protein complexes found in different regions of the muscle cell (Peters et al., 1997a). Thus, α1- and β1-syntrophin bind to the dystrophin protein complex at the sarcolemma and neuromuscular junction, whereas β1- and β2-syntrophins bind to the utrophin protein complex at the neuromuscular junction (Peters et al., 1994, 1997a). Each of the syntrophins contains two pleckstrin homology (PH) domains (implicated in signal-dependent membrane-association of proteins), a PDZ domain (implicated in a growing number of protein-protein interactions), and a unique C-terminal domain. The PDZ domain is responsible for the role of syntrophins in recruiting the neuronal form of nitric oxide synthase (nNOS) (Brenman et al., 1996; Grozdanovic and Baumgarten, 1999; Tochio et al., 1999) to the sarcolemma as part of the dystrophin-associated protein complex (Brenman et al., 1995). In addition, other poten-

tially interesting syntrophin interactions have been demonstrated, including interaction with voltage-gated sodium channels both in muscle and in brain (Gee et al., 1998; Schultz et al., 1998).

• The dystrobrevins are a family of dystrophin-related and dystrophin-associated proteins, containing three domains. Two of these domains are homologous to the cysteine-rich and C-terminal domains of dystrophin. The N-terminal domain is unique to dystrobrevin. α-dystrobrevin is expressed in muscle, heart, and brain (Blake et al., 1996a; Ambrose et al., 1997). Its primary transcript is expressed from multiple promoters (Holzfeind et al., 1999) and is subject to alternative splicing (Blake et al., 1996a), generating several different transcripts that give rise to diverse protein isoforms. In muscle, the full-length α-dystrobrevin-1 isoform is located mainly at the cytoplasmic face of the neuromuscular junction as part of the utrophin protein complex. The C-terminal truncated α-dystrobrevin-2 is located at the cytoplasmic face of the sarcolemma, associated with dystrophin (Peters et al., 1998). α-Dystrobrevin interacts with dystrophin via coiled-coil motifs in the C-terminal domain of both proteins (Sadoulet Puccio et al., 1997), and with syntrophins. β-Dystrobrevin is expressed in a number of tissues, but not muscle (Peters et al., 1997b; Blake et al., 1998). Its genomic structure is similar to that of α-dystrobrevin (Loh et al., 1998). In the brain, β-dystrobrevin and α-dystrobrevin are associated with different specific dystrophin protein complexes (Blake et al., 1999).

Absence of dystrophin from muscle results in loss of the DAPC from the sarcolemma (Ervasti et al., 1990; Matsumura et al., 1992; Matsumura and Campbell, 1994) implying that dystrophin is required to maintain the localization of the DAPC. This prompted the hypothesis that the function of dystrophin was to maintain the correct localization of the DAPC in the sarcolemma, and that loss of the DAPC was responsible for muscle degeneration in DMD. However, restoration of the DAPC is not adequate alone to correct the abnormal phenotype arising from dystrophin deficiency. Relocalization of the DAPC to the sarcolemma occurs in Dp71 overexpressing transgenic mice that do not express full-length dystrophin. These mice still undergo muscle degeneration, indicating that dystrophin has functions other than stabilization of the DAPC (Cox et al., 1994; Greenberg et al., 1994).

Mutations and Null Alleles of DAPC Genes

The crucial role of the DAPC in muscle is demonstrated by the muscle degeneration that results from mutation of its components. Several human diseases have been attributed to mutations in the genes encoding DAPC proteins; murine null mutants of many of the DAPC genes have been generated and several of these show progressive muscle-wasting phenotypes.

Limb Girdle Muscular Dystrophies (LGMD)

Limb girdle dystrophies are a genetically and clinically heterogeneous group of conditions characterized by muscle wasting and weakness that initially affects the muscles of the pelvic and pectoral girdles. Three types of autosomal dominant and eight types of autosomal recessive LGMD have been distinguished by linkage analysis. The phenotype of at least some of these shows resemblance to that of DMD. Following demonstration and characterization of the DAPC, the genes encoding its components became biological candidates for LGMD genes. Four of the recessive LGMDs are now known to arise from loss of function mutations in the genes encoding α-, β-, γ-, and δ-sarcoglycan (Lim et al., 1995; Bonnemann et al., 1996; McNally et al., 1996; Nigro et al., 1996a; Vainzof et al., 1996; Duggan et al., 1997). Various types of mutation have been described in each of the genes, including missense mutations that seem to cluster in the sequence encoding the immediate extracellular portion of each protein (reviewed in Bushby, 1999).

No human disease has yet been attributed to mutations in the DG gene, possibly on account of embryonic lethality (*see* below). Likewise, disease arising from syntrophin or dystrobrevin mutations has not been described.

Murine Null Mutants

Mouse gene "knockouts" have been generated for each of the sarcoglycans, sarcospan, dystroglycan, α1-syntrophin, and α-dystrobrevin.

- Inactivation of the DG gene in mice results in an embryonic lethal phenotype (Williamson et al., 1997). Embryoid bodies fail to form basement membranes (Henry and Campbell, 1998), and embryos fail to form a crucial extra-embryonic structure, Reichert's membrane (Williamson et al., 1997). However, mice chimeric for the null allele have now been generated. These animals express no DG in skeletal muscle. Interestingly, they show normal basement membrane formation, but progressive muscle degeneration and disruption of the ultrastructure of the neuromuscular junction (Cote et al., 1999). In addition, the SG complex and in some fibers, dystrophin, were absent from the sarcolemma. This indicates that assembly of the DAPC is dependent on the presence of DG.
- Inactivation of the α- (Duclos et al., 1998), β- (Araishi et al., 1999), γ- (Hack et al., 1998, 1999) and δ- (Coral Vazquez et al., 1999) SG genes results in a dystrophic phenotype, reflecting the pathology seen in naturally occurring human loss-of-function mutations of the same genes. In general, the whole SG subcomplex is missing from the sarcolemma in null mutants of any of its components. However, partial preservation of α-SG in the γ-SG knockout has led to speculation that β- and δ-SG are more closely associated with γ-SG than the α- or ε-SG (Hack et al., 1998). The preservation and correct localization of dystrophin and DG in SG knockout mice argues that disruption of the link from the cytoskeleton to the extracellular matrix is not essential for the development of a dystrophic phenotype. There is some evidence, however, that loss of the SG complex in the β-SG null mouse de-stabilizes the interaction between β-DG and γ-SG (Araishi et al., 1999). In addition to skeletal myopathy, γ- and δ-SG null mice develop a lethal cardiomyopathy, as does the naturally occurring δ-SG null hamster (Straub et al., 1998). In contrast, human patients with SG mutations have an unpredictable degree of cardiac dysfunction that is unrelated to which SG gene is mutated (Melacini et al., 1999). In the δ-SG null mouse, the pathogenic mechanism of the cardiomyopathy appears to be ischemia secondary to disorganization of the vascular smooth-muscle DAPC and consequent blood-vessel irregularity (Coral Vazquez et al., 1999).
- Inactivation of sarcospan does not affect the localization of the other components of the DAPC and does not result in a muscle-wasting phenotype (Lebakken et al., 2000).
- Inactivation of the α1-syntrophin gene causes loss of localization of nNOS at the sarcolemma, but does not cause dystrophic features or alter the mechanical properties of muscle. nNOS accumulates in the cytoplasm, and the myofibers remain responsive to nNOS substrates and inhibitors. The remainder of the DPC is unaffected by the loss of α1-syntrophin (Kameya et al., 1999).
- Inactivation of α-dystrobrevin results in relatively mild skeletal and cardiac myopathies (Grady et al., 1999). The remainder of the DPC is intact, but nNOS localization to the sarcolemma is lost. Functional deficits in the generation of cGMP by muscle in response to activity, and the coupling of muscle contraction to blood flow are demonstrable, although their relevance is not clear (Grady et al., 1999).

The relevance of these models to understanding the mechanisms of muscle degeneration in DMD is discussed in subsequent sections.

Loss of Basement Membrane and Associated Proteins

Dystrophin participates in a molecular link between the cytoskeleton and the exterior of the muscle cell. As described above, mutations in the genes encoding other intracellular and transmembrane components of this link result in muscle-wasting phenotypes. Interestingly, mutations in genes encoding components of the muscle-cell basement membrane and associated proteins have been implicated in some congenital muscular-dystrophy (CDM) syndromes. These syndromes differ from the Duchenne/Becker and limb girdle phenotypes described above; affected infants often manifest neuromuscular deficits from birth and there may be prominent CNS involvement. A detailed exposition of the large heterogeneous group of congenital muscular dystrophies is beyond the scope of this chapter, but the following reported mutations give rise to muscular dystrophy and are of interest to the present discussion:

- Absence of merosin, the basement membrane component that interacts with α-dystroglycan, causes a severe recessive form of CMD that may be associated with peripheral neuropathy (Shorer et al., 1995) and defects in CNS myelination (Philpot et al., 1999). Loss-of-function mutations in the gene encoding laminin α2 chain (a component of merosin) result in this syndrome (Helbling et al., 1995; Pegoraro et al., 1998). Loss of merosin is also responsible for the phenotype of the *dy/dy* mouse (Xu et al., 1994).
- Bethlem myopathy is a variable, autosomal dominant form of congenital muscular dystrophy caused by mutations in the COL6A1 gene that result in loss of collagen VI from muscle (Jobsis et al., 1996, 1999). The disease runs a benign course, may be present in childhood, and is associated with early development of contractures. Collagen VI is thought to be involved in the attachment of cells to basement membrane collagen.
- Integrin α7 is a muscle membrane receptor for components of the basal lamina; loss of function mutations in the gene encoding this integrin have been associated with a form of congenital muscular dystrophy (Hayashi et al., 1998).

Thus, loss of extracellular binding partners of either the DAPC or of other muscle proteins can result in a dystrophic phenotype.

ANIMAL MODELS OF DMD

Adequately testing pathophysiological hypotheses often involves experimental interventions that cannot be carried out in human subjects. This gives rise to a demand for representative in vivo models of DMD. There are several animal models of DMD (Ytterberg, 1991; Gorospe et al., 1997; Partridge, 1997). Although all are good genetic models of DMD, in that the affected animals lack dystrophin, the resemblance of the muscle phenotype of these animals to that seen in DMD patients is variable.

Dogs

There are several breeds of dog that exhibit an X-linked muscular dystrophy (Valentine et al., 1986; Gorospe et al., 1997; Schatzberg et al., 1999). The first of these to be biochemically characterized and shown to be associated with dystrophin deficiency was a golden retriever (Cooper et al., 1988). This was subsequently shown to be caused by a splice acceptor mutation in intron 6 of the dystrophin gene that leads to aberrant splicing and a translational frame shift (Sharp et al., 1992). The dogs suffer a rapidly progressive muscular dystrophy. Clinically, the onset of weakness is at 6–9 wk of age, with myopathic features progressing over the next few months (Valentine et al., 1988). Some pups succumb at birth, for reasons that are not clear. Progression after the first few months is variable, with dogs from some litters apparently becoming stable (Valentine et al., 1988). Many of these animals, however, eventually develop cardiac failure. Histologically, the muscles of affected dogs exhibit necrosis, regeneration, endomysial and perimysial fibrosis, and fat infiltration (Valentine et al., 1992). As such, the microscopic features are virtually indistinguishable from those of the human disease. Some of the dogs show a high percentage of revertant fibers, which may account for the arrest in disease progression. Despite the phenotypic resemblance to the human disease, these animals have not been exhaustively used in studies of the pathogenesis of DMD, perhaps because of the logistical, financial, and emotive issues that accompany breeding dogs for experimentation (Partridge, 1997).

Cats

The first dystrophin-deficient cats to be described were two male siblings from a litter of four kittens, consistent with X-linked inheritance (Carpenter et al., 1989). Unfortunately, the animals were neutered and then euthanized soon after identification, precluding detailed clinical or genetic study. Further litters of dystrophin-deficient cats have been identified, however, and the genetic lesion in these animals appears to be a large deletion of both the muscle- and Purkinje-cell-specific promoters of dystrophin (Winand et al., 1994). This model, however, displays several features that are atypical of DMD. First, there is a paucity of overt muscle-fiber necrosis and almost complete absence of fibrosis. Second, there are multiple deposits of extracellular calcium (Gorospe et al., 1997). Third, there is striking muscle hypertrophy, which may be clinically problematic (one animal died from esophageal obstruction secondary to diaphragmatic hypertrophy). For this reason, the name hypertrophic feline muscular dystrophy (HFMD) has been proposed (Gaschen et al., 1992).

Mice

The *mdx* mouse is a naturally occurring mutant that was isolated by virtue of its raised serum CK activity during a search for inbred strains with red-blood-cell abnormalities. Because of its normal life span, and the absence of overt or progressive weakness, the *mdx* mouse was initially rejected as a model for DMD. It was subsequently shown that these mice do not express any full-length dystrophin; a point mutation introduces a premature translational termination codon into exon 23 of the dystrophin gene (Sicinski et al., 1989). The resulting protein is unstable and rapidly degraded. Since this discovery, several other dystrophin-null mutant mice have been generated by ethyl-nitroso-urea mutagenesis (Cox et al., 1993; Im et al., 1996). These animals have mutations in different parts of the dystrophin gene, and have been generated on different genetic backgrounds. Their phenotype is similar to the *mdx* mouse, indicating that the mild clinical features arise from species-specific factors rather than idiosyncratic consequences of the type of mutation or the background strain.

As already stated, *mdx* mice have a relatively mild clinical picture. The life span of the mice is normal, and they remain ambulant throughout (Carnwath and Shotton, 1987). However, detailed analysis shows clear evidence of a myopathic process. Histo-

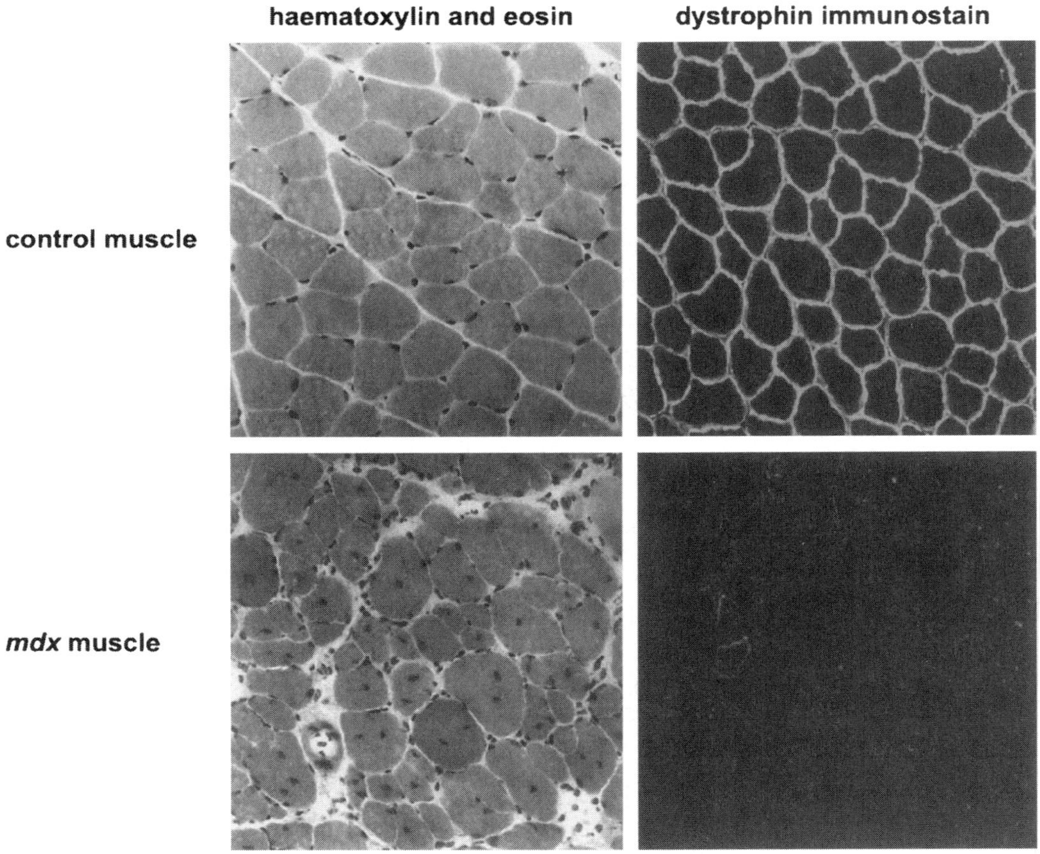

haematoxylin and eosin   **dystrophin immunostain**

control muscle

mdx muscle

Fig. 5. Features of dystrophin-deficient muscle. Transverse sections from the diaphragms of a 12-wk-old *mdx* mouse and age/sex-matched normal control are shown. The hematoxylin and eosin stained section of *mdx* muscle demonstrates the presence of centrally nucleated myofibers (indicative of ongoing muscle regeneration), muscle necrosis, myofiber size variability, mononuclear cell infiltrate, and early fibrosis. Immunofluorescence micrography using an anti-dystrophin primary antibody shows the physiological sarcolemmal localization of dystrophin in control muscle and its absence in *mdx* muscle.

logically, the muscles undergo repeated cycles of degeneration and regeneration, similar to DMD muscle (Dangain and Vrbova, 1984; Carnwath and Shotton, 1987). This is dramatic and extensive at 3–4 wk of age, with the appearance of features—for example, grouped necrosis and hyaline fibers—typical of DMD muscle. However, the myopathic process appears to abate thereafter. Although there is continuing evidence of muscle dysfunction and necrosis, sustained regeneration is apparently able to compensate for the dystrophic process. Consequently, there is little loss of muscle fiber mass. In addition, there is much less fibro-fatty infiltration than seen in DMD (Carnwath and Shotton, 1987). With time, however, the diaphragm becomes weakened and infiltrated with fibrous tissue (Stedman et al., 1991) (Fig. 5), and there is evidence of limb muscle deterioration (Pastoret and Sebille, 1995). In addition, several physiological abnormalities are demonstrable in *mdx* muscle (discussed below). Many of these are similar to those seen in DMD muscle, and several are easily measured allowing quantification of the dystrophic process in response to therapeutic interventions.

The reasons for the phenotypic differences between *mdx* mouse and DMD muscle are not entirely clear. Some have speculated that the *mdx* mouse may be a good model of the initial biochemical changes in muscle fibers that give rise to degeneration, but that the secondary consequences of such dysfunction are different to those seen in DMD (Gorospe et al., 1997; Partridge, 1997). This may be due to species-specific factors relating to the regenerative capacity of muscle and the response to tissue injury. Indeed, ongoing muscle cell necrosis is revealed in irradiated *mdx* muscle that has lost the ability to regenerate (Pagel and Partridge, 1999). An alternative hypothesis is that the dystrophin homolog, utrophin, is able to compensate for the loss of dystrophin in *mdx* mouse muscle and protect myofibers from necrosis. The onset of necrosis in *mdx* muscle occurs around the same age at which sarcolemmal utrophin disappears (Khurana et al., 1991; Pons et al., 1994b). In contrast, utrophin disappears from the sarcolemma of developing human muscle *in utero*, and human muscle would not be afforded protection by utrophin (Clerk et al., 1993). Evidence from the recent generation of utrophin–dystrophin double-null mutant mice lends support to the utrophin hypothesis. Utrophin-null mutants have a normal clinical phenotype and subtle abnormalities relating to the ultrastructure and physiology of the neuromuscular junction (Deconinck et al., 1997a). In contrast, dystrophin–utrophin double-null mutants have a severe, rapidly progressive myopathy that is fatal, similar to that seen in DMD (Deconinck et al., 1997b). The most economical explanation for these data is that utrophin-mediated protection of *mdx* muscle fibers is lost from the double mutants. An alternative mechanism might be that utrophin is pivotal in the murine myofiber regenerative program that allows *mdx* muscle to retain its morphology and function in the face of ongoing necrosis. This seems less likely, however, as the proportion of recently regenerated myofibers in *mdx* and double-null mutant muscle is similar (Deconinck et al., 1997b). A recent review hypothesized that the ability of mouse muscle to arrest myogenesis at a premature stage in dystrophic muscle is pivotal in giving rise to a mild phenotype in *mdx* mice (Infante and Huszagh, 1999). Perhaps protection of the immature myofibers is afforded by prolonged expression of utrophin at the sarcolemma.

Despite the potential objections to use of the *mdx* mouse, it remains a valuable genetic and biochemical model. Mice are genetically manipulated with ease, have a short generation time, and provide easy access to tissues for study. This has enabled the generation of much data on the pathophysiological changes occurring in dystrophin-null muscle from mice. The applicability of these findings to the secondary changes that are observed following the onset of muscle necrosis is less certain.

PROCESSES RESULTING
IN DEGENERATION OF DYSTROPHIN-NULL MUSCLE

Although the molecular basis of mutations that give rise to DMD is now well understood, the mechanisms giving rise to the resulting muscle cell degeneration remain enigmatic. Understanding the physiological role of dystrophin in muscle may be fundamental to comprehending the consequences of its absence. Any model that seeks to describe and elucidate the series of events that intervene between the primary biochemical abnormality and the clinical phenotype must provide cogent explanations for a number of observations. First, dystrophin is absent from conception, yet the phenotype only manifests in early childhood. Second, myofiber necrosis is grouped and segmental, even though dystrophin is absent from the entire length of all skeletal muscle fibers. Third, there is substantial interspecies variation in the manifestations of dystrophin deficiency.

The association of dystrophin with a protein complex in the plasma membrane, and its role in maintaining a link between the cytoskeleton and ECM has attracted much attention to the possibility that muscle degeneration in dystrophin-deficient myofibers is caused by changes in the properties of this link. Earlier work, published prior to the cloning of dystrophin, had provided evidence suggesting that the sarcolemma was intimately involved in the pathogenic process:

- The serum levels of creatine kinase and other muscle cytosolic enzymes are raised in DMD patients, and in *mdx* mice and other animal models (Shaw et al., 1967; Zellweger and Antonik, 1975). This suggests that the membrane is abnormally "leaky" in the absence of dystrophin and allows diffusion of these molecules into the circulation. The objection to this argument is that CK levels can be induced to increase by any process that causes muscle necrosis, including inflammatory myopathy and muscle ischemia where dystrophin expression is not altered. However, this objection seems untenable in view of the striking elevations in CK levels that are often seen in neonatal DMD patients in the presence of relatively subtle muscle histology (Bradley et al., 1972; Emery, 1977).
- Electron microscopic studies showed evidence of focal breaches in the continuity of the sarcolemma, in areas of muscle where a small segment of subsarcolemmal cytoplasm appeared abnormal (Mokri and Engel, 1975). These "delta lesions" are not found in control muscle, and it was hypothesized that they may represent a primary pathological change in muscle destined to become dystrophic.

Much of the later work on the pathophysiology of dystrophin deficiency has focused on the role of sarcolemmal dysfunction in the dystrophic process. Essentially, the majority of current hypotheses concerning the physiological role of dystrophin and the pathophysiological consequences of its absence fall into two categories. The first category consists of hypotheses that attribute a major structural role to dystrophin, envisaging a mechanical function in force transmission across the sarcolemma, or maintaining sarcolemmal stability in the face of local forces generated during contraction. The second category contains models in which dystrophin acts to organize important biologically active molecules in the sarcolemma or subsarcolemmal compartment.

Despite the strength of the ongoing debate, it is worth reflecting that the two categories of hypothesis are not mutually exclusive; dystrophin may have multiple functional roles in normal muscle. It follows that the cascade of events leading to muscle degeneration in DMD may be complex and not fully explicable in terms of a single hypothetical model. Indeed, data from transgenic and mouse null mutant models suggest that dystrophic changes may result from absence of components of the DAPC in the presence of dystrophin, or from absence of full-length dystrophin in the presence of the DAPC. It follows that functional deficits attributable to loss of full-length dystrophin may be superimposed on those resulting from loss of the DAPC. The resulting pathophysiological cascades are likely to be complex and multiple.

In the following sections, we highlight differences that have been demonstrated between normal and dystrophin-deficient muscles or myofibers, and discuss how these differences may help elucidate the normal role of dystrophin in muscle and the pathophysiology resulting from its deficiency.

Transmission of Contractile Force Across the Sarcolemma

The mechanisms of force transduction from the contractile apparatus to the tendon may be affected by loss of a mechanical link from the interior of the muscle cell to its exterior. It is known that both dystrophin and utrophin accumulate at the myotendinous junction (Samitt and Bonilla, 1990), the primary site of longitudinal force transmission. In addition, painstaking morphometric study shows that many skeletal muscle fibers decrease dramatically in cross-sectional area toward their extremities; many do not terminate at a myotendinous junction, but seem to end in the perimysium (Monti et al., 1999). In these fibers, a component (or all) of the contractile force vector must be transmitted across the sarcolemma outside the region of the myotendinous junction. This is thought to occur at specialized attachments between the Z-discs of the contractile apparatus and the sarcolemma. These attachments, called costameres, contain assemblies of cytoskeletal proteins, including talin, vinculin, spectrin, desmin, and dystrophin (Minetti et al., 1992; Ridge et al., 1994; Maconochie et al., 1996; Ehmer et al., 1997; Gossrau, 1998). It is conceivable that the actin/dystrophin/DAPC/laminin link is involved in trans-sarcolemmal force transmission at both the costamere and the MTJ, although other proteins at these positions are located suitably to carry out this function.

Electron microscopic studies have shown differences in the ultrastructure of the myotendinous junction and costameres of muscle between normal and *mdx* mice (Williams and Bloch, 1999). These changes occurred at all time points and were thought to be independent of muscle necrosis. The sarcolemma at the MTJ showed less infolding and the digit-like processes of the MTJ were less numerous. The costameric markings on the outside surface of the muscle fiber were less distinct, and there was a reduction in the lateral association of bundles of thin filaments with the sarcolemma (Law and Tidball, 1993).

The mechanics of muscle failure under extreme stretch are quantitatively similar between *mdx* and wild-type muscle. Tetanically stimulated muscles of both types fail at stresses of around 6×10^5 N m^{-2}, although the site of mechanical disruption differs. Wild-type muscle failed at an A-band, whereas *mdx* muscle most often failed in the reticular lamina of the tendon, immediately external to the myotendinous junction. There is no dystrophin at this location in normal muscle, and it is assumed that the altered site of mechanical failure is secondary to changes in the way that dystrophin deficient muscle distributes tensile stress, although no detailed analysis has been published (Law et al., 1995).

Mechanical failure of both wild-type and *mdx* muscle under similar extreme circumstances argues against a simple failure of force transmission and/or tensile strength as a mechanism for dystrophic change. In addition, it is difficult to attribute abnormalities of muscle contractile strength to this mechanism, as the forces generated by maximal contraction are much less than those at which the muscles mechanically fail. A substantial loss in maximal isometric force per unit cross-sectional area of muscle has been documented in *mdx* diaphragm (Stedman et al., 1991; Petrof et al., 1993b) and limb muscle (Tinsley et al., 1998) compared with control muscle. Furthermore, the force generated falls sharply following a series of eccentric contractions under tetanic stimulation (Petrof et al., 1993a). It is difficult to attribute these observations directly to mechanical consequences of dystrophin deficiency, as other factors may influence the specific force generation of fibers.

- First, the muscle contractile apparatus may behave in an altered manner secondary to changes in the intracellular environment, consequent to signaling effects or alterations in ion flux. Abnormalities of the stability or permeability of the sarcolemma may manifest in this way by allowing inappropriate influx or efflux of ions and molecules. Regardless of mechanism, it is now clear that the contractile apparatus itself is dysfunctional in *mdx* myofibers (Coirault et al., 1999); it seems likely that this is a secondary phenomenon in view of the localization of dystrophin at the periphery of the muscle cell.
- Second, the composition of the muscle, in terms of contractile fiber type or amount of connective tissue, is altered in chronically dystrophic/regenerating muscle (Stedman et al., 1991). Even young *mdx* mice that have little evidence of fibrosis show a switch in expression profile of myosin heavy-chain isoforms in actively regenerating muscle that complicates the interpretation of force generation data (Petrof et al., 1993b).

To summarize, there is no conclusive evidence that the absence of dystrophin results in a functional deficit in the transmission of contractile forces across the sarcolemma.

*Biomechanical Properties of the Sarcolemma in DMD/*mdx *Muscle*

The early electron microscopic observations of sarcolemmal discontinuities in dystrophic muscle (Mokri and Engel, 1975) have been confirmed in a number of subsequent studies of dystrophin-null tissue (Carpenter and Karpati, 1979). These observations have given rise to hypotheses concerning the structural integrity of the sarcolemma at rest and during contraction. The previous section dealt with the transmission of longitudinal contractile force across the sarcolemma. The present section deals with the response of the sarcolemma to both the longitudinal forces discussed above, and the radial forces that are generated when a muscle fiber changes length (Cecchi et al., 1990; Bagni et al., 1994).

Electron microscopic data suggest that the sarcolemma is closely attached to the subsarcolammeal cytoskeleton at the costameres. During muscle contraction, the cell membrane is seen to bulge out from the costameres, giving rise to a characteristic "festooned" appearance (Petrof, 1998). The association of dystrophin with the sarcolemma at the costameres has led to investigation of the response of the sarcolemma to radial displacement. The underlying hypothesis tested in these experiments is that the membrane biomechanical properties are altered in the absence of dystrophin in a way that gives rise to the local breaches of integrity observed by electron microscopy. The data are apparently conflicting. In the original series of experiments, the "tensile strength" of the sarcolemma was measured during studies of membrane ionic conductance by applying suction through the patch pipet until the elastic limit of the membrane was reached. No significant difference between normal or *mdx* membrane was seen in cultured myotubes (Franco and Lansman, 1990). This finding was later confirmed by others in a more extensive study of isolated *mdx* myofibers, in which tensile strength was estimated by the patch-suction technique in addition to a technique involving suction of membrane vesicles into a parallel bore pipet (Hutter et al., 1991). A more recent report documents a difference in the deformability of the sarcolemma between of *mdx*

and control myotubes in vitro. A small glass micropipet was used to indent the sarcolemma of myotubes by approx 1 μm, and the force per unit displacement of the membrane was measured. The sarcolemma of *mdx* myotubes showed approx one-quarter of
the stiffness of wild-type sarcolemma. The authors argued that the tensile strength of
the membrane did not contribute significantly to the forces measured by such small
deformations in the presence of considerable membrane folding. Rather, the radial
strength of the supporting cytoskeleton was assayed (Pasternak et al., 1995). This
assertion has been questioned on account of the large difference in sarcolemmal redundancy between mature myofibers and the myotubes in which the experiments were
done (Brown and Lucy, 1997).

Experiments examining the osmotic fragility of cells demonstrate that *mdx* myofibers
are less able to withstand a hypo-osmotic challenge that normal myofibers. The lower
threshold of osmotic stress necessary to irreversibly damage *mdx* compared with wild-
type muscle fibers (Menke and Jockusch, 1991, 1995) has been taken as evidence of
sarcolemmal fragility of dystrophin null myocytes (Gillis, 1996; Brown and Lucy,
1997). Indeed, it is possible to observe changes in the calibre of myofibers subject to
osmotic challenges of this nature, consistent with the production of sarcolemmal stress
(Deconinck et al., 1997). However, it has been argued that this conclusion is incompatible with the absence of any demonstrable difference in the tensile strength of the sarcolemma. Differences in the shape and morphology of the sarcolemma were postulated
to be responsible for the observed disparity between normal and *mdx* fibers in these
tests (Hutter et al., 1991). Recent separate observations on the distribution and density
of aquaporin 4 (Frigeri et al., 1998; Liu et al., 1999), and on the genetic expression of
aquaporin 1 (Tkatchenko et al., 2000), in *mdx* and wild-type myocytes add further
complexity to the problem. Aquaporins are a family of membrane water-transport proteins essential for osmotic homeostasis. Both AQ1 and AQ4 are downregulated in
dystrophin-deficient muscle. The predicted response to loss of these channels might be
blunting of osmotic sensitivity rather than the observed increase, but the demonstration
of changes in muscle components essential for the establishment of osmotic equilib-
rium in this experimental model demands caution in the interpretation of data obtained
by osmotic manipulations.

Studies of the movement of large molecules across the muscle-cell membrane are
perhaps the most persuasive evidence that the sarcolemma is susceptible to disruption
by mechanical stimuli in dystrophin-null myofibers. Procion Orange (McArdle et al.,
1994; Deconinck, N. et al., 1997; Rafael et al., 1998; Tinsley et al., 1998; Pagel and
Partridge, 1999) and Evans Blue (Matsuda et al., 1995; Brussee et al., 1997; Straub et
al., 1997) are low-molecular-weight dyes that are unable to cross the plasma membrane
under normal circumstances. However, degenerating muscle fibers in *mdx* mice take
up Evans Blue (Matsuda et al., 1995) and Procion Orange enters the cytoplasm of
degenerating DMD (Bradley and Fulthorpe, 1978) and *mdx* (McArdle et al., 1994)
muscle. These observations presumably arise from impaired membrane integrity. Membrane permeability to these, and other, agents can be augmented dramatically in non-
necrotic dystrophin-deficient myofibers by forced lengthening. Eccentric muscle
contraction is predicted to stress the sarcolemma, by both longitudinal and radial forces
(Cecchi et al., 1990; Petrof et al., 1993a; Bagni et al., 1994). The proportion of muscle
fibers containing IgG, presumably owing to increased membrane permeability, was

higher in *mdx* than normal tibialis anterior muscle following eccentric contraction in vivo (Weller et al., 1990). It was subsequently shown that eccentric contraction of the diaphragm in vitro caused myofibers from *mdx* mice to become permeable to Procion Orange dye in much higher proportions than seen in identically treated control diaphragm (Petrof et al., 1993b). The proportion of fibers becoming permeable to the dye increased with the peak force generated during the contraction, suggesting a relationship between mechanical stress and sarcolemmal damage. Similar results were obtained in limb muscle in the same study (Petrof et al., 1993b). Data obtained from these unphysiological stimuli have been corroborated recently by examining the limb muscles of *mdx* mice following an exercise protocol involving physiological eccentric contraction. Wild-type and *mdx* mice were made to run downhill on a treadmill. The proportion of limb muscle fibers that were permeable to Evan's Blue increased dramatically more in the *mdx* mice than in the normal mice following the exercise protocol (Brussee et al., 1997). In a series of complementary experiments *mdx*-transgenic and non-*mdx* transgenic mice expressing β-galactosidase under the control of a muscle promoter were exercised under a similar protocol. The efflux of β-gal and CK from muscle into the circulation following exercise was significantly higher in *mdx* mice than controls (Vilquin et al., 1998).

In summary, there is evidence to support the hypothesis that mechanical stresses are less well-tolerated by the sarcolemma of dystrophin-deficient muscle cells than controls. It is unclear whether this is a direct effect of the absence of a crucial mechanical function of dystrophin or a secondary change. In addition, it is not clear if sarcolemmal fragility is important in dictating subsequent myofiber degeneration, and if so, by what mechanism.

Metabolism of Calcium in DMD/mdx Muscle

Dysregulation of intracellular ionic calcium metabolism with secondary activation of calcium-dependent proteases has been suggested as a link between loss of dystrophin from the sarcolemma and myofiber degeneration. This mechanism has been invoked as an explanation for the deleterious consequences of losing sarcolemmal integrity. In addition, alterations in calcium homeostasis have been implicated in putative dystrophic mechanisms that do not involve loss of sarcolemmal integrity.

This section examines data concerning calcium levels, transmembrane calcium conductance and putative calcium-dependent effector mechanisms in dystrophin-deficient muscle.

Ca^{2+} Content of Dystrophin-Deficient Muscle

Early work showed that the proportion of myofibers on DMD biopsy specimens positive for a histochemical stain for calcium was elevated (Bodensteiner and Engel, 1978). Levels of total calcium were reported to be elevated in DMD muscle biopsy specimens by X-ray fluorescence spectrometry (Maunder Sewry et al., 1980) and by atomic absorption spectrometry (Jackson et al., 1985). The findings were confirmed in *mdx* heart and muscle, at all ages and prior to the onset of muscle necrosis (Dunn and Radda, 1991). The significance of these findings is not clear, as the contribution of extracellular calcium and sequestered intracellular calcium to the quantitative data is uncertain.

The availability of calcium-sensitive fluorescent indicators enabled the question of intracellular free-calcium concentration to be directly addressed. The technique relies on the detection of a fluorescent signal that is altered when calcium is bound to an indicator. Most of the data have been acquired using ratiometric dyes that display separate spectral peaks in the Ca^{2+}-bound and unbound forms. This enables calculation of $[Ca^{2+}]$ from the ratio of the fluorescence excited at two separate wavelengths, and is independent of intracellular dye concentration and cell morphology. Absolute determination of Ca^{2+} concentration, however, demands a stringent calibration procedure that has been the subject of controversy on account of debated differences in the way that normal and dystrophic cells handle the dye (Gailly et al., 1993; Gillis, 1996; Hopf et al., 1996). The type of fluorescent dye used and the technique for its introduction are additional variables, as are the tissue subjected to measurement, and the procedures used in its preparation. These factors may, in part, account for the large body of seemingly contradictory literature on the subject.

The first report described measurement of $[Ca^{2+}]_i$ in isolated myofibers dissected from the flexor digitorum brevis muscle of normal and *mdx* mice (Turner et al., 1988). The dye Fura-2 was introduced by microinjection. Twofold increases in $[Ca^{2+}]_i$ of *mdx* vs normal muscle were demonstrated under a range of $[Ca^{2+}]_o$. Using tyrosine release as a surrogate marker for protein degradation, it was found that the rate of proteolysis in *mdx* muscle was elevated with respect to controls, and could be varied by altering the $[Ca^{2+}]_o$. The finding of elevated $[Ca^{2+}]_i$ was confirmed in cultured DMD and *mdx* myotubes shortly afterwards by the same group (Fong et al., 1990), and in *mdx* myotubes in vitro by a different group (Bakker et al., 1993).

Others, however, were unable to demonstrate similar differences in calcium homeostasis between *mdx* and control muscle. Myofibers from *mdx* and control FDB were shown to have indistinguishable $[Ca^{2+}]_i$, which correlated with the resting membrane potential (Head, 1993). Marked differences were observed in the calibration of Fura-2 dye between control and *mdx* muscle (Gailly et al., 1993); no difference was found between the $[Ca^{2+}]_i$ of *mdx* and control tissue when the behavior of the dye in the different tissues was taken into account. Both of these studies have been criticized for the use of enzymatic isolation of muscle fibers that may theoretically have damaged a critical extracellular link of dystrophin in the control fibers (Carlson, 1998). In vitro data are equally controversial. One study demonstrated raised $[Ca^{2+}]_i$ in DMD myoblasts prior to the expression of dystrophin, but not in myotubes (Rivet Bastide et al., 1993). No difference between $[Ca^{2+}]_i$ in *mdx* and DMD myotubes vs controls was observed by another group (Pressmar et al., 1994). A further independent study showed that the $[Ca^{2+}]_i$ of *mdx* myotubes was only elevated with respect to controls in response to external hypercalcaemic or hypoosmotic stress (Leijendekker et al., 1996).

A subsequent study sought to address some of the criticisms leveled at the original work (Hopf et al., 1996). A compartmentalization-resistant ratiometric dye was introduced into cultured myotubes by microinjection and stringent calibration procedures undertaken. $[Ca^{2+}]_i$ was significantly raised in *mdx* myotubes compared with controls. Inhibition of myotube contractile activity with tetrodotoxin abolished the difference, confirming other reports of the activity-dependence of intracellular calcium overload in dystrophin-deficient muscle (Imbert et al., 1995, 1996). A further recent paper describes small but significant increases in $[Ca^{2+}]_i$ between enzymatically dissociated normal and *mdx* myofibers (Tutdibi et al., 1999).

The debate is not over, however. The most recent publication on the subject described measurements of $[Ca^{2+}]_i$ in enzymatically prepared myofibers that were voltage-clamped to examine the contribution of the resting membrane potential to the $[Ca^{2+}]_i$. No difference was demonstrated between normal and *mdx* muscle (Collet et al., 1999). In addition, alterations of $[Ca^{2+}]_i$ in response to depolarization pulses were almost identical between the two groups, both in terms of time-course and magnitude.

In summary, there is evidence that differences between $[Ca^{2+}]_i$ of normal and dystrophin-null myogenic cells may be detected under some circumstances. It is not clear, however, whether this represents an experimental artifact, or whether methodological considerations account for some studies failing to detect a real difference. It has been argued that the elaborate intracellular machinery for buffering $[Ca^{2+}]_i$ is enhanced in *mdx* cells (Gillis, 1996). No study has yet attempted to address whether any change in $[Ca^{2+}]_i$ is localized to specific areas or regions of the cytosol.

Abnormal Ca²⁺ Currents in Dystrophin-Deficient Muscle

The development of patch-clamping techniques allowed the comparison of membrane Ca^{2+} conductances in normal and dystrophin-deficient muscle. Two initial studies identified different abnormal Ca^{2+} conductances in dystrophin-null membranes:

- Ca^{2+} leak channels are voltage-independent and have conductances of 10–15 pS. Single channel studies were undertaken comparing *mdx* and DMD myotubes with controls. Ca^{2+} leak channels had similar conductance and open times in normal and dystrophin-deficient myotubes, but the closed times were greatly reduced in *mdx* and DMD membrane. The probability of channels opening was estimated to be threefold higher in the *mdx* and DMD samples compared with controls (Fong et al., 1990). This type of conductance was enhanced by nifedipine, treatment with which produced an elevation in $[Ca^{2+}]_i$ in normal myotubes.
- Channels with conductance of 17-25 pS, which were highly active in resting membrane were described in *mdx* myotubes (Franco and Lansman, 1990). Unlike the mechanosensitive channels present in normal myotubes, which increase in activity when the membrane is stretched, application of suction to the patch pipet reduced the opening of the highly active channels found in *mdx* membrane. These findings were later confirmed in *mdx* myofibers. Channel activity was found to reduce with age in wild-type but not *mdx* myofibers (Haws and Lansman, 1991).

The existence of these two types of current has been debated. The authors of the first report were unable to demonstrate any stretch-inactivated channels in their preparation (Fong et al., 1990). The authors of the second report were unable to show evidence of leak channels in *mdx* myotubes or myofiber, leading them to conclude that leak-channel activity may have been a result of muscle degeneration (Franco Obregon and Lansman, 1994). Leak-type activity has, however, been recorded by another group (Carlson and Officer, 1996).

The origin of these activities is controversial, in that the channel proteins responsible for the observed conductances have not been isolated. It was hypothesized that abnormal local forces at the sarcolemma in the absence of dystrophin accounted for the resting open state of stretch-inactivated channels (Fong et al., 1990), although the mechanism by which these channels gained their unusual mechanosensitive properties was not discussed. The expression of calcium leak conductance in *mdx* myotubes was prevented by incubation with leupeptin, an inhibitor of Ca^{2+}-dependent proteases (Fong et al., 1990). This led to speculation that Ca^{2+} influx through mechanically induced sarcolemmal rupture in *mdx* myoblasts was adequate to activate proteases that modified transmembrane channels, which in turn increased the membrane permeability to Ca^{2+} and established a Ca^{2+}-dependent positive feedback loop (Fong et al., 1990). An alternative hypothesis has been proposed (Carlson, 1998). This is based on observations of the relationships between Ca^{2+} leak currents and acetylcholine receptor activity in membrane patches. Essentially, patches from normal and *mdx* myotubes could be shown to exhibit both AChR and Ca^{2+} leak activity, the latter more frequent in *mdx* patches and independent of the presence of ACh in the pipet. At constant $[ACh]_o$, there was an inverse and discontinuous relationship between the frequency of AChR and Ca^{2+} leak events in individual patches. *mdx* cell-attached patches that showed only Ca^{2+} leak activity could be induced to show only AChR activity by acquisition of an inside-out patch. This gave rise to the hypothesis that AChRs may be responsible for abnormal Ca^{2+} leak currents by changes in their properties induced by association with the disorganized dystrophin-deficient cytoskeleton. This could occur in the absence of a structural defect in *mdx* membrane (Carlson and Officer, 1996; Carlson, 1998, 1999; Williams and Bloch, 1999).

Consequences of Ca²⁺ Entry

The inhibition of Ca^{2+} leak activity by leupeptin, discussed in the previous section, formed a core argument in favor of the Ca^{2+}-dependent protease hypothesis that would account for the secondary consequences of Ca^{2+} entry into dystrophin-deficient muscle cells (Turner et al., 1993). Other evidence included elevations of protein degradation in *mdx* myotubes in response to calcium that was blocked by thiol protease inhibitors (MacLennan et al., 1991), and evidence of specific increases in the expression of calpains in *mdx* muscle (Combaret et al., 1996). Recent evidence supports this hypothesis in that intramuscuar injection of the calpain inhibitor leupeptin causes some amelioration of the histological phenotype of *mdx* mice after 30 d of treatment (Badalamente and Stracher, 2000).

There is not universal agreement on this point, however. Although it was possible to demonstrate that *mdx* mouse muscle contained a higher concentration of calpains by immunoblotting, and that the proteins detected were supposedly activated forms (Spencer et al., 1995), paradoxically, the Ca^{2+}-dependent protease activity from *mdx* mice was lower than that measured from controls (Spencer and Tidball, 1992). In addition, it has been argued that the levels of Ca^{2+} found in *mdx* and DMD muscle cells are not adequate to reach the threshold at which calpain activation occurs (Gillis, 1996), even if the controversial data regarding $[Ca^{2+}]_i$ are accepted. In addition, the mutations in caplain 3 associated with LGMD2A show that muscle wasting can be produced by loss of function of calpains, in addition to the putative activation that occurs in the Ca^{2+} theory of DMD pathogenesis (Bushby, 1999).

Abnormal Signaling
and Other Biochemical Abnormalities in DMD/mdx Muscle

The association of dystrophin with proteins implicated in signal transduction at the sarcolemma has led to speculation about the role of disorganized signaling in the pathogenesis of DMD.

Nitric oxide (NO) signaling was the first such pathway to be investigated in any depth. The neuronal form of NO is expressed in skeletal muscle, where it is localized at the sarcolemma and enriched at the neuromuscular junction. In both places it interacts with the α1-syntrophin component of the DAPC, and is missing from both the NMJ and sarcolemma in DMD and the *mdx* mouse (Brenman et al., 1995, 1996; Christova et al., 1997; Peters et al., 1997a; Gee et al., 1998; Gossrau, 1998; Grozdanovic and Baumgarten, 1999). This prompted speculation that the loss or redistribution of nNOS may cause muscle degeneration. It is unlikely that loss of nNOS from the sarcolemma results in muscle degeneration, as nNOS null mice do not have an abnormal muscle phenotype (Huang et al., 1993). In addition, restoration of components of the DAPC that localize nNOS to the sarcolemma is not alone sufficient to prevent the muscle degeneration occurring in the absence of dystrophin (Cox et al., 1994). An alternative hypothesis was that the redistribution of nNOS to the cytoplasm in the *mdx* myofiber led to production of toxic species within the cell. This also seems unlikely, both on account of the persistence of the muscle phenotype in Dp71 transgenic *mdx* mice, and because inactivation of the nNOS gene in *mdx* mice does not improve the muscle-wasting phenotype (Chao et al., 1998; Crosbie et al., 1998). In addition, there is relocalization of nNOS to the cytoplasm in the α1-syntrophin null mouse, but no muscular dystrophy (Kameya et al., 1999).

There is evidence, however, for a nonmechanical function of dystrophin and the DAPC in the development of muscular dystrophy.

- Antibody-mediated blockade of the interaction of α-dystroglycan with α2-laminin in primary muscle-cell cultures induces changes in the myofibrillar organization and size of the myofibers, in addition to alterations in the distribution of AChR (Brown et al., 1999). It is possible that this is mediated by a signaling function of α-dystroglycan.
- The recently reported α-dystrobrevin null mutant mouse lacks sarcolemmal nNOS, but retains dystrophin and other components of the DAPC. The resulting muscle phenotype has been attributed to loss or re-distribution of nNOS or another signaling protein from the sarcolemma (Grady et al., 1999). Involvement of nNOS seems unlikely in view of the evidence discussed above regarding nNOS in other types of mutant. However, the possibility that an unidentified absent signaling component is responsible for muscle degeneration is intriguing.
- The γ-SG null mouse has a severe muscle wasting phenotype. In this model of LGMD-2C, animals have reduced sarcolemmal localization of β- and δ-SG, but retain dystrophin, laminin, and DG localization. Muscle from these animals exhibits normal mechanical properties and normal responses to eccentric contraction protocols (Hack et al., 1999). No evidence of sarcolemmal damage was found by Procion Orange staining.

As γ-SG and α-DB are both absent from the sarcolemma of DMD/*mdx* muscle, it is possible that in addition to any mechanical effects on membrane stability, dystrophin deficiency has nonmechanical consequences. The nature of other signaling or biochemical disturbances is largely speculative at present. Defects in the inositol trisphosphate signaling pathway (Liberona et al., 1998) and in sodium regulation (Dunn et al., 1993) have been described, but their relevance is unclear.

The Role of Extracellular Influences

Although the primary biochemical deficit that makes DMD and *mdx* cells vulnerable to necrosis is undoubtedly intracellular, various contributions to the dystrophic process have been postulated to arise from the extracellular environment. There are two observations that suggest involvement of extracellular factors in the pathogenesis of the muscle phenotype arising from dystrophin deficiency:

- Early pathological changes include necrosis that is segmental and grouped. As the absence of dystrophin within myofibers is universal, it is possible that the extracellular environment can locally influence the phenotypic manifestations of the absence of dystrophin in a cluster of myofibers.
- Late pathological changes include extensive fibrosis in DMD muscle and *mdx* diaphragm, but not in *mdx* limb muscle. The fibrosis arises from extracellular factors and modifies the extracellular environment. It is possible that the fibrosis affects the regenerative capacity of muscle and thereby alters the outcome of the primary deficit.

There are three main theories of grouped necrosis. (Gorospe et al., 1997). The neurogenic hypothesis invokes a role for motor neurons influencing degeneration of myofibers that they innervate. This hypothesis is untenable, as the pattern of grouped necrosis does not mirror the topographical distribution of myofibers in a motor unit. In addition, there is no clinical or investigative evidence of neurogenic dysfunction of the type that would be necessary to give rise to re-innervation and type grouping. The vascular hypothesis suggests that regional areas of necrosis could arise from ischemia, acting on partially compromised dystrophin-deficient cells. It is known that dystrophin is expressed in vascular smooth muscle and absent in DMD, although there is no evidence that the blood vessels may be overtly dysfunctional sufficient to cause ischemia (Boland et al., 1995; Hoffman et al., 1988b; Miyatake et al., 1989). It has been hypothesized that vasoactive substances might be released from mast cells or other infiltrating cells that are visible in dystrophic muscle (Gorospe et al., 1994a, 1994b, 1996). There is no direct evidence at present for this model; indeed, *mdx* mice that are genetically deficient for mast cells have a skeletal myopathy indistinguishable from that of normal *mdx* mice (Gorospe et al., 1997). It is of interest, however, that it is possible to demonstrate abnormalities in *mdx* mice of the nNOS signaling that links muscle activity to local vasodilatation (Lau et al., 1998). It is feasible that aberrant contraction/perfusion coupling interacts with other factors to trigger areas of necrosis in vulnerable cells, but no evidence exists to support this point (Gudrun et al., 1975). Finally, it has been suggested that inflammatory cells in the perimysium and endomysium may influence the survival of dystrophin-deficient myofibers (Rafael et al., 1998). The inflammatory infiltrate in dystrophin-deficient muscle is not marked, however. In addition, the muscle-wasting phenotype of *mdx* mice that are genetically deficient for either mast cells or cells of the macrophage lineage is identical to that of normal *mdx* mice (Gorospe et al., 1997).

The functional results of fibrosis in muscle are not clear. It has been hypothesized that fibrosis inhibits myogenesis, and that the extensive fibrosis seen in DMD muscle may be causally related to the failure of ongoing muscle regeneration. No data exist that substantiate or refute this hypothesis. The origin of fibrosis in DMD muscle is not clear. The fibrosis of the *mdx* diaphragm correlates with an increased expression of the procollagen I and III genes (Goldspink et al., 1994). The stimulus that initiates expression of these genes is not known, and various signaling molecules have been investigated. The detection of basic fibroblast growth factor (bFGF) in the serum of DMD patients (D'Amore et al., 1994) and bound to the basement membrane of muscle is

probably related to its release from damaged myofibers. Striking expression of transforming growth factor-β-1 was seen in DMD muscle, but not in muscle biopsies from
patients with other diagnoses in one study (Yamazaki et al., 1994). Another study demonstrated a correlation between intramuscular expression of mRNA for TGF-β-1 and
the extent of fibrosis in DMD, BMD, and other biopsy specimens (Bernasconi et al.,
1995). There was no correlation between the expression of TGF-β-2 and fibrosis in
biopsy specimens examined in another study (Murakami et al., 1999). Finally, some
insight into factors that promote muscle regeneration has come from the study of mice
that have a null allele for fibroblast growth factor 6. FGF-6 belongs to a family of
cytokines that have been implicated in the control of cellular proliferation and differentiation. Expression of FGF-6 is mainly restricted to myogenic cells; normally, FGF-6 is
upregulated after skeletal-muscle injury. FGF-6 null mice have a severe defect in
muscle regeneration after injury, and *mdx*/FGF-6 double mutants show a progressive
myopathy with regenerative failure and fibrosis (Floss et al., 1997). This indicates that
FGF-6 has a crucial role in signaling the need for repair of muscle damage in dystrophic muscle tissue; further clarification of its role may help elucidate the factors that
give rise to different clinical phenotypes in the *mdx* mouse and DMD muscle.

CONCLUSIONS: PROSPECTS FOR THERAPY

The molecular basis of DMD has been established over the last 15 years by genetic
and protein studies, but the pathophysiology linking the primary molecular defect to
the clinical phenotype is incompletely understood.

The cloning of dystrophin provided some hope that the fundamental defect in this
disorder might be addressed by gene therapy. Expression of dystrophin and improvement of the muscle phenotype after direct adenoviral delivery of a dystrophin transgene
have been demonstrated in isolated muscles of experimental animals (Ragot et al., 1993;
Acsadi et al., 1996; Deconinck et al., 1996; Yang et al., 1998; Yuasa et al., 1998).
However, long-term expression has been limited owing to immune responses both to
dystrophin and to viral antigens (Yang et al., 1994; Howell et al., 1998; Ohtsuka et al.,
1998; Petrof et al., 1999). Recent advances in the field include the use of "gutted"
adenovirus vectors that have all viral genes deleted (Kochanek et al., 1996; Chen et al.,
1997). These vectors may persist long-term in skeletal muscle and have adequate insert
capacity to accommodate the entire dystrophin coding sequence. The delivery and long-
term expression of a therapeutic dystrophin transgene in all muscle cells of the body
remains a formidable challenge, which has yet to be accomplished. Gene therapy has,
however, shown some promise for the treatment of other types of muscular dystrophy,
using adeno-associated virus vectors and arterial delivery via a permeabilized vascular
bed (Xiao et al., 1996; Greelish et al., 1999). Unfortunately, these vectors are not able
to accommodate the large dystrophin gene. Consequently this approach is unlikely to
be helpful in DMD.

For these reasons, alternative approaches to the treatment of DMD seem attractive.
We have concentrated our efforts on upregulating utrophin, the autosomal homolog of
dystrophin, in skeletal muscle, in view of the functional redundancy between the two
proteins (Tinsley and Davies, 1993; Dennis et al., 1996; Tinsley et al., 1996, 1998;
Burton et al., 1999). Others have attempted to manipulate splicing events in dystrophic
myonuclei using antisense technology (Dunckley et al., 1998).

Perhaps a better understanding of the pathophysiology of DMD would help identify further alternative targets for therapeutic intervention that may be amenable to pharmacological manipulation.

REFERENCES

Acsadi, G., Lochmuller, H., Jani, A., Huard, J., Massie, B., Prescott, S., et al. (1996) Dystrophin expression in muscles of mdx mice after adenovirus-mediated in vivo gene transfer. *Hum. Gene Ther.* **7,** 129–140.

Adams, M. E., Butler, M. H., Dwyer, T. M., Peters, M. F., Murnane, A. A., and Froehner, S. C. (1993) Two forms of mouse syntrophin, a 58 kd dystrophin-associated protein, differ in primary structure and tissue distribution. *Neuron* **11,** 531–540.

Adams, M. E., Dwyer, T. M., Dowler, L. L., White, R. A., and Froehner, S. C. (1995) Mouse alpha 1- and beta 2-syntrophin gene structure, chromosome localization, and homology with a discs large domain. *J. Biol. Chem.* **270,** 25,859–25,865.

Ahn, A. H., Freener, C. A., Gussoni, E., Yoshida, M., Ozawa, E., and Kunkel, L. M. (1996) The three human syntrophin genes are expressed in diverse tissues, have distinct chromosomal locations, and each bind to dystrophin and its relatives. *J. Biol. Chem.* **271,** 2724–2730.

Ahn, A. H. and Kunkel, L. M. (1995) Syntrophin binds to an alternatively spliced exon of dystrophin. *J. Cell Biol.* **128,** 363–371.

Ambrose, H. J., Blake, D. J., Nawrotzki, R. A., and Davies, K. E. (1997) Genomic organization of the mouse dystrobrevin gene: comparative analysis with the dystrophin gene. *Genomics* **39,** 359–369.

Arahata, K. and Engel, A. G. (1984) Monoclonal antibody analysis of mononuclear cells in myopathies. I: Quantitation of subsets according to diagnosis and sites of accumulation and demonstration and counts of muscle fibres invaded by T cells. *Ann. Neurol.* **16,** 193–208.

Arahata, K., Hoffman, E. P., Kunkel, L. M., Ishiura, S., Tsukahara, T., Ishihara, T., et al. (1989) Dystrophin diagnosis: comparison of dystrophin abnormalities by immunofluorescence and immunoblot analyses. *Proc. Natl. Acad. Sci. USA* **86,** 7154–7158.

Araishi, K., Sasaoka, T., Imamura, M., Noguchi, S., Hama, H., Wakabayashi, E., et al. (1999) Loss of the sarcoglycan complex and sarcospan leads to muscular dystrophy in beta-sarcoglycan-deficient mice. *Hum. Mol. Genet.* **8,** 1589–1598.

Badalamente, M. A. and Stracher, A. (2000) Delay of muscle degeneration and necrosis in mdx mice by calpain inhibition. *Muscle Nerve* **23,** 106–111.

Bagni, M. A., Cecchi, G., Griffiths, P. J., Maeda, Y., Rapp, G., and Ashley, C. C. (1994) Lattice spacing changes accompanying isometric tension development in intact single muscle fibres. *Biophys. J.* **67,** 1965–1975.

Bakker, A. J., Head, S. I., Williams, D. A., and Stephenson, D. G. (1993) Ca2+ levels in myotubes grown from the skeletal muscle of dystrophic (mdx) and normal mice. *J. Physiol.* **460,** 1–13.

Barnea, E., Zuk, D., Simantov, R., Nudel, U., and Yaffe, D. (1990) Specificity of expression of the muscle and brain dystrophin gene promoters in muscle and brain cells. *Neuron* **5,** 881–888.

Bartolo, C., Papp, A. C., Snyder, P. J., Sedra, M. S., Burghes, A. H., Hall, C. D., et al. (1996) A novel splice site mutation in a Becker muscular dystrophy patient. *J. Med. Genet.* **33,** 324–327.

Baumbach, L. L., Chamberlain, J. S., Ward, P. A., Farwell, N. J., and Caskey, C. T. (1989) Molecular and clinical correlations of deletions leading to Duchenne and Becker muscular dystrophies. *Neurology* **39,** 465–474.

Becker, E. and Keiner, F. (1955) Eine neue X-chromosomale muskeldystrophie. *Archiv für psychiatric und nervenkrankheiten* **193,** 427–428 (reviewed in Emery, 1993).

Beggs, A. H., Koenig, M., Boyce, F. M., and Kunkel, L. M. (1990) Detection of 98% of DMD/BMD gene deletions by polymerase chain reaction. *Hum. Genet.* **86,** 45–48.

Bell, C. D. and Conen, P. E. (1968) Histopathological changes in Duchenne muscular dystrophy. *J. Neurol. Sci.* **7,** 529–544.

Bernasconi, P., Torchiana, E., Confalonieri, P., Brugnoni, R., Barresi, R., Mora, M., et al. (1995) Expression of transforming growth factor-beta 1 in dystrophic patient muscles correlates with fibrosis. Pathogenetic role of a fibrogenic cytokine. *J. Clin. Invest.* **96,** 1137–1144.

Bies, R. D., Friedman, D., Roberts, R., Perryman, M. B., and Caskey, C. T. (1992) Expression and localization of dystrophin in human cardiac Purkinje fibres. *Circulation* **86,** 147–153.

Blake, D. J., Hawkes, R., Benson, M. A., and Beesley, P. W. (1999) Different dystrophin-like complexes are expressed in neurons and glia. *J. Cell Biol.* **147,** 645–658.

Blake, D. J., Nawrotski, R., Peters, M. F., Froehner, S. C., and Davies, K. E. (1996a) Isoform diversity of dystrobrevin, the murine 87-kDa postsynaptic protein. *J. Biol. Chem.* **271,** 7802–7810.

Blake, D. J., Nawrotzki, R., Loh, N. Y., Gorecki, D. C., and Davies, K. E. (1998) Beta-dystrobrevin, a member of the dystrophin-related protein family. *Proc. Natl. Acad. Sci. USA* **95,** 241–246.

Blake, D. J., Schofield, J. N., Zuellig, R. A., Gorecki, D. C., Phelps, S. R., Barnard, E. A., et al. (1995a) G-utrophin, the autosomal homologue of dystrophin Dp116, is expressed in sensory ganglia and brain. *Proc. Natl. Acad. Sci. USA* **92,** 3697–3701.

Blake, D. J., Tinsley, J. M., and Davies, K. E. (1996b) Utrophin: a structural and functional comparison to dystrophin. *Brain Pathol.* **6,** 37–47.

Blake, D. J., Tinsley, J. M., Davies, K. E., Knight, A. E., Winder, S. J., and Kendrick Jones, J. (1995b) Coiled-coil regions in the carboxy-terminal domains of dystrophin and related proteins: potentials for protein-protein interactions. *Trends Biochem. Sci.* **20,** 133–135.

Bodensteiner, J. B. and Engel, A. G. (1978) Intracellular calcium accumulation in Duchenne dystrophy and other myopathies: a study of 567,000 muscle fibres in 114 biopsies. *Neurology* **28,** 439–446.

Boland, B., Himpens, B., Denef, J. F., and Gillis, J. M. (1995) Site-dependent pathological differences in smooth muscles and skeletal muscles of the adult mdx mouse. *Muscle Nerve* **18,** 649–657.

Boland, B. J., Silbert, P. L., Groover, R. V., Wollan, P. C., and Silverstein, M. D. (1996) Skeletal, cardiac, and smooth muscle failure in Duchenne muscular dystrophy. *Pediatr. Neurol.* **14,** 7–12.

Bonilla, E., Samitt, C. E., Miranda, A. F., Hays, A. P., Salviati, G., DiMauro, S., et al. (1988) Duchenne muscular dystrophy: deficiency of dystrophin at the muscle cell surface. *Cell* **54,** 447–452.

Bonnemann, C. G., Passos Bueno, M. R., McNally, E. M., Vainzof, M., de Sa Moreira, E., Marie, S. K., et al. (1996) Genomic screening for beta-sarcoglycan gene mutations: missense mutations may cause severe limb-girdle muscular dystrophy type 2E (LGMD 2E). *Hum. Mol. Genet.* **5,** 1953–1961.

Boyce, F. M., Beggs, A. H., Feener, C., and Kunkel, L. M. (1991) Dystrophin is transcribed in brain from a distant upstream promoter. *Proc. Natl. Acad. Sci. USA* **88,** 1276–1280.

Bradley, W. G. and Fulthorpe, J. J. (1978) Studies of sarcolemmal integrity in myopathic muscle. *Neurology* **28,** 670–677.

Bradley, W. G., Hudgson, P., Larson, P. F., Papapetropoulos, T. A., and Jenkison, M. (1972) Structural changes in the early stages of Duchenne muscular dystrophy. *J. Neurol. Neurosurg. Psychiatry* **35,** 451–455.

Brenman, J. E., Chao, D. S., Gee, S. H., McGee, A. W., Craven, S. E., Santillano, D. R., et al. (1996) Interaction of nitric oxide synthase with the postsynaptic density protein PSD-95 and alpha1-syntrophin mediated by PDZ domains. *Cell* **84,** 757–767.

Brenman, J. E., Chao, D. S., Xia, H., Aldape, K., and Bredt, D. S. (1995) Nitric oxide synthase complexed with dystrophin and absent from skeletal muscle sarcolemma in Duchenne muscular dystrophy. *Cell* **82**, 743–752.

Brown, S. C., Fassati, A., Popplewell, L., Page, A. M., Henry, M. D., Campbell, K. P., and Dickson, G. (1999) Dystrophic phenotype induced in vitro by antibody blockade of muscle alpha-dystroglycan-laminin interaction. *J. Cell Sci.* **112**, 209–216.

Brown, S. C. and Lucy, J. A. (1997) Functions of Dystrophin, in *Dystrophin: Gene, Protein and Cell Biology* (Brown, S. C. and Lucy, J. A., eds.), Cambridge University Press, Cambridge, pp. 163–200.

Brussee, V., Tardif, F., and Tremblay, J. P. (1997) Muscle fibres of mdx mice are more vulnerable to exercise than those of normal mice. *Neuromuscul. Disord.* **7**, 487–492.

Buckle, V. J., Guenet, J. L., Simon Chazottes, D., Love, D. R., and Davies, K. E. (1990) Localization of a dystrophin-related autosomal gene to 6q24 in man, and to mouse chromosome 10 in the region of the dystrophia muscularis (dy) locus. *Hum. Genet.* **85**, 324–326.

Burton, E. A., Tinsley, J. M., Holzfiend, P., Rodrigues, N., and Davies, K. E. (1999) A second promoter provides an alternative target for therapeutic upregulation of utrophin in Duchenne muscular dystrophy. *Proc. Natl. Acad. Sci. USA* **96**, 14,025–14,030.

Bushby, K. M. (1999) Making sense of the limb-girdle muscular dystrophies. *Brain* **122**, 1403–1420.

Bushby, K. M., Appleton, R., Anderson, L. V., Welch, J. L., Kelly, P., and Gardner Medwin, D. (1995) Deletion status and intellectual impairment in Duchenne muscular dystrophy. *Dev. Med. Child Neurol.* **37**, 260–269.

Bushby, K. M., Cleghorn, N. J., Curtis, A., Haggerty, I. D., Nicholson, L. V., Johnson, M. A., et al. (1991) Identification of a mutation in the promoter region of the dystrophin gene in a patient with atypical Becker muscular dystrophy. *Hum. Genet.* **88**, 195–199.

Byers, T. J., Kunkel, L. M., and Watkins, S. C. (1991) The subcellular distribution of dystrophin in mouse skeletal, cardiac, and smooth muscle. *J. Cell Biol.* **115**, 411–421.

Byers, T. J., Lidov, H. G., and Kunkel, L. M. (1993) An alternative dystrophin transcript specific to peripheral nerve. *Nat. Genet.* **4**, 77–81.

Campbell, K. P. and Kahl, S. D. (1989) Association of dystrophin and an integral membrane glycoprotein. *Nature* **338**, 259–262.

Carlson, C. G. (1998) The dystrophinopathies: an alternative to the structural hypothesis. *Neurobiol. Dis.* **5**, 3–15.

Carlson, C. G. (1999) Spontaneous changes in acetylcholine receptor and calcium leakage activity in cell-attached patches from cultured dystrophic myotubes. *Pflugers Arch.* **437**, 371–380.

Carlson, C. G. and Officer, T. (1996) Single channel evidence for a cytoskeletal defect involving acetylcholine receptors and calcium influx in cultured dystrophic (mdx) myotubes. *Muscle Nerve* **19**, 1116–1126.

Carnwath, J. W. and Shotton, D. M. (1987) Muscular dystrophy in the mdx mouse: histopathology of the soleus and extensor digitorum longus muscles. *J. Neurol. Sci.* **80**, 39–54.

Carpenter, J. L., Hoffman, E. P., Romanul, F. C., Kunkel, L. M., Rosales, R. K., Ma, N. S., et al. (1989) Feline muscular dystrophy with dystrophin deficiency. *Am. J. Pathol.* **135**, 909–919.

Carpenter, S. and Karpati, G. (1979) Duchenne muscular dystrophy: plasma membrane loss initiates muscle cell necrosis unless it is repaired. *Brain* **102**, 147–161.

Cecchi, G., Bagni, M. A., Griffiths, P. J., Ashley, C. C., and Maeda, Y. (1990) Detection of radial crossbridge force by lattice spacing changes in intact single muscle fibres. *Science* **250**, 1409–1411.

Chao, D. S., Silvagno, F., and Bredt, D. S. (1998) Muscular dystrophy in mdx mice despite lack of neuronal nitric oxide synthase. *J. Neurochem.* **71**, 784–789.

Chelly, J., Hamard, G., Koulakoff, A., Kaplan, J. C., Kahn, A., and Berwald Netter, Y. (1990a) Dystrophin gene transcribed from different promoters in neuronal and glial cells. *Nature* **344,** 64–65.

Chelly, J., Gilgenkrantz, H., Lambert, M., Hamard, G., Chafey, P., Recan, D., et al. (1990b) Effect of dystrophin gene deletions on mRNA levels and processing in Duchenne and Becker muscular dystrophies. *Cell* **63,** 1239–1248.

Chen, H. H., Mack, L. M., Kelly, R., Ontell, M., Kochanek, S., and Clemens, P. R. (1997) Persistence in muscle of an adenoviral vector that lacks all viral genes. *Proc. Natl. Acad. Sci. USA* **94,** 1645–1650.

Christova, T., Grozdanovic, Z., and Gossrau, R. (1997) Nitric oxide synthase (NOS) I during postnatal development in rat and mouse skeletal muscle. *Acta Histochem.* **99,** 311–324.

Clerk, A., Morris, G. E., Dubowitz, V., Davies, K. E., and Sewry, C. A. (1993) Dystrophin-related protein, utrophin, in normal and dystrophic human fetal skeletal muscle. *Histochem. J.* **25,** 554–561.

Coffey, A. J., Roberts, R. G., Green, E. D., Cole, C. G., Butler, R., Anand, R., et al. (1992) Construction of a 2. 6-Mb contig in yeast artificial chromosomes spanning the human dystrophin gene using an STS-based approach. *Genomics* **12,** 474–484.

Coirault, C., Lambert, F., Marchand Adam, S., Attal, P., Chemla, D., and Lecarpentier, Y. (1999) Myosin molecular motor dysfunction in dystrophic mouse diaphragm. *Am. J. Physiol.* **277,** C1170–C1176.

Collet, C., Allard, B., Tourneur, Y., and Jacquemond, V. (1999) Intracellular calcium signals measured with indo-1 in isolated skeletal muscle fibres from control and mdx mice. *J. Physiol.* **2,** 417–429.

Combaret, L., Taillandier, D., Voisin, L., Samuels, S. E., Boespflug Tanguy, O., and Attaix, D. (1996) No alteration in gene expression of components of the ubiquitin-proteasome proteolytic pathway in dystrophin-deficient muscles. *FEBS Lett.* **393,** 292–296.

Cooper, B. J., Winand, N. J., Stedman, H., Valentine, B. A., Hoffman, E. P., Kunkel, L. M., et al. (1988) The homologue of the Duchenne locus is defective in X-linked muscular dystrophy of dogs. *Nature* **334,** 154–156.

Coral Vazquez, R., Cohn, R. D., Moore, S. A., Hill, J. A., Weiss, R. M., Davisson, R. L., et al. (1999) Disruption of the sarcoglycan-sarcospan complex in vascular smooth muscle: a novel mechanism for cardiomyopathy and muscular dystrophy. *Cell* **98,** 465–474.

Cote, P. D., Moukhles, H., Lindenbaum, M., and Carbonetto, S. (1999) Chimaeric mice deficient in dystroglycans develop muscular dystrophy and have disrupted myoneural synapses. *Nat. Genet.* **23,** 338–342.

Cox, G. A., Phelps, S. F., Chapman, V. M., and Chamberlain, J. S. (1993) New mdx mutation disrupts expression of muscle and nonmuscle isoforms of dystrophin. *Nat. Genet.* **4,** 87–93.

Cox, G. A., Sunada, Y., Campbell, K. P., and Chamberlain, J. S. (1994) Dp71 can restore the dystrophin-associated glycoprotein complex in muscle but fails to prevent dystrophy. *Nat. Genet.* **8,** 333–339.

Crosbie, R. H., Heighway, J., Venzke, D. P., Lee, J. C., and Campbell, K. P. (1997) Sarcospan, the 25-kDa transmembrane component of the dystrophin-glycoprotein complex. *J. Biol. Chem.* **272,** 31,221–31,224.

Crosbie, R. H., Lebakken, C. S., Holt, K. H., Venzke, D. P., Straub, V., Lee, J. C., et al. (1999) Membrane targeting and stabilization of sarcospan is mediated by the sarcoglycan subcomplex. *J. Cell Biol.* **145,** 153–165.

Crosbie, R. H., Straub, V., Yun, H. Y., Lee, J. C., Rafael, J. A., Chamberlain, J. S., et al. (1998) mdx muscle pathology is independent of nNOS perturbation. *Hum. Mol. Genet.* **7,** 823–829.

Cullen, M. J., Walsh, J., Nicholson, L. V., and Harris, J. B. (1990) Ultrastructural localization of dystrophin in human muscle by using gold immunolabelling. *Proc. R. Soc. Lond. B. Biol. Sci.* **240,** 197–210.

D'Amore, P. A., Brown, R. H., Jr., Ku, P. T., Hoffman, E. P., Watanabe, H., Arahata, K., et al. (1994) Elevated basic fibroblast growth factor in the serum of patients with Duchenne muscular dystrophy. *Ann. Neurol.* **35,** 362–365.

Dangain, J. and Vrbova, G. (1984) Muscle development in mdx mutant mice. *Muscle Nerve* **7,** 700–704.

de Kermadec, J. M., Becane, H. M., Chenard, A., Tertrain, F., and Weiss, Y. (1994) Prevalence of left ventricular systolic dysfunction in Duchenne muscular dystrophy: an echocardiographic study. *Am. Heart J.* **127,** 618–623.

Deconinck, A. E., Potter, A. C., Tinsley, J. M., Wood, S. J., Vater, R., Young, C., et al. (1997a) Postsynaptic abnormalities at the neuromuscular junctions of utrophin-deficient mice. *J. Cell Biol.* **136,** 883–894.

Deconinck, A. E., Rafael, J. A., Skinner, J. A., Brown, S. C., Potter, A. C., Metzinger, L., et al. (1997b) Utrophin-dystrophin-deficient mice as a model for Duchenne muscular dystrophy. *Cell* **90,** 717–727.

Deconinck, N., Ragot, T., Marechal, G., Perricaudet, M., and Gillis, J. M. (1996) Functional protection of dystrophic mouse (mdx) muscles after adenovirus-mediated transfer of a dystrophin minigene. *Proc. Natl. Acad. Sci. USA* **93,** 3570–3574.

Deconinck, N., Tinsley, J., De Backer, F., Fisher, R., Kahn, D., Phelps, S., et al. (1997) Expression of truncated utrophin leads to major functional improvements in dystrophin-deficient muscles of mice. *Nat. Med.* **3,** 1216–1221.

Dennis, C. L., Tinsley, J. M., Deconinck, A. E., and Davies, K. E. (1996) Molecular and functional analysis of the utrophin promoter. *Nucleic Acids Res.* **24,** 1646–1652.

Dixon, A. K., Tait, T. M., Campbell, E. A., Bobrow, M., Roberts, R. G., and Freeman, T. C. (1997) Expression of the dystrophin-related protein 2 (Drp2) transcript in the mouse. *J. Mol. Biol.* **270,** 551–558.

D'Souza, V. N., Nguyen, T. M., Morris, G. E., Karges, W., Pillers, D. A., and Ray, P. N. (1995) A novel dystrophin isoform is required for normal retinal electrophysiology. *Hum. Mol. Genet.* **4,** 837–842.

Duchenne, G. B. A. (1868) Recherches sur la paralysie myosclerotique. *Archives Generales de Medicine* **11,** 5–25, 179–209, 305–321, 421–444, 552–588 (Translated by Brody and Wilkins (1968) *Arch. Neurol.* **19,** 629–636).

Duclos, F., Straub, V., Moore, S. A., Venzke, D. P., Hrstka, R. F., Crosbie, R. H., et al. (1998) Progressive muscular dystrophy in alpha-sarcoglycan-deficient mice. *J. Cell Biol.* **142,** 1461–1471.

Duggan, D. J., Gorospe, J. R., Fanin, M., Hoffman, E. P., and Angelini, C. (1997) Mutations in the sarcoglycan genes in patients with myopathy. *N. Engl. J. Med.* **336,** 618–624.

Dunckley, M. G., Manoharan, M., Villiet, P., Eperon, I. C., and Dickson, G. (1998) Modification of splicing in the dystrophin gene in cultured Mdx muscle cells by antisense oligoribonucleotides. *Hum. Mol. Genet.* **7,** 1083–1090.

Dunn, J. F., Bannister, N., Kemp, G. J., and Publicover, S. J. (1993) Sodium is elevated in mdx muscles: ionic interactions in dystrophic cells. *J. Neurol. Sci.* **114,** 76–80.

Dunn, J. F. and Radda, G. K. (1991) Total ion content of skeletal and cardiac muscle in the mdx mouse dystrophy: Ca2+ is elevated at all ages. *J. Neurol. Sci.* **103,** 226–231.

Durbeej, M. and Campbell, K. P. (1999) Biochemical characterization of the epithelial dystroglycan complex. *J. Biol. Chem.* **274,** 26,609–26,616.

Ehmer, S., Herrmann, R., Bittner, R., and Voit, T. (1997) Spatial distribution of beta-spectrin in normal and dystrophic human skeletal muscle. *Acta Neuropathol.* **94,** 240–246.

Emery, A. (1993) *Duchenne Muscular Dystrophy, 2nd ed. Oxford Monographs on Medical Genetics, vol. 24,* Oxford University Press, Oxford.

Emery, A. E. (1977) Muscle histology and creatine kinase levels in the foetus in Duchenne muscular dystrophy. *Nature* **266,** 472–473.

Engel, A. G. and Arahata, K. (1986) Mononuclear cells in myopathies: quantitation of functionally distinct subsets, recognition of antigen-specific cell-mediated cytotoxicity in some diseases, and implications for the pathogenesis of the different inflammatory myopathies. *Hum. Pathol.* **17**, 704–721.

England, S. B., Nicholson, L. V., Johnson, M. A., Forrest, S. M., Love, D. R., Zubrzycka Gaarn, E. E., et al. (1990) Very mild muscular dystrophy associated with the deletion of 46% of dystrophin. *Nature* **343**, 180–182.

Ervasti, J. M. and Campbell, K. P. (1991) Membrane organization of the dystrophin-glycoprotein complex. *Cell* **66**, 1121–1131.

Ervasti, J. M., Ohlendieck, K., Kahl, S. D., Gaver, M. G., and Campbell, K. P. (1990) Deficiency of a glycoprotein component of the dystrophin complex in dystrophic muscle. *Nature* **345**, 315–319.

Ettinger, A. J., Feng, G., and Sanes, J. R. (1997) epsilon-Sarcoglycan, a broadly expressed homologue of the gene mutated in limb-girdle muscular dystrophy 2D. *J. Biol. Chem.* **272**, 32,534–32,538.

Ferlini, A., Sewry, C., Melis, M., Mateddu, A., and Mutoni, F. (1999) X-linked dilated cardiomyopathy and the dystrophin gene. *Neuromuscul. Disord* **9**, 339–346.

Floss, T., Arnold, H. H., and Braun, T. (1997) A role for FGF-6 in skeletal muscle regeneration. *Genes Dev.* **11**, 2040–2051.

Fong, P. Y., Turner, P. R., Denetclaw, W. F., and Steinhardt, R. A. (1990) Increased activity of calcium leak channels in myotubes of Duchenne human and mdx mouse origin. *Science* **250**, 673–676.

Franco, A., Jr. and Lansman, J. B. (1990) Calcium entry through stretch-inactivated ion channels in mdx myotubes. *Nature* **344**, 670–673.

Franco Obregon, A., Jr. and Lansman, J. B. (1994) Mechanosensitive ion channels in skeletal muscle from normal and dystrophic mice. *J. Physiol.* **481**, 299–309.

Frigeri, A., Nicchia, G. P., Verbavatz, J. M., Valenti, G., and Svelto, M. (1998) Expression of aquaporin-4 in fast-twitch fibres of mammalian skeletal muscle. *J. Clin. Invest.* **102**, 695–703.

Gailly, P., Boland, B., Himpens, B., Casteels, R., and Gillis, J. M. (1993) Critical evaluation of cytosolic calcium determination in resting muscle fibres from normal and dystrophic (mdx) mice. *Cell Calcium* **14**, 473–83.

Gaschen, F. P., Hoffman, E. P., Gorospe, J. R., Uhl, E. W., Senior, D. F., Cardinet, G. H. D., and Pearce, L. K. (1992) Dystrophin deficiency causes lethal muscle hypertrophy in cats. *J. Neurol. Sci.* **110**, 149–159.

Gee, S. H., Madhavan, R., Levinson, S. R., Caldwell, J. H., Sealock, R., and Froehner, S. C. (1998) Interaction of muscle and brain sodium channels with multiple members of the syntrophin family of dystrophin-associated proteins. *J. Neurosci.* **18**, 128–137.

Gillis, J. M. (1996) Membrane abnormalities and Ca homeostasis in muscles of the mdx mouse, an animal model of the Duchenne muscular dystrophy: a review. *Acta Physiol. Scand.* **156**, 397–406.

Goldberg, L. R., Hausmanowa Petrusewicz, I., Fidzianska, A., Duggan, D. J., Steinberg, L. S., and Hoffman, E. P. (1998) A dystrophin missense mutation showing persistence of dystrophin and dystrophin-associated proteins yet a severe phenotype. *Ann. Neurol.* **44**, 971–976.

Goldspink, G., Fernandes, K., Williams, P. E., and Wells, D. J. (1994) Age-related changes in collagen gene expression in the muscles of mdx dystrophic and normal mice. *Neuromuscul. Disord.* **4**, 183–191.

Gorecki, D. C., Monaco, A. P., Derry, J. M., Walker, A. P., Barnard, E. A., and Barnard, P. J. (1992) Expression of four alternative dystrophin transcripts in brain regions regulated by different promoters. *Hum. Mol. Genet.* **1**, 505–510.

Gorospe, J. R. M., Nishikawa, B. K., and Hoffman, E. P. (1997) Pathophysiology of dystrophin deficiency, in *Dystrophin: Gene, Protein and Cell Biology* (Brown, S. C. and Lucy, J. A., eds.), Cambridge University Press, Cambridge, pp. 201–232.

Gorospe, J. R. M., Nishikawa, B. K., and Hoffman, E. P. (1996) Recruitment of mast cells to muscle after mild damage. *J. Neurol. Sci.* **135,** 10–17.

Gorospe, J. R. M., Tharp, M. D., Hinckley, J., Kornegay, J. N., and Hoffman, E. P. (1994a) A role for mast cells in the progression of Duchenne muscular dystrophy? Correlations in dystrophin-deficient humans, dogs, and mice. *J. Neurol. Sci.* **122,** 44–56.

Gorospe, J. R. M., Tharp, M., Demitsu, T., and Hoffman, E. P. (1994b) Dystrophin-deficient myofibers are vulnerable to mast cell granule-induced necrosis. *Neuromuscul. Disord.* **4,** 325–333.

Gossrau, R. (1998) Nitric oxide synthase I (NOS I) is a costameric enzyme in rat skeletal muscle. *Acta Histochem.* **100,** 451–462.

Gowers, W. (1879) Clinical lecture on pseudo-hypertrophic muscular paralysis. *Lancet* **2,** 1–2, 37–39, 73–75, 113–116.

Grady, R. M., Grange, R. W., Lau, K. S., Maimone, M. M., Nichol, M. C., Stull, J. T., and Sanes, J. R. (1999) Role for alpha-dystrobrevin in the pathogenesis of dystrophin-dependent muscular dystrophies. *Nat. Cell Biol.* **1,** 215–220.

Greelish, J. P., Su, L. T., Lankford, E. B., Burkman, J. M., Chen, H., Konig, S. K., et al. (1999) Stable restoration of the sarcoglycan complex in dystrophic muscle perfused with histamine and a recombinant adeno-associated viral vector. *Nat. Med.* **5,** 439–443.

Greenberg, D. S., Schatz, Y., Levy, Z., Pizzo, P., Yaffe, D., and Nudel, U. (1996) Reduced levels of dystrophin associated proteins in the brains of mice deficient for Dp71. *Hum. Mol. Genet.* **5,** 1299–1303.

Greenberg, D. S., Sunada, Y., Campbell, K. P., Yaffe, D., and Nudel, U. (1994) Exogenous Dp71 restores the levels of dystrophin associated proteins but does not alleviate muscle damage in mdx mice. *Nat. Genet.* **8,** 340–344.

Grozdanovic, Z. and Baumgarten, H. G. (1999) Nitric oxide synthase in skeletal muscle fibres: a signaling component of the dystrophin-glycoprotein complex. *Histol. Histopathol.* **14,** 243–256.

Gudrun, B., Andrew, G. E., Boysen, G., and Engel, A. G. (1975) Effects of microembolization on the skeletal muscle blood flow. A critique of the microvascular occlusion model of Duchenne dystrophy. *Acta Neurol. Scand.* **52,** 71–80.

Hack, A. A., Cordier, L., Shoturma, D. I., Lam, M. Y., Sweeney, H. L., and McNally, E. M. (1999) Muscle degeneration without mechanical injury in sarcoglycan deficiency. *Proc. Natl. Acad. Sci. USA* **96,** 10,723–10,728.

Hack, A. A., Ly, C. T., Jiang, F., Clendenin, C. J., Sigrist, K. S., Wollmann, R. L., and McNally, E. M. (1998) Gamma-sarcoglycan deficiency leads to muscle membrane defects and apoptosis independent of dystrophin. *J. Cell Biol.* **142,** 1279–1287.

Hagiwara, Y., Nishio, H., Kitoh, Y., Takeshima, Y., Narita, N., Wada, H., et al. (1994) A novel point mutation (G-1 to T) in a 5′ splice donor site of intron 13 of the dystrophin gene results in exon skipping and is responsible for Becker muscular dystrophy. *Am. J. Hum. Genet.* **54,** 53–61.

Haws, C. M. and Lansman, J. B. (1991) Developmental regulation of mechanosensitive calcium channels in skeletal muscle from normal and mdx mice. *Proc. R. Soc. Lond. B Biol. Sci.* **245,** 173–177.

Hayashi, Y. K., Chou, F. L., Engvall, E., Ogawa, M., Matsuda, C., Hirabayashi, S., et al. (1998) Mutations in the integrin alpha7 gene cause congenital myopathy. *Nat. Genet.* **19,** 94–97.

Head, S. I. (1993) Membrane potential, resting calcium and calcium transients in isolated muscle fibres from normal and dystrophic mice. *J. Physiol.* **469,** 11–19.

Helbling Leclerc, A., Zhang, X., Topaloglu, H., Cruaud, C., Tesson, F., Weissenbach, J., et al. (1995) Mutations in the laminin alpha 2-chain gene (LAMA2) cause merosin-deficient congenital muscular dystrophy. *Nat. Genet.* **11,** 216–218.

Helliwell, T. R., Man, N. T., Morris, G. E., and Davies, K. E. (1992) The dystrophin-related protein, utrophin, is expressed on the sarcolemma of regenerating human skeletal muscle fibres in dystrophies and inflammatory myopathies. *Neuromuscul. Disord.* **2**, 177–184.

Hemmings, L., Kuhlman, P. A., and Critchley, D. R. (1992) Analysis of the actin-binding domain of alpha-actinin by mutagenesis and demonstration that dystrophin contains a functionally homologous domain. *J. Cell Biol.* **116**, 1369–1380.

Henry, M. D. and Campbell, K. P. (1998) A role for dystroglycan in basement membrane assembly. *Cell* **95**, 859–870.

Hodgson, S. V., Abbs, S., Clark, S., Manzur, A., Heckmatt, J. Z., Dubowitz, V., and Bobrow, M. (1992) Correlation of clinical and deletion data in Duchenne and Becker muscular dystrophy, with special reference to mental ability. *Neuromuscul. Disord.* **2**, 269–276.

Hoffman, E. P., Brown, R. H., Jr., and Kunkel, L. M. (1987) Dystrophin: the protein product of the Duchenne muscular dystrophy locus. *Cell* **51**, 919–928.

Hoffman, E. P., Fischbeck, K. H., Brown, R. H., Johnson, M., Medori, R., Loike, J. D., et al. (1988a) Characterization of dystrophin in muscle-biopsy specimens from patients with Duchenne's or Becker's muscular dystrophy. *N. Engl. J. Med.* **318**, 1363–1368.

Hoffman, E. P., Hudecki, M. S., Rosenberg, P. A., Pollina, C. M., and Kunkel, L. M. (1988b) Cell and fiber-type distribution of dystrophin. *Neuron* **1**, 411–420.

Holder, E., Maeda, M., and Bies, R. D. (1996) Expression and regulation of the dystrophin Purkinje promoter in human skeletal muscle, heart and brain. *Hum. Genet.* **97**, 232–239.

Holzfeind, P. J., Ambrose, H. J., Newey, S. E., Nawrotzki, R. A., Blake, D. J., and Davies, K. E. (1999) Tissue-selective expression of alpha-dystrobrevin is determined by multiple promoters. *J. Biol. Chem.* **274**, 6250–6258.

Hoogerwaard, E. M., Bakker, E., Ippel, P. F., Oosterwijk, J. C., Majoor Krakauer, D. F., Leschot, N. J., et al. (1999) Signs and symptoms of Duchenne muscular dystrophy and Becker muscular dystrophy among carriers in The Netherlands: a cohort study. *Lancet* **353**, 2116–2119.

Hopf, F. W., Turner, P. R., Denetclaw, W. F., Jr., Reddy, P., and Steinhardt, R. A. (1996) A critical evaluation of resting intracellular free calcium regulation in dystrophic mdx muscle. *Am. J. Physiol.* **271**, C1325–C1339.

Howard, P. L., Dally, G. Y., Ditta, S. D., Austin, R. C., Worton, R. G., Klamut, H. J., and Ray, P. N. (1999) Dystrophin isoforms DP71 and DP427 have distinct roles in myogenic cells. *Muscle Nerve* **22**, 16–27.

Howard, P. L., Dally, G. Y., Wong, M. H., Ho, A., Weleber, R. G., Pillers, D. A., and Ray, P. N. (1998) Localization of dystrophin isoform Dp71 to the inner limiting membrane of the retina suggests a unique functional contribution of Dp71 in the retina. *Hum. Mol. Genet.* **7**, 1385–1391.

Howell, J. M., Lochmuller, H., O'Hara, A., Fletcher, S., Kakulas, B. A., Massie, B., et al. (1998) High-level dystrophin expression after adenovirus-mediated dystrophin minigene transfer to skeletal muscle of dystrophic dogs: prolongation of expression with immunosuppression. *Hum. Gene Ther.* **9**, 629–634.

Huang, P. L., Dawson, T. M., Bredt, D. S., Snyder, S. H., and Fishman, M. C. (1993) Targeted disruption of the neuronal nitric oxide synthase gene. *Cell* **75**, 1273–1286.

Hutter, O. F., Burton, F. L., and Bovell, D. L. (1991) Mechanical properties of normal and mdx mouse sarcolemma: bearing on function of dystrophin. *J. Muscle Res. Cell Motil.* **12**, 585–589.

Ibraghimov Beskrovnaya, O., Ervasti, J. M., Leveille, C. J., Slaughter, C. A., Sernett, S. W., and Campbell, K. P. (1992) Primary structure of dystrophin-associated glycoproteins linking dystrophin to the extracellular matrix. *Nature* **355**, 696–702.

Im, W. B., Phelps, S. F., Copen, E. H., Adams, E. G., Slightom, J. L., and Chamberlain, J. S. (1996) Differential expression of dystrophin isoforms in strains of mdx mice with different mutations. *Hum. Mol. Genet.* **5**, 1149–1153.

Imbert, N., Cognard, C., Duport, G., Guillou, C., and Raymond, G. (1995) Abnormal calcium homeostasis in Duchenne muscular dystrophy myotubes contracting in vitro. *Cell Calcium* **18,** 177–186.

Imbert, N., Vandebrouck, C., Constantin, B., Duport, G., Guillou, C., Cognard, C., and Raymond, G. (1996) Hypoosmotic shocks induce elevation of resting calcium level in Duchenne muscular dystrophy myotubes contracting in vitro. *Neuromuscul. Disord.* **6,** 351–360.

Infante, J. P. and Huszagh, V. A. (1999) Mechanisms of resistance to pathogenesis in muscular dystrophies. *Mol Cell Biochem.* **195,** 155–167.

Jackson, M. J., Jones, D. A., and Edwards, R. H. (1985) Measurements of calcium and other elements in muscle biopsy samples from patients with Duchenne muscular dystrophy. *Clin. Chim. Acta* **147,** 215–221.

James, M., Nguyen, T. M., Wise, C. J., Jones, G. E., and Morris, G. E. (1996) Utrophin-dystroglycan complex in membranes of adherent cultured cells. *Cell Motil. Cytoskeleton* **33,** 163–174.

James, M., Simmons, C., Wise, C. J., Jones, G. E., and Morris, G. E. (1995) Evidence for a utrophin-glycoprotein complex in cultured cell lines and a possible role in cell adhesion. *Biochem. Soc. Trans.* **23,** 398s.

Jobsis, G. J., Boers, J. M., Barth, P. G., and de Visser, M. (1999) Bethlem myopathy: a slowly progressive congenital muscular dystrophy with contractures. *Brain* **122,** 649–655.

Jobsis, G. J., Keizers, H., Vreijling, J. P., de Visser, M., Speer, M. C., Wolterman, R. A., et al. (1996) Type VI collagen mutations in Bethlem myopathy, an autosomal dominant myopathy with contractures. *Nat. Genet.* **14,** 113–115.

Jung, D., Duclos, F., Apostol, B., Straub, V., Lee, J. C., Allamand, V., et al. (1996) Characterization of delta-sarcoglycan, a novel component of the oligomeric sarcoglycan complex involved in limb-girdle muscular dystrophy. *J. Biol. Chem.* **271,** 32,321–32,329.

Kameya, S., Araki, E., Katsuki, M., Mizota, A., Adachi, E., Nakahara, K., et al. (1997) Dp260 disrupted mice revealed prolonged implicit time of the b-wave in ERG and loss of accumulation of beta-dystroglycan in the outer plexiform layer of the retina. *Hum. Mol. Genet.* **6,** 2195–2203.

Kameya, S., Miyagoe, Y., Nonaka, I., Ikemoto, T., Endo, M., Hanaoka, K., et al. (1999) Alpha1-syntrophin gene disruption results in the absence of neuronal-type nitric-oxide synthase at the sarcolemma but does not induce muscle degeneration. *J. Biol. Chem.* **274,** 2193–2200.

Karpati, G., Carpenter, S., Morris, G. E., Davies, K. E., Guerin, C., and Holland, P. (1993) Localization and quantitation of the chromosome 6-encoded dystrophin-related protein in normal and pathological human muscle. *J. Neuropathol. Exp. Neurol.* **52,** 119–128.

Khurana, T., Hoffman, E., and Kunkel, L. (1990) Identification of a chromosome 6-encoded Dystrophin-related protein. *J. Biol. Chem.* **265,** 16,717–16,720.

Khurana, T. S., Watkins, S. C., Chafey, P., Chelly, J., Tome, F. M. S., Fardeau, M., et al. (1991) Immunolocalization and developmental expression of dystrophin related protein in skeletal muscle. *Neuromuscular Disorders* **1,** 185–194.

Khurana, T. S., Watkins, S. C., and Kunkel, L. M. (1992) The subcellular distribution of chromosome 6-encoded dystrophin-related protein in the brain. *J. Cell Biol.* **119,** 357–366.

Klamut, H., Gangopadhyay, S., Worton, R., and Ray, P. (1990) Molecular and functional analysis of the muscle-specific promoter region of the Duchenne muscular dystrophy gene. *Mol. Cell. Biol.* **10,** 193–205.

Kochanek, S., Clemens, P. R., Mitani, K., Chen, H. H., Chan, S., and Caskey, C. T. (1996) A new adenoviral vector: replacement of all viral coding sequences with 28 kb of DNA independently expressing both full-length dystrophin and beta-galactosidase. *Proc. Natl. Acad. Sci. USA* **93,** 5731–5736.

Koenig, M., Beggs, A. H., Moyer, M., Scherpf, S., Heindrich, K., Bettecken, T., et al. (1989) The molecular basis for Duchenne versus Becker muscular dystrophy: correlation of severity with type of deletion. *Am. J. Hum. Genet.* **45,** 498–506.

Koenig, M., Hoffman, E. P., Bertelson, C. J., Monaco, A. P., Feener, C., and Kunkel, L. M. (1987) Complete cloning of the Duchenne muscular dystrophy (DMD) cDNA and preliminary genomic organization of the DMD gene in normal and affected individuals. *Cell* **50,** 509–517.

Koenig, M. and Kunkel, L. M. (1990) Detailed analysis of the repeat domain of dystrophin reveals four potential hinge segments that may confer flexibility. *J. Biol. Chem.* **265,** 4560–4566.

Koenig, M., Monaco, A. P., and Kunkel, L. M. (1988) The complete sequence of dystrophin predicts a rod-shaped cytoskeletal protein. *Cell* **53,** 219–226.

Lau, K. S., Grange, R. W., Chang, W. J., Kamm, K. E., Sarelius, I., and Stull, J. T. (1998) Skeletal muscle contractions stimulate cGMP formation and attenuate vascular smooth muscle myosin phosphorylation via nitric oxide. *FEBS Lett.* **431,** 71–74.

Law, D. J., Caputo, A., and Tidball, J. G. (1995) Site and mechanics of failure in normal and dystrophin-deficient skeletal muscle. *Muscle Nerve* **18,** 216–223.

Law, D. J. and Tidball, J. G. (1993) Dystrophin deficiency is associated with myotendinous junction defects in prenecrotic and fully regenerated skeletal muscle. *Am. J. Pathol.* **142,** 1513–1523.

Lebakken, C. S., Venzke, D. P., Hrstka, R. F., Consolino, C. M., Faulkner, J. A., Williamson, R. A., and Campbell, K. P. (2000) Sarcospan-deficient mice maintain normal muscle function. *Mol. Cell. Biol.* **20,** 1669–1677.

Lederfein, D., Yaffe, D., and Nudel, U. (1993) A housekeeping type promoter, located in the 3' region of the Duchenne muscular dystrophy gene, controls the expression of Dp71, a major product of the gene. *Hum. Mol. Genet.* **2,** 1883–1888.

Leijendekker, W. J., Passaquin, A. C., Metzinger, L., and Ruegg, U. T. (1996) Regulation of cytosolic calcium in skeletal muscle cells of the mdx mouse under conditions of stress. *Br. J. Pharmacol.* **118,** 611–616.

Lenk, U., Hanke, R., Thiele, H., and Speer, A. (1993) Point mutations at the carboxy terminus of the human dystrophin gene: implications for an association with mental retardation in DMD patients. *Hum. Mol. Genet.* **2,** 1877–1881.

Lenk, U., Oexle, K., Voit, T., Ancker, U., Hellner, K. A., Speer, A., and Hubner, C. (1996) A cysteine 3340 substitution in the dystroglycan-binding domain of dystrophin associated with Duchenne muscular dystrophy, mental retardation and absence of the ERG b-wave. *Hum. Mol. Genet.* **5,** 973–975.

Liberona, J. L., Powell, J. A., Shenoi, S., Petherbridge, L., Caviedes, R., and Jaimovich, E. (1998) Differences in both inositol 1,4,5-trisphosphate mass and inositol 1,4,5-trisphosphate receptors between normal and dystrophic skeletal muscle cell lines. *Muscle Nerve* **21,** 902–909.

Lidov, H. G., Byers, T. J., Watkins, S. C., and Kunkel, L. M. (1990) Localization of dystrophin to postsynaptic regions of central nervous system cortical neurons. *Nature* **348,** 725–728.

Lidov, H. G., Selig, S., and Kunkel, L. M. (1995) Dp140: a novel 140 kDa CNS transcript from the dystrophin locus. *Hum. Mol. Genet.* **4,** 329–335.

Liechti Gallati, S., Koenig, M., Kunkel, L. M., Frey, D., Boltshauser, E., Schneider, V., et al. (1989) Molecular deletion patterns in Duchenne and Becker type muscular dystrophy. *Hum. Genet.* **81,** 343–348.

Lim, L. E., Duclos, F., Broux, O., Bourg, N., Sunada, Y., Allamand, V., Meyer, J., et al. (1995) Beta-sarcoglycan: characterization and role in limb-girdle muscular dystrophy linked to 4q12. *Nat. Genet.* **11,** 257–265.

Liu, J. W., Wakayama, Y., Inoue, M., Shibuya, S., Kojima, H., Jimi, T., and Oniki, H. (1999) Immunocytochemical studies of aquaporin 4 in the skeletal muscle of mdx mouse. *J. Neurol. Sci.* **164,** 24–28.

Liu, L. A. and Engvall, E. (1999) Sarcoglycan isoforms in skeletal muscle. *J. Biol. Chem.* **274,** 38,171–38,176.

Loh, N. Y., Ambrose, H. J., Guay Woodford, L. M., DasGupta, S., Nawrotzki, R. A., Blake, D. J., and Davies, K. E. (1998) Genomic organization and refined mapping of the mouse beta-dystrobrevin gene. *Mamm. Genome* **9,** 857–862.

Lotz, B. P. and Engel, A. G. (1987) Are hypercontracted muscle fibres artifacts and do they cause rupture of the plasma membrane? *Neurology* **37,** 1466–1475.

Love, D. R., Flint, T. J., Genet, S. A., Middleton Price, H. R., and Davies, K. E. (1991a) Becker muscular dystrophy patient with a large intragenic dystrophin deletion: implications for functional minigenes and gene therapy. *J. Med. Genet.* **28,** 860–864.

Love, D. R., Flint, T. J., Marsden, R. F., Bloomfield, J. F., Daniels, R. J., Forrest, S. M., et al. (1990) Characterization of deletions in the dystrophin gene giving mild phenotypes. *Am. J. Med. Genet.* **37,** 136–142.

Love, D. R., Hill, D. F., Dickson, G., Spurr, N. K., Byth, B. C., Marsden, R. F., et al. (1989) An autosomal transcript in skeletal muscle with homology to dystrophin. *Nature* **339,** 55–58.

Love, D. R., Morris, G. E., Ellis, J. M., Fairbrother, U., Marsden, R. F., Bloomfield, J. F., et al. (1991b) Tissue distribution of the dystrophin-related gene product and expression in the mdx and dy mouse. *Proc. Natl. Acad. Sci. USA* **88,** 3243–3247.

MacLennan, P. A., McArdle, A., and Edwards, R. H. (1991) Effects of calcium on protein turnover of incubated muscles from mdx mice. *Am. J. Physiol.* **260,** E594–598.

Maconochie, M. K., Simpkins, A. H., Damien, E., Coulton, G., Greenfield, A. J., and Brown, S. D. (1996) The cysteine-rich and C-terminal domains of dystrophin are not required for normal costameric localization in the mouse. *Transgenic Res.* **5,** 123–130.

Makover, A., Zuk, D., Breakstone, J., Yaffe, D., and Nudel, U. (1991) Brain-type and muscle-type promoters of the dystrophin gene differ greatly in structure. *Neuromusc. Disord.* **1,** 39–45.

Malhotra, S. B., Hart, K. A., Klamut, H. J., Thomas, N. S., Bodrug, S. E., Burghes, A. H., et al. (1988) Frame-shift deletions in patients with Duchenne and Becker muscular dystrophy. *Science* **242,** 755–759.

Matsuda, R., Nishikawa, A., and Tanaka, H. (1995) Visualization of dystrophic muscle fibres in mdx mouse by vital staining with Evans blue: evidence of apoptosis in dystrophin-deficient muscle. *J. Biochem.* **118,** 959–964.

Matsumura, K. and Campbell, K. (1994) Dystrophin-glycoprotein complex—its role in the molecular pathogenesis of muscular dystrophies. *Muscle Nerve* **17,** 2–15.

Matsumura, K., Ervasti, J. M., Ohlendieck, K., Kahl, S. D., and Campbell, K. P. (1992) Association of dystrophin-related protein with dystrophin-associated proteins in mdx mouse muscle. *Nature* **360,** 588–591.

Maunder Sewry, C. A., Gorodetsky, R., Yarom, R., and Dubowitz, V. (1980) Element analysis of skeletal muscle in Duchenne muscular dystrophy using x-ray fluorescence spectrometry. *Muscle Nerve* **3,** 502–508.

McArdle, A., Edwards, R. H., and Jackson, M. J. (1994) Time course of changes in plasma membrane permeability in the dystrophin-deficient mdx mouse. *Muscle Nerve* **17,** 1378–1384.

McDouall, R. M., Dunn, M. J., and Dubowitz, V. (1990) Nature of the mononuclear infiltrate and the mechanism of muscle damage in juvenile dermatomyositis and Duchenne muscular dystrophy. *J. Neurol. Sci.* **99,** 199–217.

McNally, E. M., Duggan, D., Gorospe, J. R., Bonnemann, C. G., Fanin, M., Pegoraro, E., et al. (1996) Mutations that disrupt the carboxyl-terminus of gamma-sarcoglycan cause muscular dystrophy. *Hum. Mol. Genet.* **5,** 1841–1847.

McNally, E. M., Ly, C. T., and Kunkel, L. M. (1998) Human epsilon-sarcoglycan is highly related to alpha-sarcoglycan (adhalin), the limb girdle muscular dystrophy 2D gene. *FEBS Lett.* **422,** 27–32.

Melacini, P., Fanin, M., Duggan, D. J., Freda, M. P., Berardinelli, A., Danieli, G. A., et al. (1999) Heart involvement in muscular dystrophies due to sarcoglycan gene mutations. *Muscle Nerve* **22,** 473–479.

Melacini, P., Vianello, A., Villanova, C., Fanin, M., Miorin, M., Angelini, C., and Dalla Volta, S. (1996) Cardiac and respiratory involvement in advanced stage Duchenne muscular dystrophy. *Neuromuscl. Disord.* **6,** 367–376.

Menke, A. and Jockusch, H. (1991) Decreased osmotic stability of dystrophin-less muscle cells from the mdx mouse. *Nature* **349,** 69–71.

Menke, A. and Jockusch, H. (1995) Extent of shock-induced membrane leakage in human and mouse myotubes depends on dystrophin. *J. Cell. Sci.* **108,** 727–733.

Minetti, C., Beltrame, F., Marcenaro, G., and Bonilla, E. (1992) Dystrophin at the plasma membrane of human muscle fibres shows a costameric localization. *Neuromuscul. Disord.* **2,** 99–109.

Miyatake, M., Miike, T., Zhao, J., Yoshioka, K., Uchino, M., and Usuku, G. (1989) Possible systemic smooth muscle layer dysfunction due to a deficiency of dystrophin in Duchenne muscular dystrophy. *J. Neurol. Sci.* **93,** 11–17.

Mokri, B. and Engel, A. G. (1975) Duchenne dystrophy: electron microscopic findings pointing to a basic or early abnormality in the plasma membrane of the muscle fibre. *Neurology* **25,** 1111–1120.

Monaco, A., Walker, A., Millwood, I., Larin, Z., and Lehrach, H. (1992) A yeast artificial chromosome contig containing the complete Duchenne muscular dystrophy gene. *Genomics* **12,** 465–473.

Monaco, A. P., Bertelson, C. J., Liechti Gallati, S., Moser, H., and Kunkel, L. M. (1988) An explanation for the phenotypic differences between patients bearing partial deletions of the DMD locus. *Genomics* **2,** 90–95.

Monaco, A. P., Bertelson, C. J., Middlesworth, W., Colletti, C. A., Aldridge, J., Fischbeck, K. H., et al. (1985) Detection of deletions spanning the Duchenne muscular dystrophy locus using a tightly linked DNA segment. *Nature* **316,** 842–845.

Monaco, A. P., Neve, R. L., Colletti Feener, C., Bertelson, C. J., Kurnit, D. M., and Kunkel, L. M. (1986) Isolation of candidate cDNAs for portions of the Duchenne muscular dystrophy gene. *Nature* **323,** 646–650.

Montanaro, F., Lindenbaum, M., and Carbonetto, S. (1999) alpha-Dystroglycan is a laminin receptor involved in extracellular matrix assembly on myotubes and muscle cell viability. *J. Cell Biol.* **145,** 1325–1340.

Monti, R. J., Roy, R. R., Hodgson, J. A., and Edgerton, V. R. (1999) Transmission of forces within mammalian skeletal muscles. *J. Biomech.* **32,** 371–380.

Morris, G. E., Nguyen, T. M., Nguyen, T. N., Pereboev, A., Kendrick Jones, J., and Winder, S. J. (1999) Disruption of the utrophin-actin interaction by monoclonal antibodies and prediction of an actin-binding surface of utrophin. *Biochem. J.* **337,** 119–123.

Murakami, N., McLennan, I. S., Nonaka, I., Koishi, K., Baker, C., and Hammond Tooke, G. (1999) Transforming growth factor-beta2 is elevated in skeletal muscle disorders. *Muscle Nerve* **22,** 889–898.

Nguyen, T. M., Ellis, J. M., Love, D. R., Davies, K. E., Gatter, K. C., Dickson, G., and Morris, G. E. (1991) Localization of the DMDL gene-encoded dystrophin-related protein using a panel of nineteen monoclonal antibodies: presence at neuromuscular junctions, in the sarcolemma of dystrophic skeletal muscle, in vascular and other smooth muscles, and in proliferating brain cell lines. *J. Cell. Biol.* **115,** 1695–1700.

Nguyen, T. M., Le, T. T., Blake, D. J., Davies, K. E., and Morris, G. E. (1992) Utrophin, the autosomal homologue of dystrophin, is widely-expressed and membrane-associated in cultured cell lines. *FEBS Lett.* **313,** 19–22.

Nicholson, L. V., Johnson, M. A., Bushby, K. M., Gardner Medwin, D., Curtis, A., Ginjaar, I. B., et al. (1993) Integrated study of 100 patients with Xp21 linked muscular dystrophy using clinical, genetic, immunochemical, and histopathological data. Part 2. Correlations within individual patients. *J. Med. Genet.* **30,** 737–744.

Nigro, V., de Sa Moreira, E., Piluso, G., Vainzof, M., Belsito, A., Politano, L., et al. (1996a) Autosomal recessive limb-girdle muscular dystrophy, LGMD2F, is caused by a mutation in the delta-sarcoglycan gene. *Nat. Genet.* **14,** 195–198.

Nigro, V., Piluso, G., Belsito, A., Politano, L., Puca, A. A., Papparella, S., et al. (1996b) Identification of a novel sarcoglycan gene at 5q33 encoding a sarcolemmal 35 kDa glycoprotein. *Hum. Mol. Genet.* **5,** 1179–1186.

Noguchi, S., McNally, E. M., Ben Othmane, K., Hagiwara, Y., Mizuno, Y., Yoshida, M., et al. (1995) Mutations in the dystrophin-associated protein gamma-sarcoglycan in chromosome 13 muscular dystrophy. *Science* **270,** 819–822.

North, K. N., Miller, G., Iannaccone, S. T., Clemens, P. R., Chad, D. A., Bella, I., et al. (1996) Cognitive dysfunction as the major presenting feature of Becker's muscular dystrophy. *Neurology* **46,** 461–465.

Nudel, U., Zuk, D., Einat, P., Zeelon, E., Levy, Z. S. N., and Yaffe, D. (1989) Duchenne muscular dystrophy gene product is not identical in muscle and brain. *Nature* **337,** 76–78.

Ohtsuka, Y., Udaka, K., Yamashiro, Y., Yagita, H., and Okumura, K. (1998) Dystrophin acts as a transplantation rejection antigen in dystrophin-deficient mice: implication for gene therapy. *J. Immunol.* **160,** 4635–4640.

Pagel, C. N. and Partridge, T. A. (1999) Covert persistence of mdx mouse myopathy is revealed by acute and chronic effects of irradiation. *J. Neurol. Sci.* **164,** 103–116.

Partridge, T. A. (1997) Models of dystrophinopathy, pathological mechanisms and assessment of therapies, in *Dystrophin: Gene Protein and Cell Biology* (Brown, S. C. and Lucy, J. A., eds.), Cambridge University Press, Cambridge, pp. 310–331.

Pasternak, C., Wong, S., and Elson, E. L. (1995) Mechanical function of dystrophin in muscle cells. *J. Cell Biol.* **128,** 355–361.

Pastoret, C. and Sebille, A. (1995) mdx mice show progressive weakness and muscle deterioration with age. *J. Neurol. Sci.* **129,** 97–105.

Pearce, M., Blake, D. J., Tinsley, J. M., Byth, B. C., Campbell, L., Monaco, A. P., and Davies, K. E. (1993) The utrophin and dystrophin genes share similarities in genomic structure. *Hum. Mol. Genet.* **2,** 1765–1772.

Pegoraro, E., Marks, H., Garcia, C. A., Crawford, T., Mancias, P., Connolly, A. M., et al. (1998) Laminin alpha2 muscular dystrophy: genotype/phenotype studies of 22 patients. *Neurology* **51,** 101–110.

Peters, M. F., Adams, M. E., and Froehner, S. C. (1997a) Differential association of syntrophin pairs with the dystrophin complex. *J. Cell Biol.* **138,** 81–93.

Peters, M. F., Kramarcy, N. R., Sealock, R., and Froehner, S. C. (1994) beta 2-Syntrophin: localization at the neuromuscular junction in skeletal muscle. *Neuroreport* **5,** 1577–1580.

Peters, M. F., O'Brien, K. F., Sadoulet Puccio, H. M., Kunkel, L. M., Adams, M. E., and Froehner, S. C. (1997b) Beta-dystrobrevin, a new member of the dystrophin family. Identification, cloning, and protein associations. *J. Biol. Chem.* **272,** 31,561–31,569.

Peters, M. F., Sadoulet Puccio, H. M., Grady, M. R., Kramarcy, N. R., Kunkel, L. M., Sanes, J. R., et al. (1998) Differential membrane localization and intermolecular associations of alpha-dystrobrevin isoforms in skeletal muscle. *J. Cell Biol.* **142,** 1269–1278.

Petrof, B., Ebihara, S., Guibinga, G., Gilbert, R., Massie, B., Nalbantoglu, J., and Karpati, G. (1999) Differential effects of adenovirus-mediated dystrophin and utrophin gene transfer in dystrophic (mdx) mice. American Society of Gene Therapy, Washington, DC, abstract number 831.

Petrof, B. J. (1998) The molecular basis of activity-induced muscle injury in Duchenne muscular dystrophy. *Mol. Cell Biochem.* **179,** 111–123.

Petrof, B. J., Shrager, J. B., Stedman, H. H., Kelly, A. M., and Sweeney, H. L. (1993a) Dystrophin protects the sarcolemma from stresses developed during muscle contraction. *Proc. Natl. Acad. Sci. USA* **90,** 3710–3714.

Petrof, B. J., Stedman, H. H., Shrager, J. B., Eby, J., Sweeney, H. L., and Kelly, A. M. (1993b) Adaptations in myosin heavy chain expression and contractile function in dystrophic mouse diaphragm. *Am. J. Physiol.* **265**, C834–841.

Philpot, J., Cowan, F., Pennock, J., Sewry, C., Dubowitz, V., Bydder, G., and Muntoni, F. (1999) Merosin-deficient congenital muscular dystrophy: the spectrum of brain involvement on magnetic resonance imaging. *Neuromuscul. Disord.* **9**, 81–85.

Pons, F., Nicholson, L. V., Robert, A., Voit, T., and Leger, J. J. (1993) Dystrophin and dystrophin-related protein (utrophin) distribution in normal and dystrophin-deficient skeletal muscles. *Neuromuscul. Disord.* **3**, 507–514.

Pons, F., Robert, A., Fabbrizio, E., Hugon, G., Califano, J. C., Fehrentz, J. A., et al. (1994a) Utrophin localization in normal and dystrophin-deficient heart. *Circulation* **90**, 369–374.

Pons, F., Robert, A., Marini, J. F., and Leger, J. J. (1994b) Does utrophin expression in muscles of mdx mice during postnatal development functionally compensate for dystrophin deficiency? *J. Neurol. Sci.* **122**, 162–170.

Ponting, C. P., Blake, D. J., Davies, K. E., Kendrick Jones, J., and Winder, S. J. (1996) ZZ and TAZ: new putative zinc fingers in dystrophin and other proteins. *Trends Biochem. Sci.* **21**, 11–13.

Pressmar, J., Brinkmeier, H., Seewald, M. J., Naumann, T., and Rudel, R. (1994) Intracellular Ca2+ concentrations are not elevated in resting cultured muscle from Duchenne (DMD) patients and in MDX mouse muscle fibres. *Pflugers Arch.* **426**, 499–505.

Prior, T. W., Bartolo, C., Pearl, D. K., Papp, A. C., Snyder, P. J., Sedra, M. S., et al. (1995) Spectrum of small mutations in the dystrophin coding region. *Am. J. Hum. Genet.* **57**, 22–33.

Prior, T. W., Papp, A. C., Snyder, P. J., Burghes, A. H., Bartolo, C., Sedra, M. S., et al. (1993) A missense mutation in the dystrophin gene in a Duchenne muscular dystrophy patient. *Nat. Genet.* **4**, 357–360.

Rafael, J. A., Tinsley, J. M., Potter, A. C., Deconinck, A. E., and Davies, K. E. (1998) Skeletal muscle-specific expression of a utrophin transgene rescues utrophin-dystrophin deficient mice. *Nat. Genet.* **19**, 79–82.

Ragot, T., Vincent, N., Chafey, P., Vigne, E., Gilgenkrantz, H., Couton, D., et al. (1993) Efficient adenovirus-mediated transfer of a human minidystrophin gene to skeletal muscle of mdx mice. *Nature* **361**, 647–650.

Rapaport, D., Passos Bueno, M. R., Takata, R. I., Campiotto, S., Eggers, S., Vainzof, M., et al. (1992) A deletion including the brain promoter of the Duchenne muscular dystrophy gene is not associated with mental retardation. *Neuromuscul. Disord.* **2**, 117–120.

Ridge, J. C., Tidball, J. G., Ahl, K., Law, D. J., and Rickoll, W. L. (1994) Modifications in myotendinous junction surface morphology in dystrophin-deficient mouse muscle. *Exp. Mol. Pathol.* **61**, 58–68.

Rigoletto, C., Prelle, A., Ciscato, P., Moggio, M., Comi, G., Fortunato, F., and Scarlato, G. (1995) Utrophin expression during human fetal development. *Int. J. Dev. Neurosci.* **13**, 585–593.

Rivet Bastide, M., Imbert, N., Cognard, C., Duport, G., Rideau, Y., and Raymond, G. (1993) Changes in cytosolic resting ionized calcium level and in calcium transients during in vitro development of normal and Duchenne muscular dystrophy cultured skeletal muscle measured by laser cytofluorimetry using indo-1. *Cell Calcium* **14**, 563–571.

Roberds, S. L., Leturcq, F., Allamand, V., Piccolo, F., Jeanpierre, M., Anderson, R. D., et al. (1994) Missense mutations in the adhalin gene linked to autosomal recessive muscular dystrophy. *Cell* **78**, 625–633.

Roberts, R. G., Coffey, A. J., Bobrow, M., and Bentley, D. R. (1993) Exon structure of the human dystrophin gene. *Genomics* **16**, 536–538.

Roberts, R. G., Freeman, T. C., Kendall, E., Vetrie, D. L., Dixon, A. K., Shaw Smith, C., et al. (1996) Characterization of DRP2, a novel human dystrophin homologue. *Nat. Genet.* **13**, 223–226.

Roberts, R. G., Gardner, R. J., and Bobrow, M. (1994) Searching for the 1 in 2,400,000: a review of dystrophin gene point mutations. *Hum. Mutat.* **4,** 1–11.

Rodius, F., Claudepierre, T., Rosas Vargas, H., Cisneros, B., Montanez, C., Dreyfus, H., et al. (1997) Dystrophins in developing retina: Dp260 expression correlates with synaptic maturation. *Neuroreport* **8,** 2383–2387.

Sadoulet Puccio, H. M., Rajala, M., and Kunkel, L. M. (1997) Dystrobrevin and dystrophin: an interaction through coiled-coil motifs. *Proc. Natl. Acad. Sci. USA* **94,** 12,413–12,418.

Samitt, C. E. and Bonilla, E. (1990) Immunocytochemical study of dystrophin at the myotendinous junction. *Muscle Nerve* **13,** 493–500.

Sarig, R., Mezger Lallemand, V., Gitelman, I., Davis, C., Fuchs, O., Yaffe, D., and Nudel, U. (1999) Targeted inactivation of Dp71, the major non-muscle product of the DMD gene: differential activity of the Dp71 promoter during development. *Hum. Mol. Genet.* **8,** 1–10.

Sasaki, K., Sakata, K., Kachi, E., Hirata, S., Ishihara, T., and Ishikawa, K. (1998) Sequential changes in cardiac structure and function in patients with Duchenne type muscular dystrophy: a two-dimensional echocardiographic study. *Am. Heart J.* **135,** 937–944.

Schatzberg, S. J., Olby, N. J., Breen, M., Anderson, L. V., Langford, C. F., Dickens, H. F., et al. (1999) Molecular analysis of a spontaneous dystrophin "knockout" dog. *Neuromuscul. Disord.* **9,** 289–295.

Schmalbruch, H. (1984) Regenerated muscle fibres in Duchenne muscular dystrophy: a serial section study. *Neurology* **34,** 60–65.

Schofield, J., Houzelstein, D., Davies, K., Buckingham, M., and Edwards, Y. H. (1993) Expression of the dystrophin-related protein (utrophin) gene during mouse embryogenesis. *Dev. Dyn.* **198,** 254–264.

Schultz, J., Hoffmuller, U., Krause, G., Ashurst, J., Macias, M. J., Schmieder, P., et al. (1998) Specific interactions between the syntrophin PDZ domain and voltage-gated sodium channels. *Nat. Struct. Biol.* **5,** 19–24.

Sealock, R., Butler, M. H., Kramarcy, N. R., Gao, K. X., Murnane, A. A., Douville, K., and Froehner, S. C. (1991) Localization of dystrophin relative to acetylcholine receptor domains in electric tissue and adult and cultured skeletal muscle. *J. Cell Biol.* **113,** 1133–1144.

Sharp, N. J., Kornegay, J. N., Van Camp, S. D., Herbstreith, M. H., Secore, S. L., Kettle, S., et al. (1992) An error in dystrophin mRNA processing in golden retriever muscular dystrophy, an animal homologue of Duchenne muscular dystrophy. *Genomics* **13,** 115–121.

Shaw, R. F., Pearson, C. M., Chowdhury, S. R., and Dreifuss, F. E. (1967) Serum enzymes in sex-linked (Duchenne) muscular dystrophy. *Arch. Neurol.* **16,** 115–122.

Shiga, N., Takeshima, Y., Sakamoto, H., Inoue, K., Yokota, Y., Yokoyama, M., and Matsuo, M. (1997) Disruption of the splicing enhancer sequence within exon 27 of the dystrophin gene by a nonsense mutation induces partial skipping of the exon and is responsible for Becker muscular dystrophy. *J. Clin. Invest.* **100,** 2204–2210.

Shorer, Z., Philpot, J., Muntoni, F., Sewry, C., and Dubowitz, V. (1995) Demyelinating peripheral neuropathy in merosin-deficient congenital muscular dystrophy. *J. Child Neurol.* **10,** 472–475.

Sicinski, P., Geng, Y., Ryder Cook, A. S., Barnard, E. A., Darlison, M. G., and Barnard, P. J. (1989) The molecular basis of muscular dystrophy in the mdx mouse: a point mutation. *Science* **244,** 1578–1580.

Spencer, M. J., Croall, D. E., and Tidball, J. G. (1995) Calpains are activated in necrotic fibres from mdx dystrophic mice. *J. Biol. Chem.* **270,** 10,909–10,914.

Spencer, M. J. and Tidball, J. G. (1992) Calpain concentration is elevated although net calcium-dependent proteolysis is suppressed in dystrophin-deficient muscle. *Exp. Cell Res.* **203,** 107–114.

Stedman, H. H., Sweeney, H. L., Shrager, J. B., Maguire, H. C., Panettieri, R. A., Petrof, B., et al. (1991) The mdx mouse diaphragm reproduces the degenerative changes of Duchenne muscular dystrophy. *Nature* **352,** 536–539.

Straub, V., Duclos, F., Venzke, D. P., Lee, J. C., Cutshall, S., Leveille, C. J., and Campbell, K. P. (1998) Molecular pathogenesis of muscle degeneration in the delta-sarcoglycan-deficient hamster. *Am. J. Pathol.* **153,** 1623–1630.

Straub, V., Ettinger, A. J., Durbeej, M., Venzke, D. P., Cutshall, S., Sanes, J. R., and Campbell, K. P. (1999) Epsilon-sarcoglycan replaces alpha-sarcoglycan in smooth muscle to form a unique dystrophin-glycoprotein complex. *J. Biol. Chem.* **274,** 27,989–27,996.

Straub, V., Rafael, J. A., Chamberlain, J. S., and Campbell, K. P. (1997) Animal models for muscular dystrophy show different patterns of sarcolemmal disruption. *J. Cell Biol.* **139,** 375–385.

Tanaka, H. and Ozawa, E. (1990) Developmental expression of dystrophin on the rat myocardial cell membrane. *Histochemistry* **94,** 449–453.

Thompson, T. G., Chan, Y. M., Hack, A. A., Brosius, M., Rajala, M., Lidov, H. G., et al. (2000) Filamin 2 (FLN2). A muscle-specific sarcoglycan interacting protein. *J. Cell Biol.* **148,** 115–126.

Tinsley, J., Deconinck, N., Fisher, R., Kahn, D., Phelps, S., Gillis, J. M., and Davies, K. (1998) Expression of full-length utrophin prevents muscular dystrophy in mdx mice. *Nat. Med.* **4,** 1441–1444.

Tinsley, J. M., Blake, D. J., Roche, A., Fairbrother, U., Riss, J., Byth, B. C., et al. (1992) Primary structure of dystrophin-related protein. *Nature* **360,** 591–593.

Tinsley, J. M. and Davies, K. E. (1993) Utrophin: a potential replacement for dystrophin? *Neuromuscul. Disord.* **3,** 537–539.

Tinsley, J. M., Potter, A. C., Phelps, S. R., Fisher, R., Trickett, J. I., and Davies, K. E. (1996) Amelioration of the dystrophic phenotype of mdx mice using a truncated utrophin transgene. *Nature* **384,** 349–353.

Tkatchenko, A. V., Le Cam, G., Leger, J. J., and Dechesne, C. A. (2000) Large-scale analysis of differential gene expression in the hindlimb muscles and diaphragm of mdx mouse. *Biochim. Biophys. Acta* **3,** 17–30.

Tochio, H., Zhang, Q., Mandal, P., Li, M., and Zhang, M. (1999) Solution structure of the extended neuronal nitric oxide synthase PDZ domain complexed with an associated peptide. *Nat. Struct. Biol.* **6,** 417–421.

Todoroya, A., Bronzova, J., Miorin, M., Rosa, M., Kremensky, I., and Danieli, G. A. (1996) Mutation analysis in Duchenne and Becker muscular dystrophy patients from Bulgaria shows a peculiar distribution of breakpoints by intron. *Am. J. Med. Genet.* **65,** 40–43.

Tubiello, G., Carrera, P., Soriani, N., Morandi, L., and Ferrari, M. (1995) Mutational analysis of muscle and brain specific promoter regions of dystrophin gene in DMD/BMD Italian patients by denaturing gradient gel electrophoresis (DGGE). *Mol. Cell Probes* **9,** 441–446.

Turner, P. R., Schultz, R., Ganguly, B., and Steinhardt, R. A. (1993) Proteolysis results in altered leak channel kinetics and elevated free calcium in mdx muscle. *J. Membr. Biol.* **133,** 243–251.

Turner, P. R., Westwood, T., Regen, C. M., and Steinhardt, R. A. (1988) Increased protein degradation results from elevated free calcium levels found in muscle from mdx mice. *Nature* **335,** 735–738.

Tutdibi, O., Brinkmeier, H., Rudel, R., and Fohr, K. J. (1999) Increased calcium entry into dystrophin-deficient muscle fibres of MDX and ADR-MDX mice is reduced by ion channel blockers. *J. Physiol. (Lond.)* **515,** 859–868.

Vainzof, M., Passos Bueno, M. R., Canovas, M., Moreira, E. S., Pavanello, R. C., Marie, S. K., et al. (1996) The sarcoglycan complex in the six autosomal recessive limb-girdle muscular dystrophies. *Hum. Mol. Genet.* **5,** 1963–1969.

Valentine, B. A., Cooper, B. J., Cummings, J. F., and deLahunta, A. (1986) Progressive muscular dystrophy in a golden retriever dog: light microscope and ultrastructural features at 4 and 8 months. *Acta Neuropathol.* **71,** 301–310.

Valentine, B. A., Cooper, B. J., de Lahunta, A., O'Quinn, R., and Blue, J. T. (1988) Canine X-linked muscular dystrophy. An animal model of Duchenne muscular dystrophy: clinical studies. *J. Neurol. Sci.* **88,** 69–81.

Valentine, B. A., Winand, N. J., Pradhan, D., Moise, N. S., de Lahunta, A., Kornegay, J. N., and Cooper, B. J. (1992) Canine X-linked muscular dystrophy as an animal model of Duchenne muscular dystrophy: a review. *Am. J. Med. Genet.* **42,** 352–356.

Vilquin, J. T., Brussee, V., Asselin, I., Kinoshita, I., Gingras, M., and Tremblay, J. P. (1998) Evidence of mdx mouse skeletal muscle fragility in vivo by eccentric running exercise. *Muscle Nerve* **21,** 567–576.

Wakefield, P. M., Tinsley, J. M., Wood, M. J., Gilbert, R., Karpati, G., and Davies, K. E. (2000) Prevention of the dystrophic phenotype in dystrophin/utrophin-deficient muscle following adenovirus-mediated transfer of a utrophin minigene. *Gene Ther.* **7,** 201–204.

Watkins, S. C., Hoffman, E. P., Slayter, H. S., and Kunkel, L. M. (1988) Immunoelectron microscopic localization of dystrophin in myofibres. *Nature* **333,** 863–866.

Way, M., Pope, B., Cross, R. A., Kendrick Jones, J., and Weeds, A. G. (1992) Expression of the N-terminal domain of dystrophin in E. coli and demonstration of binding to F-actin. *FEBS Lett.* **301,** 243–245.

Weller, B., Karpati, G., and Carpenter, S. (1990) Dystrophin-deficient mdx muscle fibres are preferentially vulnerable to necrosis induced by experimental lengthening contractions. *J. Neurol. Sci.* **100,** 9–13.

Whittaker, P. A., Wood, L., Mathrubutham, M., and Anand, R. (1993) Generation of ordered phage sublibraries of YAC clones: construction of a 400-kb phage contig in the human dystrophin gene. *Genomics* **15,** 453–456.

Williams, M. W. and Bloch, R. J. (1999) Extensive but coordinated reorganization of the membrane skeleton in myofibres of dystrophic (mdx) mice. *J. Cell Biol.* **144,** 1259–1270.

Williamson, R. A., Henry, M. D., Daniels, K. J., Hrstka, R. F., Lee, J. C., Sunada, Y., et al. (1997) Dystroglycan is essential for early embryonic development: disruption of Reichert's membrane in Dag1-null mice. *Hum. Mol. Genet.* **6,** 831–841.

Wilson, J., Putt, W., Jimenez, C., and Edwards, Y. H. (1999) Up71 and up140, two novel transcripts of utrophin that are homologues of short forms of dystrophin. *Hum. Mol. Genet.* **8,** 1271–1278.

Wilson, L. A., Cooper, B. J., Dux, L., Dubowitz, V., and Sewry, C. A. (1994) Expression of utrophin (dystrophin-related protein) during regeneration and maturation of skeletal muscle in canine X-linked muscular dystrophy. *Neuropathol. Appl. Neurobiol.* **20,** 359–367.

Wilton, S. D., Johnsen, R. D., Pedretti, J. R., and Laing, N. G. (1993) Two distinct mutations in a single dystrophin gene: identification of an altered splice-site as the primary Becker muscular dystrophy mutation. *Am. J. Med. Genet.* **46,** 563–569.

Winand, N. J., Edwards, M., Pradhan, D., Berian, C. A., and Cooper, B. J. (1994) Deletion of the dystrophin muscle promoter in feline muscular dystrophy. *Neuromuscul. Disord.* **4,** 433–445.

Winder, S. J., Hemmings, L., Bolton, S. J., Maciver, S. K., Tinsley, J. M., Davies, K. E., et al. (1995a) Calmodulin regulation of utrophin actin binding. *Biochem. Soc. Trans.* **23,** 397s.

Winder, S. J., Hemmings, L., Maciver, S. K., Bolton, S. J., Tinsley, J. M., Davies, K. E., et al. (1995b) Utrophin actin binding domain: analysis of actin binding and cellular targeting. *J. Cell Sci.* **108,** 63–71.

Winder, S. J. and Kendrick Jones, J. (1995) Calcium/calmodulin-dependent regulation of the NH2-terminal F-actin binding domain of utrophin. *FEBS Lett.* **357,** 125–128.

Xiao, X., Li, J., and Samulski, R. J. (1996) Efficient long-term gene transfer into muscle tissue of immunocompetent mice by adeno-associated virus vector. *J. Virol.* **70,** 8098–8108.

Xu, H., Christmas, P., Wu, X. R., Wewer, U. M., and Engvall, E. (1994) Defective muscle basement membrane and lack of M-laminin in the dystrophic dy/dy mouse. *Proc. Natl. Acad. Sci. USA* **91,** 5572–5576.

Yamazaki, M., Minota, S., Sakurai, H., Miyazono, K., Yamada, A., Kanazawa, I., and Kawai, M. (1994) Expression of transforming growth factor-beta 1 and its relation to endomysial fibrosis in progressive muscular dystrophy. *Am. J. Pathol.* **144,** 221–226.

Yang, B., Ibraghimov Beskrovnaya, O., Moomaw, C. R., Slaughter, C. A., and Campbell, K. P. (1994) Heterogeneity of the 59-kDa dystrophin-associated protein revealed by cDNA cloning and expression. *J. Biol. Chem.* **269,** 6040–6044.

Yang, L., Lochmuller, H., Luo, J., Massie, B., Nalbantoglu, J., Karpati, G., and Petrof, B. J. (1998) Adenovirus-mediated dystrophin minigene transfer improves muscle strength in adult dystrophic (MDX) mice. *Gene Ther.* **5,** 369–379.

Yang, Y., Nunes, F. A., Berencsi, K., Furth, E. E., Gonczol, E., and Wilson, J. M. (1994) Cellular immunity to viral antigens limits E1-deleted adenoviruses for gene therapy. *Proc. Natl. Acad. Sci. USA* **91,** 4407–4411.

Yoshida, M. and Ozawa, E. (1990) Glycoprotein complex anchoring dystrophin to sarcolemma. *J. Biochem. Tokyo* **108,** 748–752.

Ytterberg, S. R. (1991) Animal models of myopathy. *Curr. Opin. Rheumatol.* **3,** 934–940.

Yuasa, K., Miyagoe, Y., Yamamoto, K., Nabeshima, Y., Dickson, G., and Takeda, S. (1998) Effective restoration of dystrophin-associated proteins in vivo by adenovirus-mediated transfer of truncated dystrophin cDNAs. *FEBS Lett.* **425,** 329–336.

Zellweger, H. and Antonik, A. (1975) Newborn screening for Duchenne muscular dystrophy. *Pediatrics* **55,** 30–34.

Zubrzycka Gaarn, E. E., Bulman, D. E., Karpati, G., Burghes, A. H., Belfall, B., Klamut, H. J., et al. (1988) The Duchenne muscular dystrophy gene product is localized in sarcolemma of human skeletal muscle. *Nature* **333,** 466–469.

Index